Deontic Logic and Normative Systems

14th International Conference,
DEON 2018,
Utrecht, The Netherlands,
3—6 July, 2018

Deontic Logic and Normative Systems

14th International Conference,
DEON 2018,
Utrecht, The Netherlands,
3—6 July, 2018

Edited by
Jan Broersen,
Cleo Condoravdi
Shyam Nair
and
Gabriella Pigozzi

© Individual authors and College Publications 2018
All rights reserved.

ISBN 978-1-84890-278-7

College Publications
Scientific Director: Dov Gabbay
Managing Director: Jane Spurr

http://www.collegepublications.co.uk

Original cover design by Laraine Welch
Printed by Lightning Source, Milton Keynes, UK

All rights reserved. No part of this publication may be reproduced, stored in a retrieval system or transmitted in any form, or by any means, electronic, mechanical, photocopying, recording or otherwise without prior permission, in writing, from the publisher.

CONTENTS

Preface 1
  Jan Broersen, Cleo Condoravdi, Shyam Nair and Gabriella Pigozzi

INVITED TALKS – ABSTRACTS

A Flexible Infrastructure for Normative Reasoning 7
  Christoph Benzmüller

Agential Free Choice 9
  Melissa Fusco

Deontic Logic and Game Theory 11
  Olivier Roy

Defeasible Conditional Imperatives 13
  Marek Sergot

ACCEPTED PAPERS

Reasoning about Conditions in STIT Logic 15
  Matthias Armgardt, Emiliano Lorini and Giovanni Sartor

A Dyadic Deontic Logic in HOL 33
  Christoph Benzmüller, Ali Farjami and Xavier Parent

Knowledge and Subjective Oughts in STIT Logic 51
  Jan Broersen and Aldo Iván Ramírez Abarca

Normative Conflicts in a Dynamic Logic of Norms and Codes 71
  Ilaria Canavotto and Alessandro Giordani

Resolving Conflicting Obligations in Mīmāṃsā: A Sequent-based Approach 91
  Agata Ciabattoni, Francesca Gulisano and Björn Lellmann

A Look at Chisholm's Ethics of Requirement 111
  Marian J. R. Gilton

Ability and Responsibility in General Action Logic 121
  Alessandro Giordani

Dialogues on Moral Theories 139
  Guido Governatori, Francesco Oliveri, Régis Riveret, Antonino Rotolo and Serena
  Villata

Epistemic Oughts in STIT Semantics (Abbreviated Version) 157
  John Horty

S5 as a Deontic Logic 177
  Fengkui Ju

Towards a Formal Ethics for Autonomous Cars 193
  Piotr Kulicki, Robert Trypuz and Michael P. Musielewicz

A Formalization of Kant's Second Formulation of the Categorical Imperative 211
  Felix Lindner and Martin Mose Bentzen

You Must! Maybe You Won't 227
  Matthew Mandelkern

Contraction of Combined Normative Sets 247
  Juliano Maranhão and Edelcio Conçalves de Souza

Toward a Systematization of Logics for Monadic and Dyadic Agency &
Ability (Preliminary Version) 263
  Paul McNamara

I/O Logics with a Consistency Check 285
  Xavier Parent and Leendert van der Torre

A Logic for reasoning about Group Norms 301
  Daniele Porello

How to Take Heroin (if at all). Holistic Detachment in Deontic Logic 317
  Frederik Van De Putte, Stef Frijters and Joke Meheus

Preface

This volume contains the proceedings of the 14th International Conference on Deontic Logic and Normative Systems (DEON 2018), hosted by the University of Utrecht, the Netherlands, and taking place on 3-6 July 2018.

Since 1991, the biennial DEON conferences have promoted interdisciplinary cooperation amongst scholars interested in linking the formal-logical study of normative concepts and normative systems with computer science, artificial intelligence, linguistics, philosophy, organization theory and law.

DEON's previous editions were held in Amsterdam, the Netherlands (1991); Oslo, Norway (1994); Sesimbra, Portugal (1996); Bologna, Italy (1998); Toulouse, France (2000); London, United Kingdom (2002); Madeira, Portugal (2004); Utrecht, the Netherlands (2006); Luxembourg, Luxembourg (2008); Fiesole, Italy (2010); Bergen, Norway (2012); Ghent, Belgium (2014); Bayreuth, Germany (2016).

For DEON 2018 the Program Committee invited contributions concerned with the following topics:

- the logical study of normative reasoning, including formal systems of deontic logic, defeasible normative reasoning, logics of action, logics of time, and other related areas of logic

- the formal analysis of normative concepts, normative systems (and their dynamics)

- the formal specification of aspects of norm-governed multi-agent systems and autonomous agents, including (but not limited to) the representation of rights, authorization, delegation, power, responsibility and liability

- the normative aspects of protocols for communication, negotiation and multi-agent decision making

- the formal analysis of the semantics and pragmatics of deontic and normative expressions in natural language

- the formal representation of legal knowledge

- the formal specification of normative systems for the management of bureaucratic processes in public or private administration applications of normative logic to the specification of database integrity constraints

- game theoretic aspects of deontic reasoning

- emergence of norms

- deontic paradoxes

- argumentation theory and normative reasoning

In addition to these general themes, DEON 2018's special focus was on *Deontic reasoning for responsible AI*. The successes of Artificial Intelligence over the last few years have brought to the fore a new and important application area for deontic logic: Responsible AI. On the one hand, this concerns systems for checking and proving responsibility characteristics of artificial intelligent agents and their designs, and on the other hand, it concerns responsible decision making and machine ethics. This DEON's special theme *Deontic reasoning for responsible AI* solicited contributions that address issues related to these two subjects.

Topics of interest in this special theme included:

- moral decision making

- norm awareness

- accountability

- explainability

- causal and probabilistic theories of responsibility

- operationalizations of ethical theories

- collective responsibility

- grades of responsibility

We received 30 submissions from a variety of research communities. Each submission was carefully reviewed by three members of the Program Committee, composed of 39 leading researchers in the field. In total, 19 papers were accepted for presentation at the conference and 18 are published in this volume.

Our four keynote speakers are: Christoph Benzmüller presenting "A Flexible Infrastructure for Normative Reasoning", Melissa Fusco talking on "Agential Free Choice", Olivier Roy presenting "Deontic Logic and Game Theory", and Marek Sergot talking on "Defeasible Conditional Imperatives".

Organizing a conference is a collective work. First of all, we wish to thank our invited speakers for accepting our request and making special arrangements in their busy schedules to ensure their presence and presenting to us their latest research. Special thanks go to all the authors who submitted state-of-the-art research papers from all over the world. We are indebted to the members of the Program

Committee for reviewing the submissions, keeping up with our tight deadlines, and providing detailed and constructive reviews. Thanks go also to EasyChair for their fine conference management system, which made it possible to take care of the submissions and of the reviewing process smoothly. We thank the Local Organizing Committee, especially Aafke van Welbergen, for creating and updating the website and taking care of all the organisational matters that a conference like this requires. We are also grateful to Leon van der Torre and Jeff Horty, respectively, Chair and Vice Chair of the DEON Steering Committee, for their guidance and commitment. Our thanks to the sponsors who financially supported DEON 2018: the Utrecht University Department of Philosophy and Religious Studies, and the ERC-2013-CoG project REINS, nr. 616512. Finally we are indebted to College Publications, especially to Dov Gabbay and Jane Spurr, for their support through the process and for ensuring that these proceedings were published on time.

May 2018
Jan Broersen
Cleo Condoravdi
Shyam Nair
Gabriella Pigozzi

Organization

Program Chairs

Jan Broersen	Utrecht University
Cleo Condoravdi	Stanford University
Shyam Nair	Arizona State University
Gabriella Pigozzi	Université Paris-Dauphine

Program Committee

Thomas Ågotnes	University of Bergen
Maria Aloni	University of Amsterdam
Albert Anglberger	Munich Center for Mathematical Philosophy
Mathieu Beirlaen	Ghent University
Jan Broersen	Utrecht University
Mark A. Brown	Syracuse University
Fabrizio Cariani	Northwestern University
Jose Carmo	University of Madeira
Roberto Ciuni	Ruhr-UniversitÃďt Bochum
Cleo Condoravdi	Stanford University
Robert Demolombe	IRIT
Janice Dowell	Syracuse University
Stephen Finlay	University of Southern California
Lou Goble	Willamette University
Guido Governatori	CSIRO
Davide Grossi	University of Groningen
Andreas Herzig	IRIT, CNRS
Jeff Horty	University of Maryland
Magdalena Kaufmann	University of Connecticut
Fenrong Liu	Tsinghua University
Emiliano Lorini	IRIT, CNRS
Paul McNamara	University of New Hampshire
Joke Meheus	Gent University
John-Jules Meyer	Utrecht University
Shyam Nair	Arizona State University
Nir Oren	University of Aberdeen
Gabriella Pigozzi	Université Paris Dauphine
Henry Prakken	Utrecht University and University of Groningen
Antonio Rotolo	University of Bologna
Olivier Roy	University of Bayreuth

Giovanni Sartor   European University Institute of Florence and CIRSFID
Audun Stolpe   Norwegian Defence Research Establishment
Christian Strasser   Ruhr University Bochum
Paolo Turrini   University of Warwick
Frederik van de Putte Ghent University
Leon van der Torre   University of Luxembourg
Peter Vranas   University of Wisconsin
Malte Willer   University of Chicago
Tomoyuki Yamada   Hokkaido University

Local Organizing Committee

Jan Broersen (Chair)
Sander Beckers
Hein Duijf
Alexandra Kuncova
Niels van Miltenburg
Allard Tamminga
Aldo Ramirez
Aafke van Welbergen

Sponsoring Institutions

The Utrecht University Department of Philosophy and Religious Studies
The ERC-2013-CoG project REINS, nr. 616512

A Flexible Infrastructure for Normative Reasoning

Christoph Benzmüller
University of Luxembourg, Luxembourg, and Freie Universität Berlin, Germany
c.benzmueller@gmail.com

If humans and intelligent machines are supposed to peacefully co-exist, appropriate forms of machine-control and human-machine-interaction are required. Intelligent machines should assess and explain their actions and decisions in a form, preferably based on explicit normative reasoning, that is accessible to human understanding. In recent AI systems, however, which put a strong focus on machine learning, this aspect appears neglected.

But what are the best formal structures to enable such advanced normative reasoning capabilities in intelligent machines? To address this question, I argue for the development of a flexible reasoning infrastructure, supporting experiments with different deontic logics and ethical theories.

The proposed framework is based on the meta-logical approach to universal logical reasoning that I have successfully applied, together with colleagues and students, in previous studies in philosophy, computer science and mathematics. A demonstrator version of this flexible reasoning engine will be presented at the conference, and exemplary instantiations of it for some prominent deontic logics will be showcased. The majority of these logics have been automated for the first time.

Agential Free Choice

Melissa Fusco
Columbia University
mf3095@columbia.edu

The Free Choice effect—whereby $\Diamond(p \text{ OR } q)$ seems to entail both $\Diamond p$ and $\Diamond q$—has long described as a phenomenon affecting the deontic "may". In this talk, I explore how to extend the theory of deontic free choice I defended in Fusco (2015) to the agentive modal "can". Getting an assist from some new experimental data, I will argue that free choice for deontic and agential phenomena, while distinct, are related in a natural way. Putting them side-by-side with respect to free choice behavior opens a new window onto the unity and diversity in natural language modality.

Deontic Logic and Game Theory

Olivier Roy
University of Bayreuth
`olivier.roy@uni-bayreuth.de`

What is the logical structure of rational recommendations in strategic interaction? After clarifying and motivating that question, I will provide a critical survey of the existing literature on the topic, and argue for one specific approach that defines obligation and permissions, respectively, in terms of necessary and sufficient conditions for rationality.

DEFEASIBLE CONDITIONAL IMPERATIVES

MAREK SERGOT
Imperial College London
m.sergot@imperial.ac.uk

An alternative to possible world semantics for deontic logic is what is sometimes called 'the imperatival tradition': formulas are interpreted not with respect to worlds but to a given set of norms or imperatives. Obligations are then the actions that best fulfil these norms. Proposals go back over many years; there has been renewed interest with the emergence of default reasoning methods, notably as mapped out by Joerg Hansen. I will look at defeasible conditional imperatives of the general form 'if F then do A!'. F is an expression representing current facts or beliefs. A is an expression specifying an action or combination of actions to be done. A itself can have a conditional structure, as when we say 'if you do A then also do/do not do B!'. Imperatives are defeasible and may conflict: a (partial) priority ordering can then be used to choose between them. I will look at the representation of such structures as logic programs (of a very general kind) and sketch the logic of imperatives that emerges. I am also interested in the use of this formalism for practical reasoning. I will provide some illustrations from legal and moral reasoning.

Reasoning about Conditions in STIT Logic

Matthias Armgardt
University of Konstanz, Germany
matthias.armgardt@uni-konstanz.de

Emiliano Lorini
CNRS-IRIT, Toulouse University, France
emiliano.lorini@irit.fr

Giovanni Sartor
University of Bologna and EUI-Florence
giovanni.sartor@unibo.it

In this paper we propose a logical formalization of the legal concepts of suspensive and resolutive conditions within the STIT approach to action. At the technical level, our proposal consists in extending the STIT language with a special operator that allows us to represent the concept of a presumption. This enables us to model the retroactive effect of conditions.

1 Introduction

This paper will provide a logical analysis of legal conditions, namely, future and uncertain events on which a legal arrangement is dependent, according to a juristic act, such as a contract or will.

First, we shall introduce notions of conditioned disposition, condition and conditioned legal arrangement. Then we shall distinguish different kinds of conditions. In particular we shall distinguish suspensive and resolutive conditions, which postpone and revoke respectively a legal arrangement. We shall also distinguish non-retroactive and retroactive conditions. The first concern a legal arrangement existing from the time of the condition and the second a legal arrangement pre-existing to the condition.

Our formalization of the notion of legal condition is based on STIT logic, a well-known logic of action introduced in philosophical logic by Belnap et al. [5]. The reason why we use STIT is that it offers a clear account of time, action and their combination. These are fundamental constituents of the notion of legal condition. In order to capture retroactive effects of conditions, we will extend the basic STIT theory by the concept of presumption. The resulting framework will allow us to represent two complementary aspects of a retroactive condition: (a) the institutional past differs depending on the realization of the retroactive condition, and (b) until a suspensive condition is realized (or a resolutive condition fails to be realized) it is presumed that the conditioned arrangement does not obtain (or does obtain).

The paper is organized as follows. Section 2 is devoted to the conceptual background of our work by discussing the relevant philosophical theories of conditions. In Section 3, the STIT framework is presented. In Section 4, it is used to formalize the notion of condition. In Section 5, we conclude.

2 Legal conditions

In this section we shall introduce the concept and the regulation of conditions, describing the phenomenon that will be formalized logically in the next section.

2.1 Conditional dispositions, conditions, conditioned positions

Our analysis will address *conditional dispositions* included in juridical acts, namely, in those performative declarations —such as contracts or wills— through which private parties establish legal arrangements, e.g., they transfer property and create or remove obligations. A *conditional disposition* makes a legal arrangement, the *conditioned arrangement*, dependent upon a future and uncertain event, the *condition*.

To avoid ambiguities, we shall reserve the term "condition" to denote the future and uncertain conditioning events. We shall explicitly speak of a "conditioned arrangement" to denote an institutional outcome whose existence is dependent upon the realization of the condition, and of a "conditional disposition" for the juristic act (or part of it) establishing a conditional arrangement.

Note that a conditional disposition is no descriptive statement, it rather is a performative one, meant to constitute the conditioned arrangement in case the condition will be happen. Consider, for instance the following example of a conditional disposition, from Roman law: "I shall give you 100 sesterces, on condition that the ship arrives from Asia" (the parties to the transport contract, agree that only if the ship arrives the fee for transport is to be paid). In this conditional disposition, the proposition "the ship arrives from Asia" is the condition, and the obligation to give you 100 sesterces is the conditioned arrangement. The conditioned arrangement and the condition can also be stated in separate statements: Statement 1: I shall give you 100 sesterces. Statement 2: Statement 1 shall have effect only on condition that the ship arrives from Asia. Or also "This contract shall only have effect if the ship arrives from Asia".

The conditioned arrangement that is constituted by this conditional disposition (the obligation of the promisor and the right of the promisee, both dependent on the ship's arrival) will (a) become pure or unconditioned, if the condition takes place or (b) terminate, if the condition becomes impossible.

Conditioned arrangements may consist in any kind of legal outcome that can be constituted by private parties: the creation, the elimination, the modification or the transfer of any legal position, such as an obligation or a property right. A conditioned legal right may be the object of a transaction, being sold, purchased, donated, etc. The effectiveness of such a transaction, however, will remain subject to the verification of the condition.

The regulation of a conditional disposition goes back to Roman law. Justinian's Institutes (Book III, Section XIV, [22]), in discussing stipulations (*stipulationes*), i.e., legally binding promises, classifies them as being pure, with a deadline, or under condition: in the first case, performance can be requested immediately; in the second, performance can only be requested at the established date; in the third case, performance can only be required if the condition obtains. The structure of the third case (which is a suspensive conditioned disposition) is as follows that "A promise is made conditionally, when the obligation is made dependent on an uncertain occurrence, so that the promise is binding if something happens or does not happen" (Book III, Section XV).

The notions of a conditional disposition can also be found in modern civil codes. For instance, Article 1335 of the Italian civil code states that "The parties to a contract can make the efficacy or the resolution of the contract, or of a single agreement, dependent upon a future and uncertain event." Similarly, according to the Draft Common Frame of Reference, a project for a Common Civil Code for the EU: "[t]he terms regulating a right, obligation or contractual relationship may provide that it is conditional upon the occurrence of an uncertain future event, so that it takes effect only if the event occurs (suspensive condition) or comes to an end if the event occurs (resolutive condition)."[21]

Also the common law addresses conditional dispositions, using different terminologies (see [12]).

2.2 The effects of conditions

Here, following the civil law tradition, we shall limit our analysis to conditional dispositions in private juridical acts. We do not address the conditional connections established by authoritative legal norms, in legislation or judicial rulings, though it may be argued that similar logical structure may also characterise the institutional outcome of authoritative conditional declarations.

The purpose of conditional dispositions in private juridical acts is to address future contingencies. The parties want to establish a legal arrangement that only fits a particular future situation (the condition), but they are uncertain on whether this situation will obtain. Therefore, they make their arrangement dependent on the existence of that situation. For instance, a person that is likely to get a job in a city can make a rental contract the legal effects of which (the landlord's obligation to provide the house, and the tenant's obligation to pay the rent for the other) are conditioned on that person's getting the job. If the prospective tenant does not get the job, neither party should be due to perform their obligations.

For a conditioned legal position to exist as such, it is essential that the condition is uncertain. If a legal effect is dependant upon an event which is certain, even though the time in which it takes place is uncertain (e.g., a person's death), then the event constitutes a deadline rather than a condition. The uncertainty of conditions is assumed to be "objective": for the condition to be uncertain, at a certain point in time, it is sufficient that it is not humanly possible to anticipate with certainty whether the event will happen or not.

2.3 Suspensive and resolutive conditions

There are two different types of conditions in all legal orders since the times of Ancient Roman Law: suspensive and resolutive ones.

A suspensive conditional disposition makes the constitution of a legal arrangement dependent on the occurrence of a condition. This means that the legal arrangement only becomes effective if the condition occurs.

Consider for instance the case in which a developer acquires a piece of land from the owner for a certain price, under the suspensive condition that within a year a building permit is issued. If the building permit is issued, then the land is transferred to the developer and the developer will be obliged to pay the price. If this is not the case, the transfer will not take place, so that the land will remain with the owner, and the developer will have no obligation to pay the price.

As another example, consider an installment sale where the transfer of the property (e.g., of a vehicle) is subject to the suspensive condition of the payment of all installments. In this case only if the payment of the last installment is completed the ownership is transferred to the buyer. If this is not the case, the property remains with the seller. This arrangement is useful for the seller in case the buyer defaults: rather than having to compete with other creditors, the seller will simply claim back what still belongs to her.

A resolutive conditional disposition constitutes a legal arrangement and makes the termination of a legal arrangement dependent on the occurrence of the condition. Consider the case in which a developer acquires a piece of land from a seller, under the resolutive condition that the developer does not obtain a building permit within a year (the sale will be cancelled if this negative condition avers). If the resolutive condition is not met (the permit is given) the developer continues to own what he has purchased. If the condition is met (the permit is not given within the deadline) the transfer it cancelled, and the seller's ownership is restored. As another example, consider the case in which a buyer purchases a property from a seller under the resolutive condition that the seller gives back the whole price paid by the buyer, plus an additional sum (an interest). In this case if the seller gives back the whole amounts due, the transfer is annulled; if he does not, the transfer remains.

2.4 Retroactivity

A condition is retroactive when the occurrence of the conditioned legal arrangement (for suspensive conditions) or its cancellation (for resolutive conditions) is assumed to take place at a time that precedes the realization of the condition (usually, at the time in which the conditional disposition was enacted).

Let us assume that at a time t_0 a contract is enacted according to which 1 sells a piece of land to 2, subject to the retroactive suspensive condition that a building permit is issued.[1] If the permit is issued at the subsequent time t_1 (e.g., one year after t_0) the transfer is assumed to have taken place at t_0. If the condition were not retroactive, the transfer would be assumed to take place only at the time in which the condition takes place, namely at t_1.

The retroactivity of a condition affects subsequent transactions dependant on the conditioned arrangement. Let us assume that 1 sells the land to 2 at t_0, under the suspensive condition that the building permit is granted, 2 unconditionally sells to 3 at t_1, and the permit is issued at t_2. The retroactivity of the suspensive condition means that 2's sale to 3 is effective, since it is assumed that 2 owned the property at the time t_1 of 2's sale to 3. On the other hand, if the permit were not granted within the deadline, 2's sale to 3 would be ineffective, since it would be assumed that 2 did not own the property at t_2.

In the case of a retroactive suspensive condition, while the condition is still pending —i.e., before the condition either is realized, or definitely fails— the law assumes that the conditioned legal arrangement has not been constituted. For instance, in our example, it is assumed that no transfer has taken place, i.e., that the seller has remained the legitimate owner. Therefore, while the condition is pending, seller 1 can exercise the rights that pertain to an owner, e.g., maintain his possession of the land and use all legal remedies against trespassers and other infringers of property rights.

If a resolutive condition is retroactive, the conditioned arrangement is cancelled from the beginning. Let us assume that at time t_0 a contract is executed according to which 1 sells a piece of land to 2, subject to the retroactive resolutive condition that the contract will be cancelled if a building permit is not provided within a year. In this case it is assumed that the transfer takes place immediately (that 2 becomes the owner at t_0). However, if the permit is not issued within a year (the condition is realized), the transfer is retroactively cancelled (it is assumed that 1 has remained the owner without interruption).

Let us assume that 1 sells to 2 under the retroactive condition at t_0, 2 unconditionally sells to 3 at t_1, and the condition is realized at t_2 (a year elapses without the building permission being issued). The retroactivity of the condition means that it is assumed that the original transfer from 1 to 2 is ineffective, so that 2 was not the

---

[1] We assume in our examples a legal regime, such as Italian law, in which the transfer of property in a sale contract does not require the delivery of the thing (as it is the case in German law).

owner of the land at t_1, when he sold it to 3. Consequently, the transfer from 2 to 3 is also ineffective.

Whether conditions are by default retroactive or not depends upon the applicable national orders. Whereas French and Italian law have retroactivity by default, German law assumes that conditions are non-retroactive. Legal systems also contain regulations that address the time in which the condition is still pending, namely, it is still uncertain whether the condition will be realized. For instance, the party that has purchased a property under a suspensive condition, may take some legal initiatives to limit the risk that the property is destroyed before the realization of the condition.

2.5 How to model retroactivity

In the history of legal thinking two main theories have been proposed for the retroactivity of legal conditions (for a historical discussion on the retroactivity of conditions, see [2]).

The first theory, that goes back to the medieval jurist Bartolus, views the retroactive effect of conditions as based on a fiction: the law makes the fictive (false but binding) assumption that the condition, once realized, was effective at the time in which the conditional disposition was enacted. For all practical purposes lawyers and citizens have to reason and behave pretending to be in the fictitious history in which the conditioned arrangement held since that time. In the example above, even if in reality 1's sale to 2 could not be effective at t_0 and so at t_1, we pretend to be in a fictitious history in which the sale from 1 to 2 took place at t_0.

The second theory —which goes back to G.W. Leibniz [14, 1]— views the retroactive effect of conditions as immediate and real. Thus, in case the condition should obtain, the conditioned arrangement holds (in suspensive condition) or does not hold (in resolutive conditions) from the time in which the conditional disposition was executed. Leibniz advanced this view in his early legal work (see [3]), when he was only 20 year old, but his approach to legal conditions is consistent with his later philosophy. This approach may indeed be linked to the principle of sufficient reason [18] that characterizes his mature metaphysics: everything that happens is determined by a chain of reasons (or causes) though such reasons may be inaccessible to human cognition, while being known to God. This is the case also for the future realization of a condition: if the condition will take place in the future, it is determined, from a divine perspective, that it will take place, and so its retroactive effect is immediate. However, the realization of the condition remains contingent, when viewed from the human stance. This idea is concisely expressed by the following famous statement in the Monadology (section 22): "And as every present state of a simple substance is a natural consequence of its preceding state, so is its present pregnant with the future." [19, 96]. In the Theodicy (section 360), the idea is developed as follows:

> It is one of the rules of my system of general harmony, that the present

is big with the future, and that he who sees all sees in that which is that which shall be. What is more, I have proved conclusively that God sees in each portion of the universe the whole universe, owing to the perfect connection of things. He is infinitely more discerning than Pythagoras, who judged the height of Hercules by the size of his footprint. There must therefore be no doubt that effects follow their causes determinately, in spite of contingency and even of freedom, which nevertheless coexist with certainty or determination. [19, 97]

A third ontology of retroactivity is also possible, namely the view that, as time goes by, the past may change. Before the realization of the retroactive suspensive condition, the legal effect does not hold in the time interval between the enactment of the suspensive conditional disposition and the verification of the condition. After the verification of the condition, the legal effect holds also in that interval. Thus, in this approach, the truth of a proposition about the past has different truth values depending on the time in which the proposition is asserted (see [9, 8], who address retroactive legislation).

Here we will shall capture retroactivity through a branching-time model corresponding to Leibniz's approach to legal conditions, i.e., with the idea of the immediate effect of retroactive conditions. In our model, the histories in which a retroactive legal disposition —making arrangement ψ dependent upon condition φ— is enacted at a time t, differ depending on whether the condition takes place. If the suspensive condition φ happens at a subsequent time t', the conditioned legal arrangement ψ holds from t. If the suspensive condition φ fails to happen, the conditioned legal arrangement ψ never holds.

Thus, a history h_1 in which φ happens is paralleled by a history h_2 in which φ fails to happen that is undivided from h_1 up to the time t' of the realization of the condition. The two histories differ on the legal arrangements that are dependent on the conditions (and on the implications of these arrangements). In this situation, there is an indeterminacy concerning the conditioned arrangement, that is only removed at the time of the realization (or failure) of the condition.

We shall complement Leibniz's idea of the immediate effect of retroactive conditions, with the following observation. While the realisation of a retroactive condition still is a contingent future event, the law addresses this indeterminacy through a presumption: regardless of whether the conditioned effect holds or does not hold (depending on the realisation of the condition), the law presumes —i.e., it assumes for all practical purposes— that the conditioned arrangement does not hold (for suspensive conditions) or holds (for resolutive conditions).

This means that while it is possible that an arrangement under suspensive condition already holds (this is the case in those histories in which the condition will be realized), at the time in which the condition is still pending, we have to presume that the arrangement does not hold.

Similarly, while it remains possible that the arrangement under resolutive condition does not really hold, at the time in which the condition is still pending we have to presume that the arrangement does hold.

Assume, for instance that at t_0, 1 sells a piece of land to 2, under the suspensive condition that a building permit is granted, and that the permit will indeed be granted, at the subsequent time t_1. At t_0 it is still uncertain whether the permit will be granted or not. In such a context, we shall say that the transfer of property (the suspended legal effect) already takes place at t_0, but that at t_0, while the condition still is uncertain, the law presumes that effect has not taken place, since it has chosen to take into consideration (to presume) only this possibility.

3 Logical framework

In this section, we present the language and the semantics of the logic STIT-P (STIT logic with presumption). STIT logic (the logic of *seeing to it that*) by Belnap et al. [5] is one of the most prominent formal accounts of agency. It is the logic of sentences of the form "the agent i (or the group of agents G) sees to it that φ is true". Two variants of STIT have been studied in the literature: 'atemporal STIT' and 'temporal STIT'. At the syntactic level, the former corresponds to the family of languages for expressing properties of individual and group agency with no temporal operators. Notable examples are the languages studied by [4, 17, 11, 10]. The latter corresponds to extensions of atemporal STIT languages by temporal operators for expressing properties of agency in connection with time such as the temporal operator 'next' of linear temporal logic [6, 16][2] as well as future and past tense operators of basic tense logic [13, 15, 23].

STIT-P consists in extending the language of the temporal STIT presented in [16] by a new modal operator for 'presumption'.

3.1 Language

Let Atm be a countable set of atomic propositions denoting atomic facts and let Agt be a finite set of agents. The language of STIT-P, denoted by $\mathcal{L}_{\mathsf{STIT\text{-}P}}(Atm, Agt)$, is defined by the following grammar in Backus-Naur form (BNF):

$$\varphi ::= p \mid \neg\varphi \mid \varphi \wedge \psi \mid \mathsf{X}\varphi \mid \mathsf{Y}\varphi \mid [i]\varphi \mid [Agt]\varphi \mid \Box\varphi \mid \mathsf{P}\varphi$$

where p ranges over Atm and i ranges over Agt. The other boolean operators are defined from negation and conjunction in the usual way.

The formulas $\mathsf{X}\varphi$, $\mathsf{Y}\varphi$, $[i]\varphi$, $[Agt]\varphi$, $\Box\varphi$ and $\mathsf{P}\varphi$ have to be read as follows:

---

[2]The main feature of the language studied by [6] is that the temporal operator 'next' and the agency operator are fused into a single operator. In the languages studied by [20, 16] they are kept separated.

- $X\varphi$: "φ will be true in the next state along the current history",
- $Y\varphi$: "φ was true in the previous state along the current history",
- $[i]\varphi$: "agent i sees to it that φ, regardless of what the others choose",
- $[Agt]\varphi$: "all agents together see to it that φ",
- $\Box\varphi$: "it is historically necessary that φ", and
- $P\varphi$: "φ is legally presumed to be true".

The duals of the operators $\Box$, $\langle i \rangle$ and $\langle Agt \rangle$ are defined as follows:

$$\Diamond\varphi \stackrel{\text{def}}{=} \neg\Box\neg\varphi$$

$$\langle i \rangle\varphi \stackrel{\text{def}}{=} [i]\varphi$$

$$\langle Agt \rangle\varphi \stackrel{\text{def}}{=} [Agt]\varphi$$

where $\Diamond\varphi$, $\langle i \rangle\varphi$ and $\langle Agt \rangle\varphi$ have to be read, respectively, "it is historically possible that φ", "φ may result from agent i's actual choice" and "φ may result from the agents' actual choices".

3.2 Semantics

Different semantics for STIT have been proposed in the literature (see [7] for a recent systematic analysis and comparison of these different semantics). The original semantics of STIT by Belnap et al. [5] is defined in terms of $BT+AC$ structures: branching-time structures (BT) augmented by agent choice functions (AC). A BT structure is made of a set of moments and a tree-like ordering over them. An AC for an agent i is a function mapping each moment m into a partition of the set of histories passing through that moment, a history h being a maximal set of linearly ordered moments and the equivalence classes of the partition being the possible choices for agent i at moment m.

Kripke-style semantics for STIT have been proposed by [17] for non-necessarily discrete time and by [16] for discrete time. We here present a semantics for STIT-P based on the following concept of agentive structure with discrete time that is well-suited to formalize the concept of condition given its explicit representation of discrete time. It turns out that this semantics and the discrete-time Kripkean semantics for temporal STIT presented in [16] are equivalent relative to the language $\mathcal{L}_{\text{STIT-P}}(Atm, Agt)$ under consideration. On the conceptual side, the main difference between this semantics for STIT and Belnap et al.'s $BT+AC$ semantics is that the former takes the concept of *history* as a primitive instead of the concept of *moment* and defines: (i) a *moment* as an equivalence class induced by a certain equivalence relation over the set of histories, and (ii) an agent i's set of *choices* at a moment as a partition of that moment.

Definition 1 (Agentive structure with discrete time). *An agentive structure with discrete time (ASDT) is a tuple*

$$M = (H, (\sim_n)_{n \in \mathbb{N}}, (\sim_{\langle n,i \rangle})_{n \in \mathbb{N}, i \in Agt}, (\sim_{\langle n,Agt \rangle})_{n \in \mathbb{N}}, \mathcal{V})$$

where:

- *H is a set of histories;*

- *all relations $\sim_n$, $\sim_{\langle n,i \rangle}$ and $\sim_{\langle n,Agt \rangle}$ are equivalence relations on H that satisfy the following conditions:*

    (C1) *for all $n \in \mathbb{N}$ and $i \in Agt$: $\sim_{\langle n,i \rangle} \subseteq \sim_n$,*

    (C2) *for all $n \in \mathbb{N}$ and $h_1, \ldots, h_n \in H$: if $h_i \sim_n h_j$ for all $i,j \in \{1, \ldots, n\}$ then $\bigcap_{i \in Agt} \sim_{\langle n,i \rangle}(h_i) \neq \emptyset$,*

    (C3) *for all $n \in \mathbb{N}$: $\sim_{\langle n,Agt \rangle} = \bigcap_{i \in Agt} \sim_{\langle n,i \rangle}$,*

    (C4) *for all $m,n \in \mathbb{N}$ and $h,h' \in H$: if $h \sim_n h'$ and $m < n$ then $h \sim_{\langle m,Agt \rangle} h'$, and*

- *$\mathcal{V} : \mathbb{N} \times H \longrightarrow 2^{Atm}$ is a valuation function for atomic propositions.*

An ASDT is defined by a set of histories H. The truth value of an atomic proposition depends on the time point n along a given history h. In particular, proposition p is true at time point n along the history h if and only if $p \in \mathcal{V}(n,h)$. The equivalence relation $\sim_n$ defines the historical alternatives of a history at the time point n. Specifically, if $h \sim_n h'$ then, at time point n, h' is a historic alternative of h.

The equivalence relations $\sim_n$ also define the set of moments Mom. In particular, $Mom = \bigcup_{n \in \mathbb{N}} H/\sim_n$ where $H/\sim_n$ is the quotient set of H by the equivalence relation $\sim_n$. Elements of Mom are denoted by $m, m', \ldots$

The equivalence relations $\sim_{\langle n,i \rangle}$ and $\sim_{\langle n,Agt \rangle}$ define, respectively, agent i's choices at the time point n and the collective choices of all agents at time point n. Specifically, if $h \sim_{\langle n,i \rangle} h'$ then histories h and h' belong to the same choice of agent i at time n. If $h \sim_{\langle n,i \rangle} h'$ then histories h and h' belong to the same collective choice of all agents at time n.

Constraint C1 just means that an agent can only choose among possible alternatives. This constraint ensures that, for every history h, the equivalence relation $\sim_{\langle n,i \rangle}$ induces a partition of the moment $\sim_n(h)$. An element of this partition is a choice that is possible (or available) for agent i at moment $\sim_n(h)$. Constraint C2 expresses the so-called assumption of independence of agents or independence of choices. Intuitively, this means that agents can never be deprived of choices due to the choices made by other agents. Constraint C3 just says that the collective choice of the grand coalition Agt is equal to the intersection of the choices of all individuals.

Finally, Constraint C4 corresponds to the so-called property of no choice between undivided histories. It captures the idea that if two histories come together in some future moment then, in the present, each agent does not have a choice between these two histories. This implies that if an agent can choose between two histories at a later stage, then she does not have a choice between them in the present.

The STIT semantics assumes that agents' choices may have non-deterministic effects, that is to say, it is not necessarily the case that the consequences of the agents' current choices are unequivocally determined. In formal terms, it could be the case that $h \sim_{\langle n, Agt \rangle} h'$ and $h \not\sim_{n+1} h'$. The latter means that h and h' belong to the same collective choice at time n but do not belong to the same moment at time $n+1$.

The following definition extends the concept of ASDT by the concept of presumption.

Definition 2 (Agentive structure with discrete time and presumption). *An agentive structure with discrete time and presumption (ASDTP) is a tuple $M = (H, (\sim_n)_{n \in \mathbb{N}}, (\sim_{\langle n,i \rangle})_{n \in \mathbb{N}, i \in Agt}, (\sim_{\langle n, Agt \rangle})_{n \in \mathbb{N}}, \mathcal{P}, \mathcal{V})$ where*

- $M = (H, (\sim_n)_{n \in \mathbb{N}}, (\sim_{\langle n,i \rangle})_{n \in \mathbb{N}, i \in Agt}, (\sim_{\langle n, Agt \rangle})_{n \in \mathbb{N}}, \mathcal{V})$ *is an ASDT and*

- $\mathcal{P}$ *is a legal presumption function mapping every moment $m \in Mom$ to a non-empty set of histories $\mathcal{P}(m) \subseteq m$ passing through it.*

The idea is that the law, when making a presumption, requires legal reasoners —in particular legal decision-makers— to take into account only a subset of the possible histories (those passing though a given moment), i.e., it requires them to consider only the histories that are compatible with what is presumed. The remaining histories passing through the given moment are excluded from the law's perspective.

Thus, intuitively, $\mathcal{P}(m)$ is the set of possible histories that the law takes into consideration at moment m, i.e., the set of histories that are compatible with what law presumes at moment m. Conversely, $m \setminus \mathcal{P}(m)$ are the histories that are excluded from the law's consideration, being incompatible with what law presumes at moment m.

Accordingly, we can define satisfaction as a relation between formulas of the language $\mathcal{L}_{\mathsf{STIT\text{-}P}}(Atm, Agt)$ and pointed ASDTPs.

In particular, let $M = (H, (\sim_n)_{n \in \mathbb{N}}, (\sim_{\langle n,i \rangle})_{n \in \mathbb{N}, i \in Agt}, (\sim_{\langle n, Agt \rangle})_{n \in \mathbb{N}}, \mathcal{V})$ be a AS-

DTP and let $(n,h) \in \mathbb{N} \times H$, then:

$$M, (n,h) \models p \iff p \in \mathcal{V}(n,h)$$
$$M, (n,h) \models \neg \varphi \iff M, (n,h) \not\models \varphi$$
$$M, (n,h) \models \varphi \wedge \psi \iff M, (n,h) \models \varphi \text{ and } M, (n,h) \models \psi$$
$$M, (n,h) \models \mathsf{X}\varphi \iff M, (n+1,h) \models \varphi$$
$$M, (n,h) \models \mathsf{Y}\varphi \iff \text{if } n > 0 \text{ then } M, (n-1,h) \models \varphi$$
$$M, (n,h) \models \Box\varphi \iff \forall h' \in H : \text{if } h \sim_n h' \text{ then } M, (n,h') \models \varphi$$
$$M, (n,h) \models [i]\varphi \iff \forall h' \in H : \text{if } h \sim_{\langle n,i \rangle} h' \text{ then } M, (n,h') \models \varphi$$
$$M, (n,h) \models [Agt]\varphi \iff \forall h' \in H : \text{if } h \sim_{\langle n,Agt \rangle} h' \text{ then } M, (n,h') \models \varphi$$
$$M, (n,h) \models \mathsf{P}\varphi \iff \forall h' \in \mathcal{P}(\sim_n(h)) : M, (n,h') \models \varphi$$

We say that a formula φ of the language $\mathcal{L}_{\mathsf{STIT\text{-}P}}(Atm, Agt)$ is satisfiable if there exists an ASDTP M and a pair (n,h) such that $M, (n,h) \models \varphi$. We say that φ is valid, denoted by $\models \varphi$, if $\neg \varphi$ is not satisfiable.

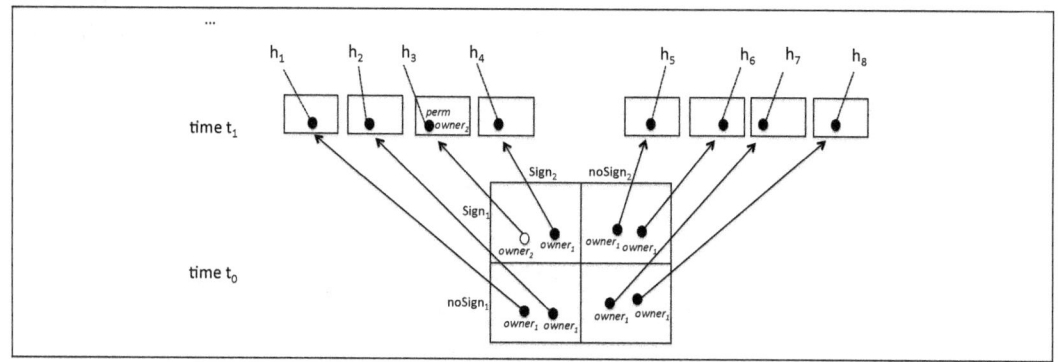

Figure 1: Example of ASDTPs.

Figure 1 illustrates the semantics of the logic STIT-P with the aid of a concrete example where two agents 1 (the seller) and 2 (the buyer) sign a contract. The contract has a retroactive suspensive condition stating that 2 will become the owner of the land ($owner_2$) only if a building permit is issued ($perm$). In the figure, rectangles correspond to moments and arrows represent histories. Black points correspond to histories that are compatible with what law presumes, while white points correspond to histories that are incompatible with what law presumes at a given moment. An agent's set of available choices at a given moment defines a partition of the moment. For instance, at moment $\sim_{t_0}(h_1)$, agents 1 and 2 have two choices available, as each of them can decide either to sign or not to sign the contract. Agent 1's choices correspond to rows (the horizontal partition), while agent 2's correspond to columns (the vertical partition).

At moment $\sim_{t_0}(h_1)$, it is uncertain whether the condition $perm$ will occur in the next state. Indeed, there are two distinct histories h_3 and h_4 passing through this moment such that the condition *perm* obtains in the former but does not obtain in the latter. In the light of this uncertainty, at moment $\sim_{t_0}(h_1)$, it is legally presumed that the condition will not occur and, consequently, that agent 2 is not yet the owner of the land. Indeed, $owner_2$ is false at all histories that are compatible with what law presumes at moment $\sim_{t_0}(h_1)$.

4 Formalization of conditions

In this section, we use the logic STIT-P to formalize the concepts of suspensive and resolutive conditions.

We focus on conditions occurring in the 'next' state. We will discuss conditions occurring before a certain deadline at the end of the section.

We say that φ is a suspensive condition for ψ (the effect of the condition), denoted by $\mathsf{Susp}(\varphi,\psi)$, if and only if the following requirements are satisfied:

- *Uncertainty requirement*: given the agents' actual choices, it is uncertain whether φ will occur in the next state.

- *Realization requirement*: given the agents' actual choices, if in the next state the condition φ is realized then, from the law's perspective, the condition φ is realized.

- *Conditional requirement*: given the agents' actual choices, the occurrence of φ in the next state is a necessary and sufficient condition for ψ to become true.

According to the uncertainty and conditional requirements, φ is a condition for ψ only if φ is a future and uncertain event on which the occurrence of ψ depends. According to the realization requirement, the law must keep track of the realization of the condition, that is to say, if the condition φ obtains then φ has to be true at all histories that are compatible with what the law presumes. The realization requirement is fundamental to capture the connection between the realization of a condition and its legal effects. Indeed, for a condition to be effective, the law must presume that it obtains and, consequently, that its legal consequences are in place.

This leads to the following formal definition of suspensive condition:

$$\mathsf{Susp}(\varphi,\psi) \stackrel{\text{def}}{=} (\langle Agt\rangle \mathsf{X}\varphi \wedge \langle Agt\rangle \neg \mathsf{X}\varphi) \wedge$$
$$([Agt]\mathsf{X}(\varphi \to \mathsf{P}\varphi)) \wedge$$
$$([Agt]\mathsf{X}(\varphi \leftrightarrow \psi))$$

where formula $\langle Agt\rangle \mathsf{X}\varphi \wedge \langle Agt\rangle \neg \mathsf{X}\varphi$ represents the uncertainty requirement, formula $[Agt]\mathsf{X}(\varphi \to \mathsf{P}\varphi)$ represents the realization requirement and formula $[Agt]\mathsf{X}(\varphi \leftrightarrow \psi)$ represents the conditional requirement.

Notice that formula $\langle Agt \rangle \mathsf{X}\varphi \wedge \langle Agt \rangle \neg \mathsf{X}\varphi$ represents *ex-post* uncertainty, i.e., uncertainty after the agents have made their choices and have publicly revealed them. This means that in the example given in Section 3.2, the fact that the building permit is issued becomes a condition for agent 2 to be the owner of the land after agents 1 and 2 have signed the contract. Similarly, $[Agt]\mathsf{X}(\varphi \leftrightarrow \psi)$ represents the fact that φ is a necessary and sufficient condition for ψ only with respect to the agents' current choices. Again in the example of Section 3.2, we assume that the issuing of the building permit is necessary and sufficient for agent 2 to become the owner of the house, under the proviso that 1 and 2 have signed the contract.

Let us consider suspensive conditions with retroactive effects. We say that φ is a suspensive condition with retroactive effect ψ (the effect of the condition), denoted by $\mathsf{RetrSusp}(\varphi,\psi)$, if and only if the following requirements are satisfied.

- *Condition requirement*: φ is a suspensive condition for making ψ true in the past.

- *Presumption requirement*: it is legally presumed that ψ is false.

According to the condition requirement, a suspensive condition with retroactive effect ψ is a special suspensive condition whose effect is $\mathsf{Y}\psi$. The presumption requirement is needed since a suspensive condition with retroactive effect ψ implies the uncertainty whether ψ is true in the present. Therefore, as explained in Section 2.4, by presuming that ψ is false the law proceeds *as if* the conditioned legal arrangement has not been constituted. In the example, of Section 3.2, before the building permit is issued it is legally presumed that agent 2 is not yet the owner of the land.

In formal terms, we have:

$$\mathsf{RetrSusp}(\varphi,\psi) \stackrel{\text{def}}{=} \mathsf{Susp}(\varphi,\mathsf{Y}\psi) \wedge \mathsf{P}\neg\psi$$

where formula $\mathsf{Susp}(\varphi,\mathsf{Y}\psi)$ represents the condition requirement and formula $\mathsf{P}\neg\psi$ represents the presumption requirement. It is easy to check that in the structure depicted in Figure 1 the following holds:

$$M,(1,h_1) \models \mathsf{Susp}(perm,owner_2) \wedge \mathsf{RetrSusp}(perm,owner_2)$$

This means that, at time 1 along the history h_1, the fact that the building permit is issued is both a suspensive condition for agent 2 to become the owner of the land in the future and a retroactive suspensive condition for agent 2 to become the owner of the land in the present.

Before considering resolutive conditions, we discuss the following properties of suspensive conditions:

$$\models \mathsf{RetrSusp}(\varphi,\psi) \to \mathsf{Susp}(\varphi,\mathsf{Y}\psi) \qquad (1)$$

$$\models (\mathsf{Susp}(\varphi,\psi) \wedge \mathsf{X}\varphi) \to \mathsf{X}(\psi \wedge \mathsf{P}\psi) \qquad (2)$$

$$\models (\mathsf{RetrSusp}(\varphi,\psi) \wedge \mathsf{X}\varphi) \to \mathsf{X}(\mathsf{Y}\psi \wedge \mathsf{PY}\psi) \qquad (3)$$

The first validity follows straightforwardly from the definition of suspensive condition with retroactive effects. The second and third validities are detachment principles for suspensive conditions: if φ is a suspensive condition for ψ and φ will occur tomorrow then, tomorrow it has to be the case that ψ and it has to be presumed that ψ. Similarly, if φ is a suspensive condition with retroactive effect ψ and φ will occur tomorrow then, tomorrow it has to be the case that ψ was true yesterday and it has to be presumed that ψ was true yesterday.

A resolutive condition is nothing but the reverse of a suspensive one. Indeed, a suspensive condition is a condition on the occurrence of which the *constitution* of a legal arrangement depends, while a resolutive condition is a condition on the occurrence of which the *termination* of a legal arrangement depends. In formal terms, we have:

$$\mathsf{Resol}(\varphi,\psi) \stackrel{\text{def}}{=} \mathsf{Susp}(\varphi,\neg\psi)$$

where $\mathsf{Resol}(\varphi,\psi)$ means that φ is a resolutive condition for ψ. Similarly, for the retroactive case, we have:

$$\mathsf{RetrResol}(\varphi,\psi) \stackrel{\text{def}}{=} \mathsf{RetrSusp}(\varphi,\neg\psi)$$

where $\mathsf{RetrResol}(\varphi,\psi)$ means that φ is a resolutive condition with retroactivity with respect to ψ.

As emphasized at the beginning of the section, the previous definitions of suspensive and resolutive conditions are relative to the 'next' state. A generalization of these notions consists in defining suspensive and regulative conditions relative to a deadline n, that is to say, a fact φ that has to occur before n steps from now to make ψ true (for suspensive condition) and to make ψ false (for resolutive condition).

Let us introduce the following inductive definition to represent the occurrence of an event in n steps from now:

$$\mathsf{X}^0 \varphi \stackrel{\text{def}}{=} \varphi$$

and for all $n > 0$:

$$\mathsf{X}^n \varphi \stackrel{\text{def}}{=} \mathsf{XX}^{n-1} \varphi$$

where $X^n\varphi$ has to be read "φ will be true in n steps from now along the current history". Furthermore, let us define:

$$X^{\exists n}\varphi \stackrel{def}{=} \bigvee_{0 \leq m \leq n} X^m \varphi$$

$$X^{\forall n}\varphi \stackrel{def}{=} \bigwedge_{0 \leq m \leq n} X^m \varphi$$

where $X^{\exists n}\varphi$ and $X^{\forall n}\varphi$ have to be read, respectively, "φ will eventually be true before the deadline n" and "φ will always be true before the deadline n".

We can now define suspensive condition and resolutive condition with deadlines:

$$\mathsf{Susp}(\varphi,\psi,n) \stackrel{def}{=} (\langle Agt\rangle X^{\exists n}\varphi \wedge \langle Agt\rangle \neg X^{\exists n}\varphi) \wedge$$
$$([Agt]X^{\forall n}(\varphi \to P\varphi)) \wedge$$
$$([Agt]X^{\forall n}(\varphi \leftrightarrow \psi))$$

$$\mathsf{Resol}(\varphi,\psi,n) \stackrel{def}{=} \mathsf{Susp}(\varphi,\neg\psi,n)$$

where $\mathsf{Susp}(\varphi,\psi,n)$ means that φ is a suspensive condition for ψ with deadline n and $\mathsf{Resol}(\varphi,\psi,n)$ means that φ is a resolutive condition for ψ with deadline n. According to the new formulation of the uncertainty requirement, uncertainty is about the occurrence of the condition φ before the deadline. Moreover, according to the new formulation of the conditional requirement, the occurrence of φ at every point before the deadline n is a necessary and sufficient condition to make ψ true there.

We have the following generalization of the detachment principle given above:

$$\models (\mathsf{Susp}(\varphi,\psi,n) \wedge X^m \varphi) \to X^m(\psi \wedge P\psi) \text{ if } 0 \leq m \leq n \qquad (4)$$

According to the previous validity, if φ is a suspensive condition for ψ with deadline n and φ will occur at time m before the deadline then, at time m it has to be the case that ψ and it has to be presumed that ψ.

5 Conclusion

We believe that our approach to conditions can be useful not only for computational models of contracts but also for legal doctrine. Our model of retroactivity —as resulting from the conditional possibility of alternative outcomes, coupled with a presumption— provides a clear understanding of the working of retroactive conditions, without the need to postulate a fiction, or to assume backward causality. It provides a better understanding of retroactivity, which may support easier legal solutions of the resulting issues.

As a follow-up of this work, we plan to provide a sound and complete axiomatization for the logic STIT-P presented in Section 3. We also plan to study complexity of model checking for this logic and to come up with a model checking algorithm for it. Indeed, we believe that our formalization of conditions can be exploited in practice via model checking to automatically verify whether, in a certain situation, a certain condition will obtain and/or a certain legal effect will be produced.

References

[1] M. Armgardt. *Das rechtslogische System der "Doctrina Conditionum" von Gottfried Wilhelm Leibniz*. Elwert, 2001.

[2] M. Armgardt. Zur Rückwirkung der Bedingung im klassischen römischen Recht und zum stoischen Determinismus. *The Legal History Review*, 78:341–349, 2010.

[3] Alberto Artosi and Giovanni Sartor. Leibniz as a jurist. In Maria Rosa Antognazza, editor, *The Oxford Handbook of Leibniz*, pages 1–29. Oxford University Press, 2016.

[4] P. Balbiani, A. Herzig, and N. Troquard. Alternative axiomatics and complexity of deliberative STIT theories. *Journal of Philosophical Logic*, 37(4):387–406, 2008.

[5] N. Belnap, M. Perloff, and M. Xu. *Facing the future: agents and choices in our indeterminist world*. Oxford University Press, New York, 2001.

[6] J. Broersen. Deontic epistemic stit logic distinguishing modes of mens rea. *Journal of Applied Logic*, 9(2):137–152, 2011.

[7] R. Ciuni and E. Lorini. Comparing semantics for temporal stit logic. *Logique & Analyse*, 96(3), 2017.

[8] G. Governatori and A. Rotolo. Changing legal systems: legal abrogations and annulments in defeasible logic. *Logic Journal of IGPL*, 18:157–94, 2010.

[9] G. Governatori, A. Rotolo, R. Riveret, M. Palmirani, and G. Sartor. Back to the future: Variants of temporal defeasible logics for modelling norm modifications. In *Proceedings of Eleventh International Conference on Artificial Intelligence and Law*, pages 155–9. ACM, 2007.

[10] D. Grossi, E. Lorini, and F. Schwarzentruber. The ceteris paribus structure of logics of game forms. *Journal of Artificial Intelligence Research*, 53:91–126, 2015.

[11] A. Herzig and F. Schwarzentruber. Properties of logics of individual and group agency. In C. Areces and R. Goldblatt, editors, *Advances in Modal Logic 7, papers from the seventh conference on "Advances in Modal Logic," held in Nancy, France, 9-12 September 2008*, pages 133–149. College Publications, 2008.

[12] Martin Hogg. *Obligations: Law and Language*. Cambridge University Press, 2017.

[13] J. F. Horty. *Agency and Deontic Logic*. Oxford Univ. Press, Oxford, 2001.

[14] G. W. Leibniz. Specimen certitudinis seu demonstrationum in jure exhibitum in doctrinam conditionum. In *Sämtliche Schriften und Briefe, sixth series, volume 1*. Akademie, [1669] 1923.

[15] E. Lorini. Temporal STIT logic and its application to normative reasoning. *Journal of Applied Non-Classical Logics*, 23(4):372–399, 2013.

[16] E. Lorini and G. Sartor. A STIT logic for reasoning about social influence. *Studia Logica*, 104(4):773–812, 2016.

[17] E. Lorini and F. Schwarzentruber. A logic for reasoning about counterfactual emotions. *Artificial Intelligence*, 175(3-4):814–847, 2011.

[18] Yitzhak Y. Melamed and Martin Lin. Principle of sufficient reason. In Edward N. Zalta, editor, *The Stanford Encyclopedia of Philosophy (Spring 2018 Edition)*. 2018.

[19] Nicholas Rescher. *G. W. Leibniz's Monadology: An Edition for Students*. University of Pittsburgh Press, 1991.

[20] F. Schwarzentruber. Complexity results of STIT fragments. *Studia Logica*, 100(5):1001–1045, 2012.

[21] Study Group on a European Civil Code, Research Group on EC Private Law (Acquis Group). *Principles, Definitions and Model Rules of European Private Law: Draft Common Frame of Reference (DCFR), Outline Edition*. Sellier, 2009.

[22] Alan Watson. *The Digest of Justinian*. University of Pennsylvania Press, 1985.

[23] S. Wölf. Propositional q-logic. *Journal of Philosophical Logic*, 31:387–414, 2002.

A Dyadic Deontic Logic in HOL

Christoph Benzmüller
University of Luxembourg, Luxembourg, and Freie Universität Berlin, Germany
c.benzmueller@gmail.com

Ali Farjami
University of Luxembourg, Luxembourg
farjami110@gmail.com

Xavier Parent
University of Luxembourg, Luxembourg
xavier.parent@uni.lu

A shallow semantical embedding of a dyadic deontic logic by Carmo and Jones in classical higher-order logic is presented. This embedding is proven sound and complete, that is, faithful.

The work presented here provides the theoretical foundation for the implementation and automation of dyadic deontic logic within off-the-shelf higher-order theorem provers and proof assistants.

Keywords: Logic of CTD conditionals by Carmo and Jones; Classical higher-order logic; Semantic embedding; Automated reasoning

1 Introduction

Dyadic deontic logic is the logic for reasoning with dyadic obligations ("it ought to be the case that ... if it is the case that ..."). A particular dyadic deontic logic, tailored to so-called contrary-to-duty (CTD) conditionals, has been proposed by Carmo and Jones [13]. We shall refer to it as DDL in the remainder. DDL comes with a neighborhood semantics and a weakly complete axiomatization over the class of finite models. The framework is immune to the well-known CTD paradoxes, like Chisholm's paradox [14, 19], and other related puzzles. However, the question of how to mechanise and automate reasoning tasks in DDL has not been studied yet.

This article adresses this challenge. We essentially devise a faithful semantical embedding of DDL in classical higher-order logic (HOL). The latter logic thereby

---

This work has been supported by the European Union's Horizon 2020 research and innovation programme under the Marie Skłodowska-Curie grant agreement No 690974.

serves as an universal meta-logic. Analogous to successful, recent work in the area of computational metaphysics (cf. [6] and the references therein), the key motivation is to mechanise and automate DDL on the computer by reusing existing theorem proving technology for meta-logic HOL. The embedding of DDL in HOL as devised in this article enables just this.

Meta-logic HOL [4], as employed in this article, was originally devised by Church [17], and further developed by Henkin [18] and Andrews [1, 3, 2]. It bases both terms and formulas on simply typed λ-terms. The use of the λ-calculus has some major advantages. For example, λ-abstractions over formulas allow the explicit naming of sets and predicates, something that is achieved in set theory via the comprehension axioms. Another advantage is, that the complex rules for quantifier instantiation at first-order and higher-order types is completely explained via the rules of λ-conversion (the so-called rules of α-, β-, and η-conversion) which were proposed earlier by Church [15, 16]. These two advantages are exploited in our embedding of DDL in HOL.

Different notions of semantics for HOL have been thoroughly studied in the literature [7, 20]. In this article we assume HOL with Henkin semantics and choice (cf. the detailed description by Benzmüller et. al. [7]). For this notion of HOL, which does not suffer from Gödel's incompleteness results, several sound and complete theorem provers have been developed in the past decades [9]. We propose to reuse these theorem provers for the mechanisation and automation of DDL. The semantical embedding as devised in this article provides both the theoretical foundation for the approach and the practical bridging technology that is enabling DDL applications within existing HOL theorem provers.

The article is structured as follows: Section 2 outlines DDL and Sec. 3 introduces HOL. The semantical embedding of DDL in HOL is then devised and studied in Sec. 4. This section also addresses soundness and completeness, but due to space restrictions the proofs can only be sketched here; for details we refer to [8]. Section 5 discusses the implementation and automation of the embedding in Isabelle/HOL [21] and Sec. 6 concludes the paper.

2 The Dyadic Deontic Logic of Carmo and Jones

This section provides a concise introduction of DDL, the dyadic deontic logic proposed by Carmo and Jones. Definitions as required for the remainder are presented. For further details we refer to the literature [13, 12].

To define the formulas of DDL we start with a countable set P of propositional symbols, and we choose $\neg$ and $\vee$ as the only primitive connectives.

The set of *DDL formulas* is given as the smallest set of formulas obeying the following conditions:

- Each $p^i \in P$ is an (atomic) DDL formula.

- Given two arbitrary DDL formulas φ and ψ, then

$\neg\varphi$	—	classical negation,
$\varphi \vee \psi$	—	classical disjunction,
$O(\psi/\varphi)$	—	dyadic deontic obligation: "it ought to be ψ, given φ",
$\Box\varphi$	—	in all worlds,
$\Box_a\varphi$	—	in all actual versions of the current world,
$\Box_p\varphi$	—	in all potential versions of the current world,
$O_a\varphi$	—	monadic deontic operator for actual obligation, and
$O_p\varphi$	—	monadic deontic operator for primary obligation

  are also DDL formulas.

Further logical connectives can be defined as usual: $\varphi \wedge \psi := \neg(\neg\varphi \vee \neg\psi)$, $\varphi \to \psi := \neg\varphi \vee \psi$, $\varphi \leftrightarrow \psi := (\varphi \to \psi) \wedge (\psi \to \varphi)$, $\Diamond\varphi := \neg\Box\neg\varphi$, $\Diamond_a\varphi := \neg\Box_a\neg\varphi$, $\Diamond_p\varphi := \neg\Box_p\neg\varphi$, $\top := \neg q \vee q$, for some propositional symbol q, $\bot := \neg\top$, and $O\varphi := O(\varphi/\top)$.

A DDL *model* is a structure $M = \langle S, av, pv, ob, V \rangle$, where S is a non empty set of items called possible worlds, V is a function assigning a set of worlds to each atomic formula, that is, $V(p^i) \subseteq S$. $av: S \to \wp(S)$, where $\wp(S)$ is the power set of S, is a function mapping worlds to sets of worlds such that $av(s) \neq \emptyset$. $av(s)$ is the set of actual versions of the world s. $pv: S \to \wp(S)$ is another, similar mapping such that $av(s) \subseteq pv(s)$ and $s \in pv(s)$. $pv(s)$ is the set of potential versions of the world s. $ob: \wp(S) \to \wp(\wp(S))$ is a function mapping sets of worlds to sets of sets of worlds. $ob(\bar{X})$ is the set of propositions that are obligatory in context $\bar{X} \subseteq S$. The following conditions hold for ob (where $\bar{X}, \bar{Y}, \bar{Z}$ designate arbitrary subsets of S):

1. $\emptyset \notin ob(\bar{X})$.

2. If $\bar{Y} \cap \bar{X} = \bar{Z} \cap \bar{X}$, then $\bar{Y} \in ob(\bar{X})$ if and only if $\bar{Z} \in ob(\bar{X})$.

3. Let $\bar{\beta} \subseteq ob(\bar{X})$ and $\bar{\beta} \neq \emptyset$. If $(\cap\bar{\beta}) \cap \bar{X} \neq \emptyset$ (where $\cap\bar{\beta} = \{s \in S \mid$ for all $\bar{Z} \in \bar{\beta}$ we have $s \in \bar{Z}\}$), then $(\cap\bar{\beta}) \in ob(\bar{X})$.

4. If $\bar{Y} \subseteq \bar{X}$ and $\bar{Y} \in ob(\bar{X})$ and $\bar{X} \subseteq \bar{Z}$, then $(\bar{Z} \setminus \bar{X}) \cup \bar{Y} \in ob(\bar{Z})$.

5. If $\bar{Y} \subseteq \bar{X}$ and $\bar{Z} \in ob(\bar{X})$ and $\bar{Y} \cap \bar{Z} \neq \emptyset$, then $\bar{Z} \in ob(\bar{Y})$.

Satisfiability of a formula φ for a model $M = \langle S, av, pv, ob, V \rangle$ and a world $s \in S$ is expressed by writing that $M, s \models \varphi$ and we define $V^M(\varphi) = \{s \in S \mid M, s \models \varphi\}$. In order to simplify the presentation, whenever the model M is obvious from context, we write $V(\varphi)$ instead of $V^M(\varphi)$. Moreover, we often use "iff" as shorthand for "if and only if".

$$
\begin{aligned}
M,s &\models p & \text{iff} \quad & s \in V(p) \\
M,s &\models \neg\varphi & \text{iff} \quad & M,s \not\models \varphi \text{ (that is, not } M,s \models \varphi) \\
M,s &\models \varphi \vee \psi & \text{iff} \quad & M,s \models \varphi \text{ or } M,s \models \psi \\
M,s &\models \Box\varphi & \text{iff} \quad & V(\varphi) = S \\
M,s &\models \Box_a\varphi & \text{iff} \quad & av(s) \subseteq V(\varphi) \\
M,s &\models \Box_p\varphi & \text{iff} \quad & pv(s) \subseteq V(\varphi) \\
M,s &\models \bigcirc(\psi/\varphi) & \text{iff} \quad & V(\psi) \in ob(V(\varphi)) \\
M,s &\models \bigcirc_a\varphi & \text{iff} \quad & V(\varphi) \in ob(av(s)) \text{ and } av(s) \cap V(\neg\varphi) \neq \emptyset \\
M,s &\models \bigcirc_p\varphi & \text{iff} \quad & V(\varphi) \in ob(pv(s)) \text{ and } pv(s) \cap V(\neg\varphi) \neq \emptyset
\end{aligned}
$$

Our evaluation rule for $\bigcirc(\_/\_)$ is a simplified version of the one used by Carmo and Jones. Given the constraints placed on ob, the two rules are equivalent (cf. [5, result II-2-2]).

As usual, a DDL formula φ is *valid in a DDL model* $M = \langle S, av, pv, ob, V \rangle$, i.e. $M \models^{DDL} \varphi$, if and only if for all worlds $s \in S$ we have $M, s \models \varphi$. A formula φ is *valid*, denoted $\models^{DDL} \varphi$, if and only if it is valid in every DDL model.

3 Classical Higher-order Logic

In this section we introduce classical higher-order logic (HOL). The presentation, which has partly been adapted from [5], is rather detailed in order to keep the article sufficiently self-contained.

3.1 Syntax of HOL

For defining the syntax of HOL, we first introduce the set T of *simple types*. We assume that T is freely generated from a set of *basic types* $BT \supseteq \{o, i\}$ using the function type constructor $\rightarrow$. Type o denotes the (bivalent) set of Booleans, and i a non-empty set of individuals.

For the definition of HOL, we start out with a family of denumerable sets of typed constant symbols $(C_\alpha)_{\alpha \in T}$, called the HOL *signature*, and a family of denumerable sets of typed variable symbols $(V_\alpha)_{\alpha \in T}$.[1] We employ Church-style typing, where each term t_α explicitly encodes its type information in subscript α.

The *language of HOL* is given as the smallest set of terms obeying the following conditions.

- Every typed constant symbol $c_\alpha \in C_\alpha$ is a HOL term of type α.

- Every typed variable symbol $X_\alpha \in V_\alpha$ is a HOL term of type α.

---

[1]For example in Section 4 we will assume constant symbols av, pv and ob with types $i \rightarrow i \rightarrow o$, $i \rightarrow i \rightarrow o$ and $(i \rightarrow o) \rightarrow (i \rightarrow o) \rightarrow o$ as part of the signature.

- If $s_{\alpha\to\beta}$ and t_α are HOL terms of types $\alpha \to \beta$ and α, respectively, then $(s_{\alpha\to\beta}\, t_\alpha)_\beta$, called *application*, is an HOL term of type β.

- If $X_\alpha \in V_\alpha$ is a typed variable symbol and s_β is an HOL term of type β, then $(\lambda X_\alpha s_\beta)_{\alpha\to\beta}$, called *abstraction*, is an HOL term of type $\alpha \to \beta$.

The above definition encompasses the simply typed λ-calculus. In order to extend this base framework into logic HOL we simply ensure that the signature $(C_\alpha)_{\alpha \in T}$ provides a sufficient selection of primitive logical connectives. Without loss of generality, we here assume the following *primitive logical connectives* to be part of the signature: $\neg_{o\to o} \in C_{o\to o}$, $\vee_{o\to o\to o} \in C_{o\to o\to o}$, $\Pi_{(\alpha\to o)\to o} \in C_{(\alpha\to o)\to o}$ and $=_{\alpha\to\alpha\to o} \in C_{\alpha\to\alpha\to o}$, abbreviated as $=^\alpha$. The symbols $\Pi_{(\alpha\to o)\to o}$ and $=_{\alpha\to\alpha\to o}$ are generally assumed for each type $\alpha \in T$. The denotation of the primitive logical connectives is fixed below according to their intended meaning. *Binder notation* $\forall X_\alpha\, s_o$ is used as an abbreviation for $\Pi_{(\alpha\to o)\to o}\lambda X_\alpha s_o$. Universal quantification in HOL is thus modeled with the help of the logical constants $\Pi_{(\alpha\to o)\to o}$ to be used in combination with lambda-abstraction. That is, the only binding mechanism provided in HOL is lambda-abstraction.

HOL is a logic of terms in the sense that the *formulas of HOL* are given as the terms of type o. In addition to the primitive logical connectives selected above, we could assume *choice operators* $\epsilon_{(\alpha\to o)\to\alpha} \in C_{(\alpha\to o)\to\alpha}$ (for each type α) in the signature. We are not pursuing this here.

Type information as well as brackets may be omitted if obvious from the context, and we may also use infix notation to improve readability. For example, we may write $(s \vee t)$ instead of $((\vee_{o\to o\to o} s_o) t_o)$.

From the selected set of primitive connectives, other logical connectives can be introduced as abbreviations.[2] For example, we may define $s \wedge t := \neg(\neg s \vee \neg t)$, $s \to t := \neg s \vee t$, $s \leftrightarrow t := (s \to t) \wedge (t \to s)$, $\top := (\lambda X_i X) = (\lambda X_i X)$, $\bot := \neg\top$ and $\exists X_\alpha s := \neg\forall X_\alpha \neg s$.

The notions of *free variables*, α-*conversion*, $\beta\eta$-*equality* (denoted as $=_{\beta\eta}$) and *substitution* of a term s_α for a variable X_α in a term t_β (denoted as $[s/X]t$) are defined as usual.

---

[2] As demonstrated by Andrews [4], we could in fact start out with only primitive equality in the signature (for all types α) and introduce all other logical connectives as abbreviations based on it. Alternatively, we could remove primitive equality from the above signature, since equality can be defined in HOL from these other logical connectives by exploiting Leibniz' principle, expressing that two objects are equal if they share the same properties. *Leibniz equality* $\doteq^\alpha$ at type α is thus defined as $s_\alpha \doteq^\alpha t_\alpha := \forall P_{\alpha\to o}(Ps \leftrightarrow Pt)$. The motivation for the redundant signature as selected here is to stay close to the the choices taken in implemented theorem provers such as LEO-II and Leo-III and also to theory paper [7], which is recommended for further details.

3.2 Semantics of HOL

The semantics of HOL is well understood and thoroughly documented. The introduction provided next focuses on the aspects as needed for this article. For more details we refer to the previously mentioned literature [7].

The semantics of choice for the remainder is Henkin semantics, i.e., we work with Henkin's general models [18]. Henkin models (and standard models) are introduced next. We start out with introducing frame structures.

A *frame* D is a collection $\{D_\alpha\}_{\alpha \in T}$ of nonempty sets D_α, such that $D_o = \{T, F\}$ (for truth and falsehood). The $D_{\alpha \to \beta}$ are collections of functions mapping D_α into D_β.

A *model* for HOL is a tuple $M = \langle D, I \rangle$, where D is a frame, and I is a family of typed interpretation functions mapping constant symbols $p_\alpha \in C_\alpha$ to appropriate elements of D_α, called the *denotation of* p_α. The logical connectives $\neg$, $\lor$, Π and $=$ are always given their expected, standard denotations:[3]

- $I(\neg_{o \to o}) = not \in D_{o \to o}$ such that $not(T) = F$ and $not(F) = T$.
- $I(\lor_{o \to o \to o}) = or \in D_{o \to o \to o}$ such that $or(a, b) = T$ iff ($a = T$ or $b = T$).
- $I(=_{\alpha \to \alpha \to o}) = id \in D_{\alpha \to \alpha \to o}$ such that for all $a, b \in D_\alpha$, $id(a, b) = T$ iff a is identical to b.
- $I(\Pi_{(\alpha \to o) \to o}) = all \in D_{(\alpha \to o) \to o}$ such that for all $s \in D_{\alpha \to o}$, $all(s) = T$ iff $s(a) = T$ for all $a \in D_\alpha$; i.e., s is the set of all objects of type α.

Variable assignments are a technical aid for the subsequent definition of an interpretation function $\|.\|^{M,g}$ for HOL terms. This interpretation function is parametric over a model M and a variable assignment g.

A *variable assignment* g maps variables X_α to elements in D_α. $g[d/W]$ denotes the assignment that is identical to g, except for variable W, which is now mapped to d.

The *denotation* $\|s_\alpha\|^{M,g}$ of an HOL term s_α on a model $M = \langle D, I \rangle$ under assignment g is an element $d \in D_\alpha$ defined in the following way:

$$\|p_\alpha\|^{M,g} = I(p_\alpha)$$
$$\|X_\alpha\|^{M,g} = g(X_\alpha)$$
$$\|(s_{\alpha \to \beta} t_\alpha)_\beta\|^{M,g} = \|s_{\alpha \to \beta}\|^{M,g}(\|t_\alpha\|^{M,g})$$
$$\|(\lambda X_\alpha s_\beta)_{\alpha \to \beta}\|^{M,g} = \text{the function } f \text{ from } D_\alpha \text{ to } D_\beta \text{ such that } f(d) = \|s_\beta\|^{M,g[d/X_\alpha]} \text{ for all } d \in D_\alpha$$

---

[3] Since $=_{\alpha \to \alpha \to o}$ (for all types α) is in the signature, it is ensured that the domains $D_{\alpha \to \alpha \to o}$ contain the respective identity relations. This addresses an issue discovered by Andrews [2]: if such identity relations did not existing in the $D_{\alpha \to \alpha \to o}$, then Leibniz equality in Henkin semantics might not denote as intended.

A model $M = \langle D, I \rangle$ is called a *standard model* if and only if for all $\alpha, \beta \in T$ we have $D_{\alpha \to \beta} = \{f \mid f : D_\alpha \longrightarrow D_\beta\}$. In a *Henkin model (general model)* function spaces are not necessarily full. Instead it is only required that for all $\alpha, \beta \in T$, $D_{\alpha \to \beta} \subseteq \{f \mid f : D_\alpha \longrightarrow D_\beta\}$. However, it is required that the valuation function $\|\cdot\|^{M,g}$ from above is total, so that every term denotes. Note that this requirement, which is called *Denotatpflicht*, ensures that the function domains $D_{\alpha \to \beta}$ never become too sparse, that is, the denotations of the lambda-abstractions as devised above are always contained in them.

Corollary 1. *For any Henkin model $M = \langle D, I \rangle$ and variable assignment g:*

1. $\|(\neg_{o \to o} s_o)_o\|^{M,g} = T$ *iff* $\|s_o\|^{M,g} = F$.

2. $\|((\vee_{o \to o \to o} s_o) t_o)_o\|^{M,g} = T$ *iff* $\|s_o\|^{M,g} = T$ *or* $\|t_o\|^{M,g} = T$.

3. $\|((\wedge_{o \to o \to o} s_o) t_o)_o\|^{M,g} = T$ *iff* $\|s_o\|^{M,g} = T$ *and* $\|t_o\|^{M,g} = T$.

4. $\|((\to_{o \to o \to o} s_o) t_o)_o\|^{M,g} = T$ *iff*   *(if $\|s_o\|^{M,g} = T$ then $\|t_o\|^{M,g} = T$).*

5. $\|((\longleftrightarrow_{o \to o \to o} s_o) t_o)_o\|^{M,g} = T$ *iff*   *($\|s_o\|^{M,g} = T$ iff $\|t_o\|^{M,g} = T$).*

6. $\|\top\|^{M,g} = T$.

7. $\|\bot\|^{M,g} = F$.

8. $\|(\forall X_\alpha s_o)_o\|^{M,g} = T$ *iff*   *for all $d \in D_\alpha$ we have $\|s_o\|^{M,g[d/X_\alpha]} = T$.*

9. $\|(\exists X_\alpha s_o)_o\|^{M,g} = T$ *iff*   *there exists $d \in D_\alpha$ such that $\|s_o\|^{M,g[d/X_\alpha]} = T$.*

Proof. We leave the proof as an exercise to the reader. □

An HOL formula s_o is *true* in an Henkin model M under assignment g if and only if $\|s_o\|^{M,g} = T$; this is also expressed by writing that $M, g \models^{\text{HOL}} s_o$. An HOL formula s_o is called *valid* in M, which is expressed by writing that $M \models^{\text{HOL}} s_o$, if and only if $M, g \models^{\text{HOL}} s_o$ for all assignments g. Moreover, a formula s_o is called *valid*, expressed by writing that $\models^{\text{HOL}} s_o$, if and only if s_o is valid in all Henkin models M. Finally, we define $\Sigma \models^{\text{HOL}} s_o$ for a set of HOL formulas Σ if and only if $M \models^{\text{HOL}} s_o$ for all Henkin models M with $M \models^{\text{HOL}} t_o$ for all $t_o \in \Sigma$.

Note that any standard model is obviously also a Henkin model. Hence, validity of a HOL formula s_o for all Henkin models, implies validity of s_o for all standard models.

4 Modeling DDL as a Fragment of HOL

This section, the core contribution of this article, presents a shallow semantical embedding of DDL in HOL and proves its soundness and completeness. In contrast to a deep logical embedding, where the syntax and semantics of logic L would be formalized in full detail (using structural induction and recursion), only the core differences in the semantics of both DDL and meta-logic HOL are explicitly encoded here.

4.1 Semantical Embedding

DDL formulas are identified in our semantical embedding with certain HOL terms (predicates) of type $i \to o$. They can be applied to terms of type i, which are assumed to denote possible worlds. That is, the HOL type i is now identified with a (non-empty) set of worlds. Type $i \to o$ is abbreviated as τ in the remainder. The HOL signature is assumed to contain the constant symbols $av_{i\to\tau}$, $pv_{i\to\tau}$ and $ob_{\tau\to\tau\to o}$. Moreover, for each propositional symbol p^i of DDL, the HOL signature must contain the corresponding constant symbol p^i_τ. Without loss of generality, we assume that besides those symbols and the primitive logical connectives of HOL, no other constant symbols are given in the signature of HOL.

The mapping $\lfloor \cdot \rfloor$ translates DDL formulas s into HOL terms $\lfloor s \rfloor$ of type τ. The mapping is recursively[4] defined:

$$\begin{aligned}
\lfloor p^i \rfloor &= p^i_\tau \\
\lfloor \neg s \rfloor &= \neg_\tau \lfloor s \rfloor \\
\lfloor s \vee t \rfloor &= \vee_{\tau\to\tau\to\tau} \lfloor s \rfloor \lfloor t \rfloor \\
\lfloor \Box s \rfloor &= \Box_{\tau\to\tau} \lfloor s \rfloor \\
\lfloor \bigcirc(t/s) \rfloor &= \bigcirc_{\tau\to\tau\to\tau} \lfloor s \rfloor \lfloor t \rfloor \\
\lfloor \Box_a s \rfloor &= \Box^a_{\tau\to\tau} \lfloor s \rfloor \\
\lfloor \Box_p s \rfloor &= \Box^p_{\tau\to\tau} \lfloor s \rfloor \\
\lfloor \bigcirc_a s \rfloor &= \bigcirc^a_{\tau\to\tau} \lfloor s \rfloor \\
\lfloor \bigcirc_p s \rfloor &= \bigcirc^p_{\tau\to\tau} \lfloor s \rfloor
\end{aligned}$$

$\neg_{\tau\to\tau}$, $\vee_{\tau\to\tau\to\tau}$, $\Box_{\tau\to\tau}$, $\bigcirc_{\tau\to\tau\to\tau}$, $\Box^a_{\tau\to\tau}$, $\Box^p_{\tau\to\tau}$, $\bigcirc^a_{\tau\to\tau}$ and $\bigcirc^p_{\tau\to\tau}$ thereby abbreviate the following HOL terms:

---

[4] A recursive definition is actually not needed in practice. By inspecting the equations below it should become clear that only the abbreviations for the logical connectives of DDL are required in combination with a type-lifting for the propositional constant symbols; cf. also Fig. 1.

$$\begin{aligned}
\neg_{\tau \to \tau} &= \lambda A_\tau \lambda X_i \neg (A\,X) \\
\vee_{\tau \to \tau \to \tau} &= \lambda A_\tau \lambda B_\tau \lambda X_i (A\,X \vee B\,X) \\
\Box_{\tau \to \tau} &= \lambda A_\tau \lambda X_i \forall Y_i (A\,Y) \\
O_{\tau \to \tau \to \tau} &= \lambda A_\tau \lambda B_\tau \lambda X_i (ob\,A\,B) \\
\Box^a_{\tau \to \tau} &= \lambda A_\tau \lambda X_i \forall Y_i (\neg(av\,X\,Y) \vee A\,Y) \\
\Box^p_{\tau \to \tau} &= \lambda A_\tau \lambda X_i \forall Y_i (\neg(pv\,X\,Y) \vee (A\,Y)) \\
O^a_{\tau \to \tau} &= \lambda A_\tau \lambda X_i ((ob\,(av\,X)\,A) \wedge \exists Y_i (av\,X\,Y \wedge \neg(A\,Y))) \\
O^p_{\tau \to \tau} &= \lambda A_\tau \lambda X_i ((ob\,(pv\,X)\,A) \wedge \exists Y_i (pv\,X\,Y \wedge \neg(A\,Y)))
\end{aligned}$$

Analyzing the truth of a translated formula $\lfloor s \rfloor$ in a world represented by term w_i corresponds to evaluating the application $(\lfloor s \rfloor \, w_i)$. In line with previous work [10], we define $\text{vld}_{\tau \to o} = \lambda A_\tau \forall S_i (A\,S)$. With this definition, validity of a DDL formula s in DDL corresponds to the validity of formula $(\text{vld}\,\lfloor s \rfloor)$ in HOL, and vice versa.

4.2 Soundness and Completeness

To prove the soundness and completeness, that is, faithfulness, of the above embedding, a mapping from DDL models into Henkin models is employed.

Definition 1 (Henkin model H^M for DDL model M). *For any DDL model $M = \langle S, av, pv, ob, V \rangle$, we define a corresponding Henkin model H^M. Thus, let a DDL model $M = \langle S, av, pv, ob, V \rangle$ be given. Moreover, assume that $p^i \in P$, for $i \geq 1$, are the only propositional symbols of DDL. Remember that our embedding requires the corresponding signature of HOL to provide constant symbols p^j_τ such that $\lfloor p^j \rfloor = p^j_\tau$ for $j = 1, \ldots, m$.*

A Henkin model $H^M = \langle \{D_\alpha\}_{\alpha \in T}, I \rangle$ for M is now defined as follows: D_i is chosen as the set of possible worlds S; all other sets $D_{\alpha \to \beta}$ are chosen as (not necessarily full) sets of functions from D_α to D_β. For all $D_{\alpha \to \beta}$ the rule that every term $t_{\alpha \to \beta}$ must have a denotation in $D_{\alpha \to \beta}$ must be obeyed (Denotatpflicht). In particular, it is required that D_τ, $D_{i \to \tau}$ and $D_{\tau \to \tau \to o}$ contain the elements Ip^j_τ, $Iav_{i \to \tau}$, $Ipv_{i \to \tau}$ and $Iob_{\tau \to \tau \to o}$. The interpretation function I of H^M is defined as follows:

1. *For $i = 1, \ldots, m$, $Ip^i_\tau \in D_\tau$ is chosen such that $Ip^i_\tau(s) = T$ iff $s \in V(p^j)$ in M.*

2. *$Iav_{i \to \tau} \in D_{i \to \tau}$ is chosen such that $Iav_{i \to \tau}(s, u) = T$ iff $u \in av(s)$ in M.*

3. *$Ipv_{i \to \tau} \in D_{i \to \tau}$ is chosen such that $Ipv_{i \to \tau}(s, u) = T$ iff $u \in pv(s)$ in M.*

4. *$Iob_{\tau \to \tau \to o} \in D_{\tau \to \tau \to o}$ is such that $Iob_{\tau \to \tau \to o}(\bar{X}, \bar{Y}) = T$ iff $\bar{Y} \in ob(\bar{X})$ in M.*

5. *For the logical connectives $\neg$, $\vee$, Π and $=$ of HOL the interpretation function I is defined as usual (see the previous section).*

Since we assume that there are no other symbols (besides the p^i, av, pv, ob and $\neg$, $\vee$, Π, and $=$) in the signature of HOL, I is a total function. Moreover, the above construction guarantees that H^M is a Henkin model: $\langle D, I \rangle$ is a frame, and the choice of I in combination with the Denotatpflicht ensures that for arbitrary assignments g, $\|.\|^{H^M,g}$ is an total evaluation function.

Lemma 1. *Let H^M be a Henkin model for a DDL model M. In H^M we have for all $s \in D_i$ and all $\bar{X}, \bar{Y}, \bar{Z} \in D_\tau$ (cf. the conditions on DDL models as stated on page 3).*[5]

(av) $Iav_{i \to \tau}(s) \neq \emptyset$.
(pv1) $Iav_{i \to \tau}(s) \subseteq Ipv_{i \to \tau}(s)$.
(pv2) $s \in Ipv_{i \to \tau}(s)$.
(ob1) $\emptyset \notin Iob_{\tau \to \tau \to o}(\bar{X})$.
(ob2) If $\bar{Y} \cap \bar{X} = \bar{Z} \cap \bar{X}$, then $(\bar{Y} \in Iob_{\tau \to \tau \to o}(\bar{X})$ iff $\bar{Z} \in Iob_{\tau \to \tau \to o}(\bar{X}))$.
(ob3) Let $\bar{\beta} \subseteq Iob_{\tau \to \tau \to o}(\bar{X})$ and $\bar{\beta} \neq \emptyset$.
       If $(\cap \bar{\beta}) \cap \bar{X} \neq \emptyset$, where $\cap \bar{\beta} = \{s \in S \mid$ for all $\bar{Z} \in \bar{\beta}$ we have $s \in \bar{Z}\}$,
       then $(\cap \bar{\beta}) \in Iob_{\tau \to \tau \to o}(\bar{X})$.
(ob4) If $\bar{Y} \subseteq \bar{X}$ and $\bar{Y} \in Iob_{\tau \to \tau \to o}(\bar{X})$ and $\bar{X} \subseteq \bar{Z}$,
       then $(\bar{Z} \setminus \bar{X}) \cup \bar{Y} \in Iob_{\tau \to \tau \to o}(\bar{Z})$.
(ob5) If $\bar{Y} \subseteq \bar{X}$ and $\bar{Z} \in Iob_{\tau \to \tau \to o}(\bar{X})$ and $\bar{Y} \cap \bar{Z} \neq \emptyset$,
       then $\bar{Z} \in Iob_{\tau \to \tau \to o}(\bar{Y})$.

Proof. Each statement follows by construction of H^M for M. □

Lemma 2. *Let $H^M = \langle \{D_\alpha\}_{\alpha \in T}, I \rangle$ be a Henkin model for a DDL model M. We have $H^M \models^{HOL} \Sigma$ for all $\Sigma \in \{AV, PV1, PV2, OB1, ..., OB5\}$, where*

---

[5] In the proof in [8] we implicitly employ curring and uncurring, and we associate sets with their characteristic functions. This analogously applies to the remainder of this article.

AV	is	$\forall W_i \exists V_i (av_{i\to\tau} W_i V_i)$
$PV1$	is	$\forall W_i \forall V_i (av_{i\to\tau} W_i V_i \to pv_{i\to\tau} W_i V_i)$
$PV2$	is	$\forall W_i (pv_{i\to\tau} W_i W_i)$
$OB1$	is	$\forall X_\tau \neg ob_{\tau\to\tau\to o} X_\tau (\lambda X_\tau \bot)$
$OB2$	is	$\forall X_\tau Y_\tau Z_\tau (\ (\forall W_i ((Y_\tau W_i \wedge X_\tau W_i) \longleftrightarrow (Z_\tau W_i \wedge X_\tau W_i)))$
		$\quad \to (ob_{\tau\to\tau\to o} X_\tau Y_\tau \longleftrightarrow ob_{\tau\to\tau\to o} X_\tau Z_\tau))$
$OB3$	is	$\forall \beta_{\tau\to\tau\to o} \forall X_\tau$
		$(\ (((\forall Z_\tau (\beta_{\tau\to\tau\to o} Z_\tau \to ob_{\tau\to\tau\to o} X_\tau Z_\tau)) \wedge \exists Z_\tau (\beta_{\tau\to\tau\to o} Z_\tau))$
		$\to (\ (\exists Y_i (((\lambda W_i \forall Z_\tau (\beta_{\tau\to\tau\to o} Z_\tau \to Z_\tau W_i)) Y_i) \wedge X_\tau Y_i))$
		$\to ob_{\tau\to\tau\to o} X_\tau (\lambda W_i \forall Z_\tau (\beta_{\tau\to\tau\to o} Z_\tau \to Z_\tau W_i))))$
$OB4$	is	$\forall X_\tau Y_\tau Z_\tau$
		$(\ (\forall W_i (Y_\tau W_i \to X_\tau W_i) \wedge ob_{\tau\to\tau\to o} X_\tau Y_\tau \wedge \forall X_\tau (X_\tau W_i \to Z_\tau W_i))$
		$\to ob_{\tau\to\tau\to o} Z_\tau (\lambda W_i ((Z_\tau W_i \wedge \neg X_\tau W_i) \vee Y_\tau W_i)))$
$OB5$	is	$\forall X_\tau Y_\tau Z_\tau$
		$(\ (\forall W_i (Y_\tau W_i \to X_\tau W_i) \wedge ob_{\tau\to\tau\to o} X_\tau Z_\tau \wedge \exists W_i (Y_\tau W_i \wedge Z_\tau W_i))$
		$\to ob_{\tau\to\tau\to o} Y_\tau Z_\tau)$

Proof. By construction of H^M for M in combination with Lemma 1. $\square$

Lemma 3. *Let H^M be a Henkin model for a DDL model M. For all DDL formulas δ, arbitrary variable assignments g and worlds s it holds:*

$$M, s \models \delta \text{ if and only if } \|\lfloor \delta \rfloor S_i\|^{H^M, g[s/S_i]} = T$$

Proof. By induction on the structure of δ. $\square$

Lemma 4. *For every Henkin model $H = \langle \{D_\alpha\}_{\alpha \in T}, I \rangle$ such that $H \models^{HOL} \Sigma$ for all $\Sigma \in \{AV, PV1, PV2, OB1,..., OB5\}$, there exists a corresponding DDL model M. Corresponding means that for all DDL formulas δ and for all assignments g and worlds s, $\|\lfloor \delta \rfloor S_i\|^{H, g[s/S_i]} = T$ if and only if $M, s \vDash \delta$.*

Proof. Suppose that $H = \langle \{D_\alpha\}_{\alpha \in T}, I \rangle$ is a Henkin model such that $H \models^{HOL} \Sigma$ for all $\Sigma \in \{AV, PV1, PV2, OB1,..,OB5\}$. Without loss of generality, we can assume that the domains of H are denumerable [18]. We construct the corresponding DDL model M as follows:

1. $S = D_i$,

2. $s \in av(u)$ for $s, u \in S$ iff $Iav_{i\to\tau}(s, u) = T$,

3. $s \in pv(u)$ for $s, u \in S$ iff $Ipv_{i\to\tau}(s, u) = T$,

4. $\bar{X} \in ob(\bar{Y})$ for $\bar{X}, \bar{Y} \in D_i \longrightarrow D_o$ iff $Iob_{\tau\to\tau\to o}(\bar{X}, \bar{Y}) = T$, and

5. $s \in V(p^j)$ iff $Ip^j_\tau(s) = T$.

Since $H \models^{HOL} \Sigma$ for all $\Sigma \in \{\text{AV, PV1, PV2, OB1, .., OB5}\}$, it is straightforward (but tedious) to verify that av, pv and ob satisfy the conditions as required for a DDL model.

Moreover, the above construction ensures that H is a Henkin model H^M for DDL model M. Hence, Lemma 3 applies. This ensures that for all DDL formulas δ, for all assignment g and all worlds s we have $\|\lfloor \delta \rfloor S_i\|^{H,g[s/S_i]} = T$ if and only if $M, s \models \delta$. □

Theorem 1 (Soundness and Completeness of the Embedding).

$$\models^{DDL} \varphi \text{ if and only if } \{\text{AV, PV1, PV2, OB1,..,OB5}\} \models^{HOL} vld \lfloor \varphi \rfloor$$

Proof. (Soundness, ←) The proof is by contraposition. Assume $\not\models^{DDL} \varphi$, that is, there is a DDL model $M = \langle S, av, pv, ob, V \rangle$, and world $s \in S$, such that $M, s \not\models \varphi$. Now let H^M be a Henkin model for DDL model M. By Lemma 3, for an arbitrary assignment g, it holds that $\|\lfloor \varphi \rfloor S_i\|^{H^M,g[s/S_i]} = F$. Thus, by definition of $\|.\|$, it holds that $\|\forall S_i (\lfloor \varphi \rfloor S)\|^{H^M,g} = \|vld \lfloor \varphi \rfloor\|^{H^M,g} = F$. Hence, $H^M \not\models^{HOL} vld \lfloor \varphi \rfloor$. Furthermore, $H^M \models^{HOL} \Sigma$ for all $\Sigma \in \{\text{AV, PV1, PV2, OB1,...,OB5}\}$ by Lemma 2. Thus, $\{\text{AV, PV1, PV2, OB1,..,OB5}\} \not\models^{HOL} vld \lfloor \varphi \rfloor$.

(Completeness, →) The proof is again by contraposition. Assume $\{\text{AV, PV1, PV2, OB1,..,OB5}\} \not\models^{HOL} vld \lfloor \varphi \rfloor$, that is, there is a Henkin model $H = \langle \{D_\alpha\}_{\alpha \in T}, I \rangle$ such that $H \models^{HOL} \Sigma$ for all $\Sigma \in \{\text{AV, PV1, PV2, OB1,..,OB5}\}$, but $\|vld \lfloor \varphi \rfloor\|^{H,g} = F$ for some assignment g. By Lemma 4, there is a DDL model M such that $M \not\models \varphi$. Hence, $\not\models^{DDL} \varphi$. □

Each DDL reasoning problem thus represents a particular HOL problem. The embedding presented in this section, which is based on simple abbreviations, tells us how the two logics are connected.

5 Implementation in Isabelle/HOL

The semantical embedding as devised in Sec. 4 has been implemented in the higher-order proof assistant Isabelle/HOL [21]. Figure 1 displays the respective encoding. Figure 2 applies this encoding to Chisholm's paradox (cf. [14]), which involves the following four statements:

1. It ought to be that Jones goes to assist his neighbors;

2. It ought to be that if Jones goes, then he tells them he is coming;

3. If Jones doesn't go, then he ought not tell them he is coming;

4. Jones doesn't go.

These statements can be given a consistent formalisation in DDL see Fig. 2. This is confirmed by the model finder Nitpick [11] integrated with Isabelle/HOL. Nitpick computes an intuitive, small model for the scenario consisting of two possible worlds i_1 and i_2. Function ob is interpreted in this model as follows:

$$ob(\{i_1, i_2\}) = \{\{i_1, i_2\}, \{i_1\}\}$$

$$ob(\{i_1\}) = \{\{i_1, i_2\}, \{i_1\}\}$$

$$ob(\{i_2\}) = \{\{i_1, i_2\}, \{i_2\}\}$$

$$ob(\emptyset) = \emptyset$$

The designated current world in the given model is i_2, in which Jones doesn't go to assist his neighbors and doesn't tell them that he is coming. In the other possible world i_1, Jones is going to assist them and he also tells them that he his coming. That is, $V(go) = V(tell) = \{i_1\}$. Also, we have $\{i_1\} \in ob(\{i_1, i_2\})$. So, $i_2 \models \bigcirc go$ by the evaluation rule for $\bigcirc$. Similarly, $\{i_1\} \in ob(\{i_1\})$ implies $i_2 \models \bigcirc(tell/go)$, and $\{i_2\} \in ob(\{i_2\})$ implies $i_2 \models \bigcirc(\neg tell/\neg go)$.

For further experiments, focusing on the automation of meta-theoretic aspects of DDL, we refer to [8, Fig. 2 and Fig. 3].

6 Conclusion

A shallow semantical embedding of Carmo and Jones's logic of contrary-to-duty conditionals in classical higher-order logic has been presented, and shown to be faithful (sound an complete). This theory work has meanwhile been implemented in the proof assistant Isabelle/HOL. This implementation constitutes the first theorem prover for the logic by Carmo and Jones that is available to date. The foundational theory for this implementation has been laid in this article.

There is much room for future work. First, experiments could investigate whether the provided implementation already supports non-trivial applications in practical normative reasoning, or whether further emendations and improvements are required. Second, the introduced framework could also be used to systematically analyse the properties of Carmo and Jones's dyadic deontic logic within Isabelle/HOL. Third, analogous to previous work in modal logic [10], the provided framework could be extended to study and support first-order and higher-order variants of the framework.

Acknowledgements

We thank the anonymous reviewers for their valuable feedback and comments.

References

[1] P.B. Andrews. Resolution in type theory. *Journal of Symbolic Logic*, 36(3):414–432, 1971.

[2] P.B. Andrews. General models and extensionality. *Journal of Symbolic Logic*, 37(2):395–397, 1972.

[3] P.B. Andrews. General models, descriptions, and choice in type theory. *Journal of Symbolic Logic*, 37(2):385–394, 1972.

[4] P.B. Andrews. Church's type theory. In E.N. Zalta, editor, *The Stanford Encyclopedia of Philosophy*. Metaphysics Research Lab, Stanford University, spring 2014 edition, 2014.

[5] C. Benzmüller. Cut-elimination for quantified conditional logic. *Journal of Philosophical Logic*, 46(3):333–353, 2017.

[6] C. Benzmüller. Recent successes with a meta-logical approach to universal logical reasoning (extended abstract). In S.A. da Costa Cavalheiro and J.L. Fiadeiro, editors, *Formal Methods: Foundations and Applications - 20th Brazilian Symposium, SBMF 2017, Recife, Brazil, November 29 - December 1, 2017, Proceedings*, volume 10623 of *Lecture Notes in Computer Science*, pages 7–11. Springer, 2017.

[7] C. Benzmüller, C. Brown, and M. Kohlhase. Higher-order semantics and extensionality. *Journal of Symbolic Logic*, 69(4):1027–1088, 2004.

[8] C. Benzmüller, A. Farjami, and X. Parent. Faithful semantical embedding of a dyadic deontic logic in HOL. CoRR, https://arxiv.org/abs/1802.08454, 2018.

[9] C. Benzmüller and D. Miller. Automation of higher-order logic. In D.M. Gabbay, J.H. Siekmann, and J. Woods, editors, *Handbook of the History of Logic, Volume 9 — Computational Logic*, pages 215–254. North Holland, Elsevier, 2014.

[10] C. Benzmüller and L.C. Paulson. Quantified multimodal logics in simple type theory. *Logica Universalis (Special Issue on Multimodal Logics)*, 7(1):7–20, 2013.

[11] J.C. Blanchette and T. Nipkow. Nitpick: A counterexample generator for higher-order logic based on a relational model finder. In *ITP 2010*, number 6172 in Lecture Notes in Computer Science, pages 131–146. Springer, 2010.

[12] J. Carmo and A.J.I. Jones. Deontic logic and contrary-to-duties. In D. M. Gabbay and F. Guenthner, editors, *Handbook of Philosophical Logic: Volume 8*, pages 265–343. Springer Netherlands, Dordrecht, 2002.

[13] J. Carmo and A.J.I. Jones. Completeness and decidability results for a logic of contrary-to-duty conditionals. *J. Log. Comput.*, 23(3):585–626, 2013.

[14] R.M. Chisholm. Contrary-to-duty imperatives and deontic logic. *Analysis*, 24:33–36, 1963.

[15] A. Church. A set of postulates for the foundation of logic. *Annals of Mathematics*, 33(3):346–366, 1932.

[16] A. Church. An unsolvable problem of elementary number theory. *American Journal of Mathematics*, 58(2):354–363, 1936.

[17] A. Church. A formulation of the simple theory of types. *Journal of Symbolic Logic*, 5(2):56–68, 1940.

[18] L. Henkin. Completeness in the theory of types. *Journal of Symbolic Logic*, 15(2):81–91,

1950.

[19] P. McNamara. Deontic logic. In E.N. Zalta, editor, *The Stanford Encyclopedia of Philosophy*. Metaphysics Research Lab, Stanford University, winter 2014 edition, 2014.

[20] R. Muskens. Intensional models for the theory of types. *Journal of Symbolic Logic*, 75(1):98–118, 2007.

[21] T. Nipkow, L.C. Paulson, and M. Wenzel. *Isabelle/HOL — A Proof Assistant for Higher-Order Logic*, volume 2283 of *Lecture Notes in Computer Science*. Springer, 2002.

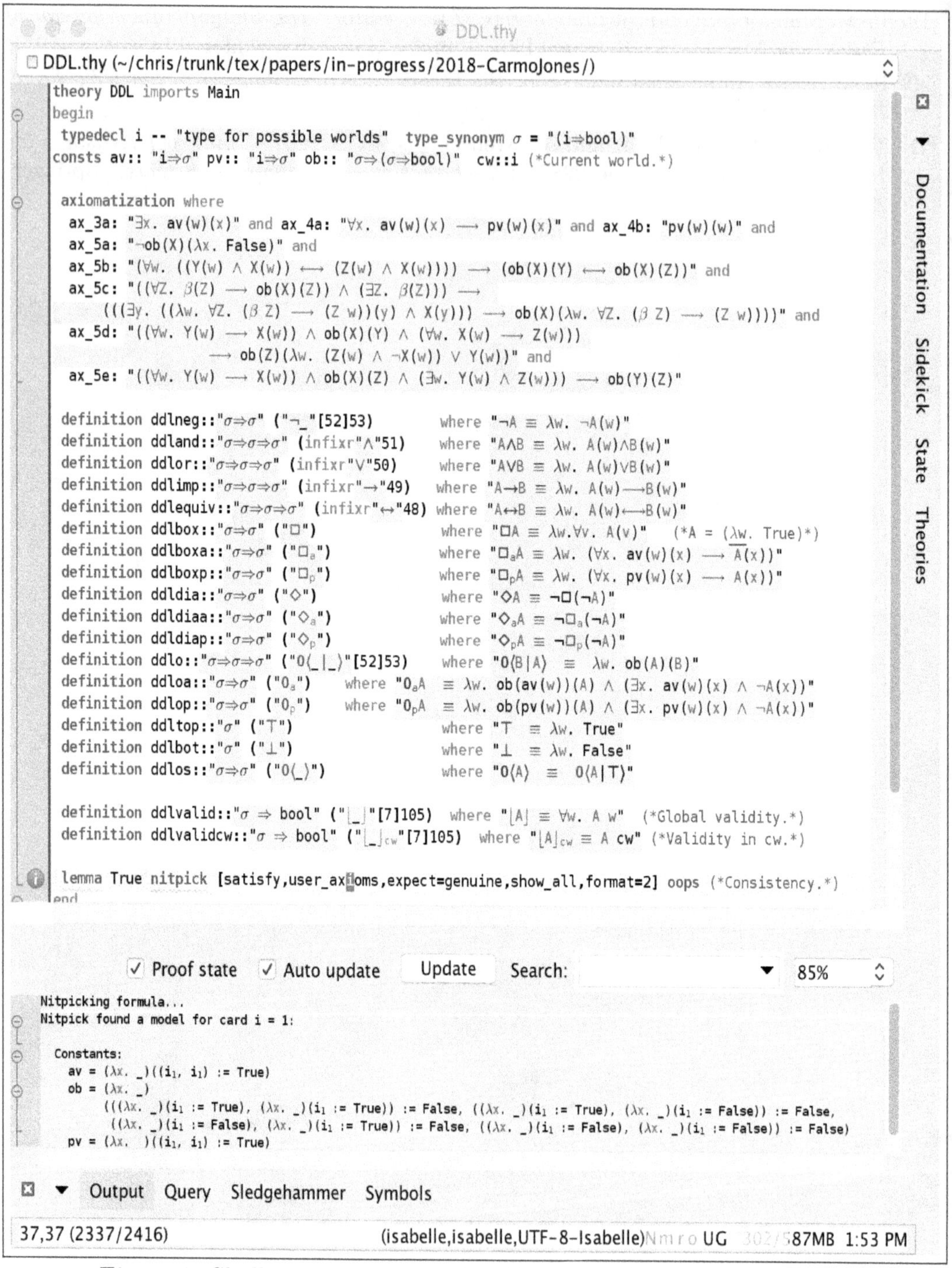

Figure 1: Shallow semantical embedding of DDL in Isabelle/HOL.

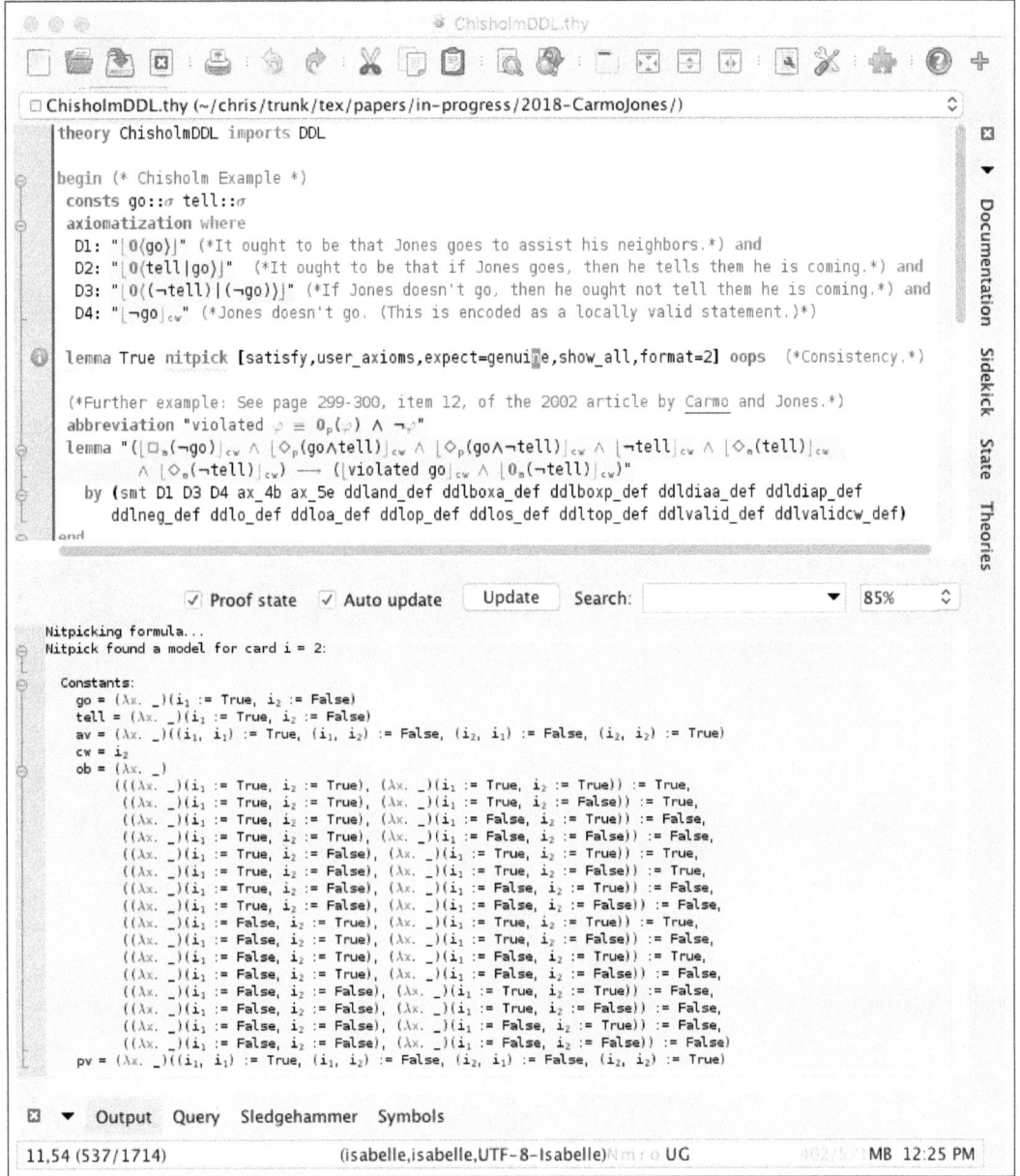

Figure 2: The Chisholm paradox scenario encoded in DDL (the shallow semantical embedding of DDL in Isabelle/HOL as displayed in Fig. 1 is imported here). Nitpick confirms consistency the encoded statements.

Knowledge and Subjective Oughts in STIT Logic

Jan Broersen
Utrecht University, The Netherlands, 3512 BL
J.M.Broersen@uu.nl

Aldo Iván Ramírez Abarca
Utrecht University, The Netherlands, 3512 BL
a.i.ramirezabarca@uu.nl

Puzzles concerning epistemic oughts, such as the ones put forward by Horty & Pacuit, can be solved without introducing action types in stit semantics. The problem of epistemic oughts is a fundamental one. Without exaggeration, we can say that the relation between agency, action, knowledge, and normativity is central to the understanding of responsibility, a main topic in Artificial Intelligence nowadays. We propose to bring together these four components of responsibility by blending (1) a theory of action and agency, (2) a theory of epistemic indistinguishability in action, and (3) deontic orderings that would allow us to distinguish right from wrong. We accomplish this by extending the standard stit language with epistemic and deontic operators. Semantically, we define subjective oughts by lifting a deontic ordering over histories to the level of 'actions that an agent can knowingly perform'. We compare the resulting definition for subjective oughts with Horty's action-type semantics. We demonstrate our findings by addressing Horty and Pacuit's coin-flip examples.

1 Introduction

With the advance of AI, questions about *obligations* and *responsibility* of intelligent systems have become increasingly important. The present work contributes to this field of research by addressing the logical modeling of agency, practical knowledge, and oughts. In particular, we show how through Kant's imperative 'ought implies can' ([6]) and the relation of 'can' and 'practical knowledge' we arrive at objective and subjective versions of the ought modality. This is done for an epistemic stit

---

The authors sincerely thank the reviewers for their careful and insightful remarks and annotations. Likewise, they gratefully acknowledge financial support from the ERC-2013-CoG Project REINS, No. 616512.

semantics with deontic operators. It is such a semantics ([3]) that allows us to present a formal theory of oughts, relative to a theory of agency and practical knowledge.

It is important to position this work correctly within the fast growing literature on issues involving knowledge, ought, and action ([2]). Our theory focuses on what we call 'knowledge-dependent oughts'. These are close to Pacuit et al.'s 'knowledge-based obligations' ([13]), but there is an important difference. We study oughts from a standpoint where agents can be excused for not meeting an obligation by claiming that they lacked the practical knowledge to do so. Pacuit et al. study how acquiring knowledge about a state can lead agents to being obliged to do certain things. Thus, our concern has got to do with 'practical' knowledge (*a doctor ought to stop the bleeding of his patient, but if he does not know how to, he is excused*), while Pacuit et al.'s regards 'theoretical' knowledge (*if a doctor knows that his patient is bleeding, he ought to stop it*).

In their recent work [9], Horty & Pacuit argue that we cannot have a logical theory of oughts, knowledge, and action without resorting to action types. This echoes similar claims by Lorrini and Herzig ([7]) and is in accordance with a broad sentiment in the DEL community and in the ATL community ([1]) –both of which use action types as basic ingredients of their models. As we see it, the difference between types and tokens is as follows. An action token is the particular performance of an action by a specific agent at a specific moment (*opening the window of my bedroom at 8 a.m. on Monday...*). Action types refer to categories or patterns of actions that are instantiated in tokens (*to open a window*). Horty & Pacuit motivate the use of types in epistemic deontic stit logic by presenting three puzzles for which standard semantics of knowledge and ought-to-do ([10]) fail. They develop a proposal that solves these puzzles, but which comes with technical disadvantages inherent to the type construction: the class of models they introduce is limited, and their logic cannot be fully axiomatized, since the constraints involving types cannot be characterized syntactically with modal axioms. In this paper, we show that we do not need action types for a formal theory of ought, knowledge, and action. We present an approach that solves the three puzzles and that is axiomatizable.

Elaborating on the matter, Horty & Pacuit favor types in order to be able to express that an agent can have the same *kinds* of available actions at separate, epistemically indistinguishable moments. This property is well known in game theory and in the epistemic version ATEL ([8]) of ATL: it is called the 'uniformity of strategies' constraint. A problem with this condition is that it does not correspond one-to-one with a property of the logic of ATEL, since types do not feature in ATEL's object language. The same thing happens in Horty and Pacuit's proposal, and what we will do is show that we do not need a uniformity-of-strategies constraint in terms of types, for its role is better played by a more general condition that *does* correspond to a syntactic property of our logic.

Our approach for modeling knowledge-dependent oughts in a stit framework is as follows: we lift a deontic ordering over choices (that itself results from lifting an ordering of histories) to an ordering over information sets. We do this to capture the idea that knowledge-dependent oughts imply practical knowledge[1]. For objective oughts, we do not have to lift to this level.

1.1 Motivating Problem: the Miners Scenario

There is a famous thought experiment that heavily inspired the formal notions that we propose in this work: the so-called *Miners Paradox*. Put forward in an unpublished paper by Derek Parfit ([14]) but made popular by Kolodny and MacFarlane ([11]), it concerns the following situation: *ten miners are trapped either in shaft A or in shaft B of a mine, and we do not know exactly in which. Water threatens to flood the shafts. However, we only have enough sandbags to block one of the shafts, not both. If one shaft is blocked, all the water will stream into the other shaft, killing all the miners in it. If we block neither shaft, both will be partially flooded, killing maximally one miner.* In a utilitarian view on morality, minimizing the number of casualties modulo the uncertainty about the location of the miners, it seems correct to say that:

(1) *We ought to block neither shaft.*

However, it also seems correct to say that:

(2) *If the miners are in shaft A, we ought to block A*

(3) *If the miners are in shaft B, we ought to block B*

As background knowledge we have that:

(4) *Either the miners are in shaft A or they are in shaft B*

Thus stated, sentences (2), (3), and (4) entail that:

(5) *Either we ought to block shaft A or we ought to block shaft B,*

which contradicts sentence (1). This conflict is perceived as a problem based on the assumption that, in a utilitarian view, sentence (1) is clearly true, while the opposite of sentence (1) is concluded from the seemingly correct premises (2), (3), and (4). We derive a contradiction from apparently valid premises and background knowledge[2].

---

[1] For this paper, we do not discuss the possibility that agents incur an obligation to learn how to do something on the basis of an obligation to do that something.

[2] For that reason, scenarios like this are often dubbed 'paradoxes', but we feel that such a name is too much of an honor for this particular scenario.

The most popular approach for solving the problem appears to be to curb the reasoning-by-cases principle underlying the derivation with sentences (2), (3), and (4) (see [16]). Sometimes humans favor reasoning by cases, and sometimes we do not. The challenge is then to come up with general criteria that would distinguish between these two contexts, and to formalize them. Solutions of this kind often come in the form of a non-monotonic theory or an update theory. Another approach, which resembles our treatment of oughts in that it involves directly the knowledge of the rescuers, addresses the conflict by arguing that (2) and (3) are not correct. For these clauses to be more adequate, they should read "If we know that the miners are in shaft A, we ought to block A", and "If we know that the miners are in shaft B, we ought to block B", respectively. To a certain extent, this approach is successful, and incorporates an epistemic dimension to the analysis. However, it is debatable whether this is an actual solution or not ([5]).

With knowledge, ought, and action all in the picture, the scenario can be easily formalized using epistemic deontic stit models for a language with both knowledge operators and Horty's famous act utilitarian ought-to-do operator from [10]. In such formalization, which we will review in detail in section 2, the saving of lives has a certain utility or payoff, and oughts are calculated as action tokens that are not weakly dominated by others, much like in game theory. This formalization faces a big problem, since one obtains that even if (2) and (3) are stated taking knowledge into consideration, (5) remains true in the situation where rescuers do *not* know where the miners are trapped.

Our framework for knowledge-dependent oughts is built on the basis that sentences (2) and (3) should be reinterpreted taking knowledge into account, and it solves the problem discussed above –the one that appears in the stit formalization of the scenario. However, we do not offer a solution in terms of prescribing a specific action for the rescuers. Our main idea stems from interpreting that the problem arises from the assumption that the oughts appearing in sentences (1), (2), (3), and (5) are all of the same kind. We believe that two senses of ought are being confused in the reasoning: an objective and a subjective one[3] (see [15] and [5]). If all oughts in the scenario are *objective*, then sentence (1) is false: there is an alternative that –from a utilitarian moral perspective– is preferable, namely blocking the shaft that contains the miners. If all oughts in the scenario are *subjective*[4], then we claim that sentences (2) and (3) are false. Subjective moral oughts demand that an agent has practical knowledge about the act under consideration (something we will elaborate

---

[3]Objective oughts can be found in legal theory, for instance, where agents sometimes are 'strictly liable' for certain outcomes, even though they could not have avoided them or knew nothing about their legal status. Subjective oughts are typically found in moral theory, where agents are blamed for having had wrong intentions. Intentions require plans that are subjectively feasible.

[4]This view is certainly supported by the fact that the truth of sentence (1) is actually never called into question, while it must be based on a form of subjective moral maximal expected utility reasoning.

on presently). However, any agent facing the miners dilemma clearly lacks such practical knowledge, since it does not know which shaft the miners are in. If the miners are in shaft A, an agent *cannot* knowingly perform the action 'rescue the miners by blocking a shaft', since it does not know which shaft to block.

Let us briefly come back to the point in the previous paragraph where we say that subjective oughts (i.e., oughts of the moral kind) require a possibility to successfully and knowingly comply with them. This is Kant's maxim of 'ought implies can', and it is probably better to present it in a contrapositive form: as 'cannot implies being excused'[5]. Kant's principle strongly binds moral responsibility of an agent to its practical knowledge, and here we see how action types might come into play, since it is generally believed that types are necessary for modeling practical knowledge. In such constraints as 'uniformity of strategies', that is exactly the role played by action types: to make sure that the same action types can be performed modulo uncertainty about the present state. We hope to show that types are not necessary to encode such a condition, or are necessary to formalize practical knowledge.

In subsection 2.1, we will discuss the three puzzles put forward by Horty & Pacuit that we mentioned above. Involving coin-flips and betting on the outcome, it turns out that one of these examples is structurally equivalent to the Miners Scenario. Therefore, in the sequel, we will treat the Miners Scenario as a variant of one of the three puzzles. Both Horty & Pacuit's and our proposal for "solving" these puzzles revolve around Horty's decision-theoretical account of ought-to-do in stit logic ([10]). Thus, we first recover the basic definitions of the logic that we will use.

2 Epistemic Act Utilitarian Stit Logic

Definition 2.1. Given a finite set Ags of agent names and a countable set of propositions P, with $p \in P$ and $\alpha \in Ags$, the grammar for the formal language $\mathcal{L}_{\mathsf{KCo}}$ is given by:

$$\varphi := p \mid \neg\varphi \mid \varphi \wedge \psi \mid \Box\varphi \mid [\alpha \ \mathtt{Cstit}]\varphi \mid K_\alpha\varphi \mid \odot[\alpha \ \mathtt{Cstit}]\varphi$$

Besides the usual propositional connectives, the syntax of $\mathcal{L}_{\mathsf{KCo}}$ comprises four modal operators. $\Box\varphi$ will express historical necessity of φ ($\Diamond\varphi$ abbreviates $\neg\Box\neg\varphi$); $[\alpha \ \mathtt{Cstit}]\varphi$ stands for 'agent α sees to it that φ'; K_α is the epistemic operator for α; and $\odot[\alpha \ \mathtt{Cstit}]$ is meant to represent what α ought to do ([10]).

Definition 2.2. Given a finite set Ags of agent names, a finite-choice epistemic utilitarian KCo-frame is a tuple $\langle T, \sqsubset, \mathbf{Choice}, \{\sim_\alpha\}_{\alpha \in Ags}, \mathbf{Value}\rangle$ such that:

- T is a non-empty set of moments. $\sqsubset$ is a strict partial ordering on T satisfying 'no backward branching': for every $m, m', m'' \in T$, if $m' \sqsubset m$ and $m'' \sqsubset m$,

---
[5]Note that an appeal to this type of excuse is actually the most popular among a wide variety of excuses available for not doing something one should.

then $m' \sqsubseteq m''$ or $m'' \sqsubseteq m'$ or $m' = m''$. Each maximal $\sqsubseteq$-chain is called a *history*, representing a way in which time might evolve. We use H to denote the set of all histories, and for each $m \in T$, we use H_m to denote the set $\{h \in H \mid m \in h\}$.

- Tuples $\langle m, h \rangle$ where $m \in T$, $h \in H$, and $m \in h$, are called *situations*. Choice is a function that maps each agent α and moment m to a *finite*[6] partition **Choice**$_\alpha^m$ of H_m. The cells of such a partition represent the possible choices $C_\alpha^m, C'^m_\alpha, C''^m_\alpha, \ldots$ of α at m. A choice profile $\langle C_{\alpha_1}^m, C_{\alpha_2}^m \ldots C_{\alpha_n}^m \rangle$ at m is a particular combination of choices $C_{\alpha_i}^m$ at m, one for each agent α_i in the system. Choice satisfies two constraints known as (NC) or 'no choice between undivided histories' and (IA) or 'independence of agents'. Formally,

  - (NC) For all $h, h' \in H_m$, if $m' \in h \cap h'$ for some $m' \sqsupset m$, then $h \in L$ iff $h' \in L$ for every $L \in$ **Choice**$_\alpha^m$.
  - (IA) For any $m \in T$, the intersection of choices in any choice profile is non-empty: $\bigcap_{\alpha_i \in Ags} C_{\alpha_i}^m \neq \emptyset$.

- **Value** is a (utilitarian) deontic function that assigns to each history $h \in H$ a real number, representing the utility of h.

- For each agent $\alpha \in Ags$, $\sim_\alpha$ is an equivalence relation on the set of situations, which is to be interpreted as the epistemic indistinguishability relation for agent α. Notice that this relation ensues between *situations*, meaning moment-history pairs. The cells of $\sim_\alpha$ are typically referred to as the *information sets* of agent α at a given situation.

Definition 2.3 (Dominance and optimality). For fixed $\alpha \in Ags$ and $m_* \in T$, we define the following constructs:

- A dominance ordering $\preceq$ on **Choice**$_\alpha^{m_*}$ such that for $L, L' \subseteq H_{m_*}$,

  $$L \preceq L' \text{ iff } \textbf{Value}(h) \leq \textbf{Value}(h') \text{ for every } h \in L, h' \in L'.$$

  We write $L \prec L'$ iff $L \preceq L'$ and $L' \not\preceq L$.

- An optimal set of actions **Optimal**$_\alpha^{m_*}$, such that

  $$\textbf{Optimal}_\alpha^{m_*} =_{def} \{L \in \textbf{Choice}_\alpha^{m_*}; \text{there is no } L' \in \textbf{Choice}_\alpha^{m_*} \text{ such that } L \prec L'\}.$$

We now define models by adding a valuation of propositional atoms to the frames of definition 2.2.

---

[6]This condition is the reason why we use the term "finite choice" in the definition of our frames.

Definition 2.4. A frame $\langle T, \sqsubset, \textbf{Choice}, \{\sim_\alpha\}_{\alpha \in Ags}, \textbf{Value}\rangle$ is extended to a model $\mathcal{M} = \langle T, \sqsubset, \textbf{Choice}, \{\sim_\alpha\}_{\alpha \in Ags}, \textbf{Value}, \mathcal{V}\rangle$ by adding a valuation function $\mathcal{V} : P \to 2^{T \times H}$ assigning to each atomic proposition the set of situations where it is true. Relative to a model $\mathcal{M} = \langle T, \sqsubset \textbf{Choice}, \{\sim_\alpha\}_{\alpha \in Ags}, \textbf{Value}, \mathcal{V}\rangle$, the semantics for the formulas of $\mathcal{L}_{\mathsf{KCo}}$ is defined recursively by the following truth conditions:

For a situation $\langle m, h \rangle$,

$\langle m, h \rangle \models p \quad \Leftrightarrow \quad \langle m, h \rangle \in \mathcal{V}(p)$
$\langle m, h \rangle \models \neg \varphi \quad \Leftrightarrow \quad \langle m, h \rangle \not\models \varphi$
$\langle m, h \rangle \models \varphi \wedge \psi \quad \Leftrightarrow \quad \langle m, h \rangle \models \varphi$ and $\langle m, h \rangle \models \psi$
$\langle m, h \rangle \models \Box \varphi \quad \Leftrightarrow \quad \forall h' \in H_m, \langle m, h' \rangle \models \varphi$
$\langle m, h \rangle \models [\alpha\ \mathtt{Cstit}]\varphi \quad \Leftrightarrow \quad \forall h' \in \textbf{Choice}_\alpha^m(h), \langle m, h' \rangle \models \varphi$
$\langle m, h \rangle \models K_\alpha \varphi \quad \Leftrightarrow \quad \forall \langle m', h' \rangle$ such that $\langle m, h \rangle \sim_\alpha \langle m', h' \rangle, \langle m', h' \rangle \models \varphi$
$\langle m, h \rangle \models \odot[\alpha\ \mathtt{Cstit}]\varphi \quad \Leftrightarrow \quad \forall L \in \textbf{Optimal}_\alpha^m, h' \in L$ implies that $\langle m, h' \rangle \models \varphi$.

Satisfiability, validity on a frame, and general validity are defined as usual.

As a convention, we will write $|\varphi|^m$ to refer to the set $\{h \in H_m \mid \langle m, h \rangle \models \varphi\}$.

2.1 Formal Account of the Puzzles

The three puzzles that Horty & Pacuit present can be summarized as follows.

Example 2.5. After a fair coin-flip, the outcome of which is still hidden, an agent can bet heads, bet tails, or refrain from betting. If the agent bets and chooses the right outcome, it wins €10. If the agent chooses the wrong outcome, it does not win anything, and if it refrains from betting, it wins €5. This is a minor variant on the Miners Scenario.

The stit diagram that represents Horty & Pacuit's interpretation of the situation is included in Figure 1. Once again, the blue dotted circle represents the epistemic class of the agent at any given situation: since the outcome of the coin-flip is unknown by the agent, it cannot distinguish whether the moment is m_1 or m_2[7].

For such an interpretation regarding the epistemic situation of the agent, a problem ensues due to the fact that for every $i \in \{1, 2\}$ and $h \in H_{m_i}$, we have that $\langle m_i, h \rangle \models K_\alpha \odot [\alpha\ \mathtt{Cstit}]G$, where G stands for the proposition of 'gambling'. This means that the agent knows that it ought to gamble, even if this is a "risky" move that could result into its having a payoff of 0. In this sense, we could say the agent's knowledge of what is objectively optimal would lead it into taking a chance and betting.

---

[7]Notice that Horty & Pacuit's formalization also yields that in none of the situations will the agent knowingly perform the available actions: it cannot epistemically distinguish between the situations in which it is 'betting heads', 'betting tails', or 'refraining'.

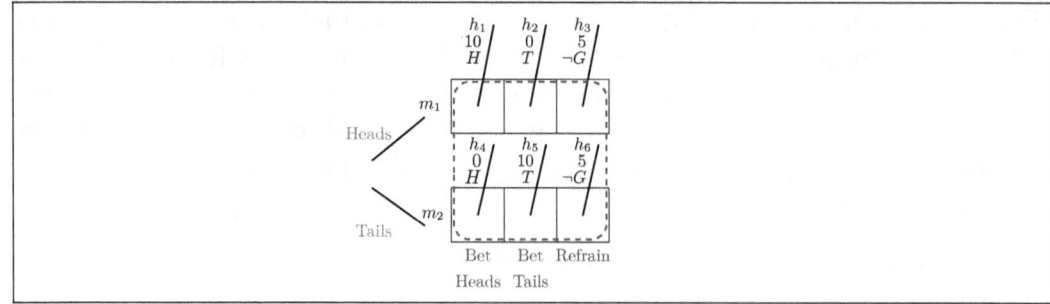

Figure 1: Coin-flip problem #1 (variant of the Miners Scenario)

Example 2.6. After a fair coin-flip, the outcome of which is still hidden, an agent can bet heads or refrain from betting. If the agent bets and chooses the right outcome, it wins €10. If the agent refrains from betting, it *also* wins €10.

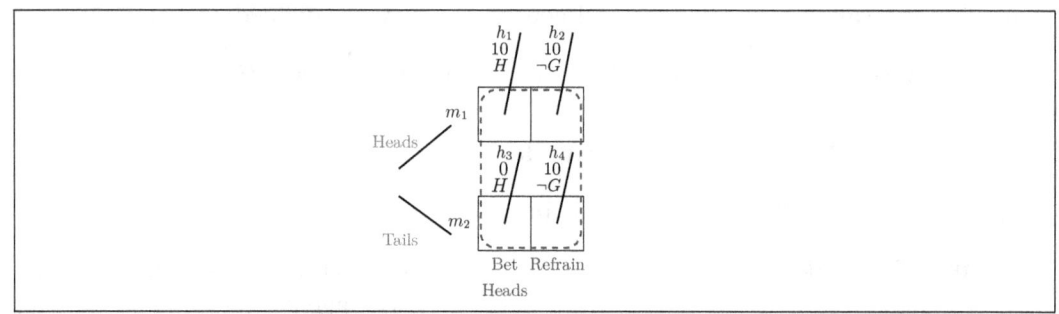

Figure 2: Coin-flip problem #2

Thus, in this scenario, the agent intuitively ought to refrain from gambling, for such action implies that the agent would win by the same amount as when betting correctly –but without engaging in an action that could possibly fail. In this case, the problem for Horty & Pacuit is that for every $i \in \{1, 2\}$ and $h \in H_{m_i}$, we have that $\langle m_i, h \rangle \not\models K_\alpha \odot [\alpha \; \texttt{Cstit}] \neg G$: the agent does not know that it ought to refrain from gambling, even if the information available seems to promote such a thing.

Example 2.7. After a fair coin-flip, the outcome of which is still hidden, an agent can bet heads or bet tails. If the agent bets and chooses the right outcome, it wins €10. If it bets incorrectly, it does not win anything.

In Horty & Pacuit's formalization, the problem is that for every $i \in \{1, 2\}$ and $h \in H_{m_i}$, $\langle m_i, h \rangle \models K_\alpha \odot [\alpha \; \texttt{Cstit}] W$. This means that the agent knows that it ought to win at any given index, but such knowledge is not *action-guiding*, meaning that it will not provide the agent with a choice to make. Though the agent knowingly ought to win, it cannot knowingly do so –it simply does not have the means due to a lack of knowledge. Thus, Kant's principle of 'ought implies can' is not satisfied ($\langle m_i, h \rangle \not\models K_\alpha \odot [\alpha \; \texttt{Cstit}] W \rightarrow \Diamond K_\alpha [\alpha \; \texttt{Cstit}] W$).

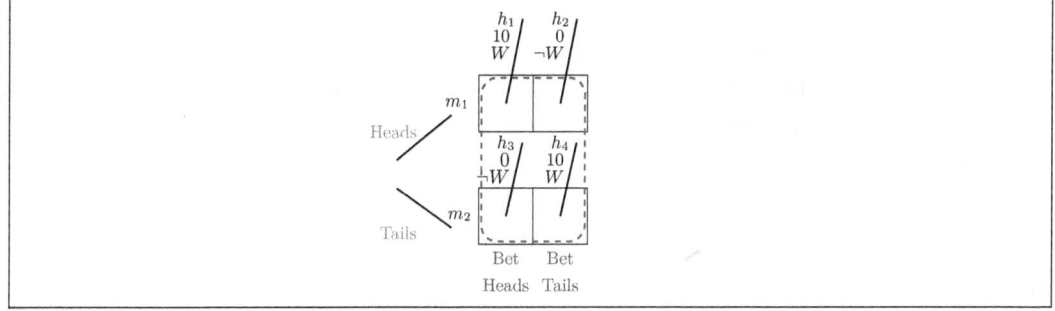

Figure 3: Coin-flip problem #3

2.2 Using Action Types to Model the Puzzles

Horty & Pacuit solve these three puzzles by introducing both syntactic and semantic addenda to epistemic deontic stit logic. They extend the language with the operator [... kstit] to encode the concept of *ex interim* knowledge, or of knowingly doing. The semantics for this operator uses *action types* to cluster histories according to the intuition that actions of the same type can lead to different outcomes. This means that Horty & Pacuit take 'betting heads' and 'betting tails' as action types. The syntactic and semantic tools added come with two unfavorable new constraints for the epistemic stit models: (1) lifting the epistemic indistinguishability relations from moment-history pairs to moments[8], which is semantically limiting, and (2) the uniformity-of-strategies constraint from ATEL, saying that indistinguishable states offer same types, which cannot be characterized syntactically. As a solution to the stated puzzles, Horty & Pacuit's approach is successful. However, we show that we can also be successful without using action types, and this comes with the advantages mentioned in the introduction: (a) technical characterizability of the epistemic constraints imposed on the structures –which is important for axiomatization–, and (b) conceptual clarity, as the role of types is replaced by the concept of 'knowingly doing'.

3 Our proposal

To define knowledge-dependent ought-to-do's, we propose to lift the dominance ordering over choices to an ordering over choices *within information sets*. This gives us the opportunity to address two different versions of the concept of ought-to-do: an *objective* ought-to-do, which is exactly the same as Horty's act utilitarian ought-to-do (faced by the possible choices relative to the current moment, the agent has to do what holds true for all weakly undominated choices), and a *subjective* ought-to-do,

---

[8]Horty & Pacuit's models satisfy the following constraint: if $\langle m, h \rangle \sim_\alpha \langle m', h' \rangle$ then $\langle m, h'' \rangle \sim_\alpha \langle m', h''' \rangle$ for every $h'' \in H_m$, $h''' \in H_{m'}$.

which selects the best candidates out of the set of actions that the agent can knowingly perform, where we determine those best candidates by looking at the *actual* alternative choices for the current moment, and by picking the ones whose epistemic equivalents are all undominated in the information sets (if this sounds too cryptic, the formalization below should clarify things).

3.1 Extension with subjective oughts

We extend the language of Definition 2.1.

Definition 3.1. Given a finite set Ags of agent names, a countable set of propositions P, with $p \in P$ and $\alpha \in Ags$, the grammar for the formal language $\mathcal{L}_{\mathsf{KCO}}$ is given by:

$$\varphi := p \mid \neg\varphi \mid \varphi \wedge \psi \mid \Box\varphi \mid [\alpha\ \mathtt{Cstit}]\varphi \mid K_\alpha\varphi \mid \odot[\alpha\ \mathtt{Cstit}]\varphi \mid \odot_\mathcal{S}[\alpha\ \mathtt{Cstit}]\varphi$$

Definition 3.2. A finite-choice epistemic utilitarian KCO-frame is a tuple $\langle T, \sqsubset, \mathbf{Choice}, \{\sim_\alpha\}_{\alpha \in Ags}, \mathbf{Value}\rangle$ such that:

- $T, \sqsubset, \mathbf{Choice}$, and $\sim_\alpha$ are defined the same way as in Definition 2.2.

- For any $\alpha \in Ags$, relation $\sim_\alpha$ and function $\mathbf{Choice}$ satisfy the following constraints:

    - (KX) For every situation $\langle m_*, h_*\rangle$, if $\langle m_*, h_*\rangle \sim_\alpha \langle m, h\rangle$ for some $\langle m, h\rangle$, then $\langle m_*, h'_*\rangle \sim_\alpha \langle m, h\rangle$ for every $h'_* \in \mathbf{Choice}_\alpha^{m_*}(h_*)$. We will refer to this constraint as the 'practical knowledge' constraint[9].

    - (US) For every situation $\langle m_*, h_*\rangle$, if $\langle m_*, h_*\rangle \sim_\alpha \langle m, h\rangle$ for some $\langle m, h\rangle$, then for every $h'_* \in H_{m_*}$, there exists $h' \in H_m$ such that $\langle m_*, h'_*\rangle \sim_\alpha \langle m, h'\rangle$. For semantics without action types, this constraint is meant to capture a notion of 'uniformity of strategies', where epistemically indistinguishable situations should offer same actions for the agent to choose upon (see [18]).

In order to present our models and the semantics for the formulas of the extended language $\mathcal{L}_{\mathsf{KCO}}$, we need some definitions first. These definitions will allow us to define an adequate semantics for $\odot_\mathcal{S}[\alpha\ \mathtt{Cstit}]\varphi$, which we will do by lifting a subjective dominance ordering to choices within information sets.

Definition 3.3 (Subjective dominance and subjective optimality). Let $\alpha \in Ags$ be a fixed agent and $m_* \in T$ be a fixed moment.

---

[9] The name comes from the fact that it is mentioned by Xu [18] as an important condition for a theory of practical knowledge. Essentially, it says that the agents should not be able to epistemically distinguish histories at which they are performing the same action (token). It yields that agents know only what they knowingly do.

- For a non-empty $L \subseteq H_{m_*}$ and $m \in T$, we define an *epistemic cluster set* of L at m:

$$[L]_\alpha^m =_{def} \{h \in H_m \mid \exists h_* \in L \text{ st } \langle m_*, h_* \rangle \sim_\alpha \langle m, h \rangle\}.$$

As a convention, we will write $m \sim_\alpha m'$ if there exist $h \in H_m$, $h' \in H_{m'}$ such that $\langle m, h \rangle \sim_\alpha \langle m', h' \rangle$[10].

- We define a subjective ordering $\preceq_s$ on $\mathbf{Choice}_\alpha^{m_*}$ such that for $L, L' \subseteq H_{m_*}$,

$$L \preceq_s L' \quad \text{iff} \quad \text{for every } m \text{ such that } m_* \sim_\alpha m, \mathbf{Value}(h) \leq \mathbf{Value}(h')$$
$$\text{for every } h \in [L]_\alpha^m, h' \in [L']_\alpha^m.$$

- We define a subjectively optimal set of actions $\mathbf{S} - \mathbf{optimal}_\alpha^{m_*}$, such that

$$\mathbf{S} - \mathbf{optimal}_\alpha^{m_*} =_{def} \{L \in \mathbf{Choice}_\alpha^{m_*}; \text{there is no } L' \in \mathbf{Choice}_\alpha^{m_*}$$
$$\text{such that } L \prec_s L'\}.$$

As expected, we will write $L \prec_s L'$ iff $L \preceq_s L'$ and $L' \not\preceq_s L$.

Definition 3.4. A frame $\langle T, \sqsubset, \mathbf{Choice}, \{\sim_\alpha\}_{\alpha \in Ags}, \mathbf{Value} \rangle$ is extended to a model $\mathcal{M} = \langle T, \sqsubset, \mathbf{Choice}, \{\sim_\alpha\}_{\alpha \in Ags}, \mathbf{Value}, \mathcal{V} \rangle$ by adding a valuation $\mathcal{V}$ of atomic propositions just as in Definition 2.4. The semantics for the formulas of $\mathcal{L}_{\mathsf{KCO}}$ is defined recursively by extending the rule given in Definition 2.4 with the following truth condition:

$$\langle m, h \rangle \models \odot_S[\alpha \,\, \mathtt{Cstit}]\varphi \,\, \Leftrightarrow \,\, \forall L \in \mathbf{S} - \mathbf{optimal}_\alpha^m, \forall m' \text{ such that}$$
$$m \sim_\alpha m', [L]_\alpha^{m'} \subseteq |\varphi|^{m'}.$$

3.2 Solution to the Epistemic Puzzles

Solution to Example 2.5

As for Example 2.5, in order to capture the assumption that the agent does not know what the outcome of the coin-flip is, we take $\sim_\alpha$ as defined by the following three equivalence classes: $\{\langle m_1, h_1 \rangle, \langle m_2, h_4 \rangle\}$, in which agent α bets heads; $\{\langle m_1, h_2 \rangle, \langle m_2, h_5 \rangle\}$, in which agent α bets tails; and $\{\langle m_1, h_3 \rangle, \langle m_2, h_6 \rangle\}$, in which agent α refrains from betting. As implied by such a layout (see Figure 4), agent α can knowingly act, in the sense that it knowingly sees to it that a given condition characterizing an action is enforced ('betting heads', 'betting tails', 'refraining from betting'). For our semantics, the problem is solved because although for every $i \in \{1, 2\}$ and $h \in H_{m_i}$, $\langle m_i, h \rangle \models K_\alpha \odot [\alpha \,\, \mathtt{Cstit}]G$, we consider this as the knowledge of an *objective* ought-to-do. *Subjectively* speaking, we do not obtain that the agent knows that it ought to gamble: for every $i \in \{1, 2\}$ and $h \in H_{m_i}$

---

[10]Notice that for a fixed moment $m_* \in M$ and $L \subseteq H_{m_*}$, $[L]_\alpha^m$ is non-empty if and only if $m_* \sim_\alpha m$ (where the "if" direction follows from (US) condition above).

$\langle m_i, h \rangle \not\models \odot_S[\alpha \text{ Cstit}]G$, which in turn implies that $\langle m_i, h \rangle \not\models K_\alpha \odot_S [\alpha \text{ Cstit}]G$ (K_α is an **S5** operator that validates the (T) axiom)[11].

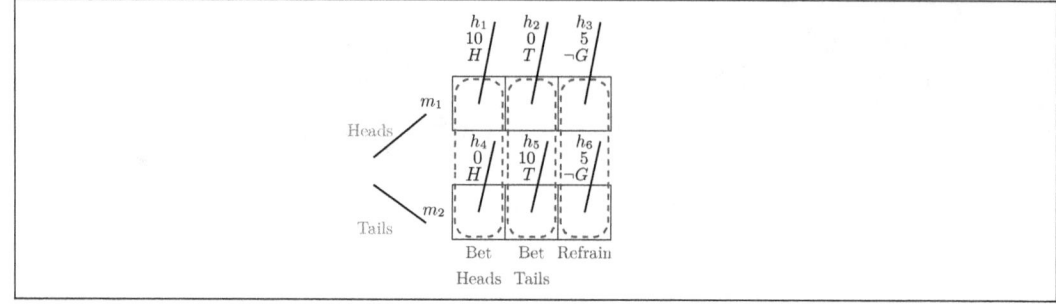

Figure 4: Coin-flip problem #1 revisited

Solution to Example 2.6

We interpret this example with $\sim_\alpha$ defined by the following two equivalence classes: $\{\langle m_1, h_1 \rangle, \langle m_2, h_3 \rangle\}$, in which agent α bets heads; and $\{\langle m_1, h_2 \rangle, \langle m_2, h_4 \rangle\}$, in which agent α refrains from betting. Therefore, once again we are working under the assumption that agent α can knowingly act at any given moment. For our semantics, the problem is solved because for every $i \in \{1, 2\}$ and $h \in H_{m_i}$, $\langle m_i, h \rangle \models \odot_S[\alpha \text{ Cstit}]\neg G$ and $\langle m_i, h \rangle \models K_\alpha \odot_S [\alpha \text{ Cstit}]\neg G$. The key argument for this comes from our definitions of the subjective ordering and of the subjectively optimal sets: **S − optimal**$_\alpha^{m_1} = \{\{h_2\}\}$, and **S − optimal**$_\alpha^{m_2} = \{\{h_4\}\}$. In this way, the agent subjectively ought to refrain from gambling, because the knowledge about the actions in this situation points it in the direction of avoiding an unnecessary risk. Notice that it still holds that for every $h \in H_{m_1}$, $\langle m_1, h \rangle \not\models K_\alpha \odot [\alpha \text{ Cstit}]\neg G$.

Solution to Example 2.7

We interpret this example with $\sim_\alpha$ defined by the following two equivalence classes: $\{\langle m_1, h_1 \rangle, \langle m_2, h_3 \rangle\}$, in which agent α bets heads; and $\{\langle m_1, h_2 \rangle, \langle m_2, h_4 \rangle\}$, in which agent α bets tails. For our semantics, the problem is solved because we get that for every $i \in \{1, 2\}$ and $h \in H_{m_i}$, $\langle m_i, h \rangle \not\models \odot_S[\alpha \text{ Cstit}]W$, which in turn implies that $\langle m_i, h \rangle \not\models K_\alpha \odot_S [\alpha \text{ Cstit}]W$ (notice that **S − optimal**$_\alpha^{m_1} = \{\{h_1\}, \{h_2\}\}$, and that **S − optimal**$_\alpha^{m_2} = \{\{h_3\}, \{h_4\}\}$). Therefore, although the agent knows that it objectively ought to win (it is always better to win than to lose), it is not the

---

[11]Notice that we can use Figure 4 to study the Miners Scenario: if we change H for A ('shaft A is blocked'), T for B ('shaft B is blocked') and $\neg G$ for N ('neither shaft is blocked'). Thus interpreted, we get that –objectively– the agent should block shaft A or should block shaft B. Moreover, we have that the agent knows that: for every $i \in \{1, 2\}$ and $h \in H_{m_i}$, $\langle m_i, h \rangle \models K_\alpha(\odot[\alpha \text{ Cstit}]A \lor \odot[\alpha \text{ Cstit}]B)$. Subjectively, things are different: for every $i \in \{1, 2\}$ and $h \in H_{m_i}$, $\langle m_i, h \rangle \not\models \odot_S[\alpha \text{ Cstit}]A$ and $\langle m_i, h \rangle \not\models \odot_S[\alpha \text{ Cstit}]B$. Notice that we do not have that $\langle m_1, h \rangle \models \odot_S[\alpha \text{ Cstit}]N$ either. Therefore, the agent is not compelled by any obligation to act somehow, and would be excused in the tragic case in which it chose the wrong shaft.

case that it subjectively ought to win, which means to say that the agent cannot knowingly guide its action towards winning according to a subjective obligation of winning.

When comparing our solutions to Horty & Pacuit's, we have to point out that our formalization is different from theirs, for we keep the indistinguishability relation between *situations*. Regardless of such difference, the solution we achieve is virtually the same; our notion of $\odot_S[\alpha\ \texttt{Cstit}]$ works as an analog of their $\odot[...\ \texttt{kstit}]$[12].

3.3 Axiomatization and some logical properties

Definition 3.5 (Axiom system). Let Λ be the axiom system defined by the following axioms and rules of inference:
(Axioms)

- All classical tautologies from propositional logic.

- The **S5** axioms for $\Box$, $[\alpha\ \texttt{Cstit}]$, K_α, and the following axioms:

$$\odot[\alpha\ \texttt{Cstit}](p \to q) \to (\odot[\alpha\ \texttt{Cstit}]p \to \odot[\alpha\ \texttt{Cstit}]q) \tag{A1}$$
$$\Box p \to [\alpha\ \texttt{Cstit}]p \land \odot[\alpha\ \texttt{Cstit}]p \tag{A2}$$
$$\Box \odot [\alpha\ \texttt{Cstit}]p \lor \Box \neg \odot [\alpha\ \texttt{Cstit}]p \tag{A3}$$
$$\odot[\alpha\ \texttt{Cstit}]p \to \odot[\alpha\ \texttt{Cstit}]([\alpha\ \texttt{Cstit}]p) \tag{A4}$$
$$\odot[\alpha\ \texttt{Cstit}]p \to \Diamond[\alpha\ \texttt{Cstit}]p \tag{Oic}$$

For $n \geq 1$ and pairwise different $\alpha_1, \ldots, \alpha_n$,

$$\bigwedge_{1 \leq k \leq n} \Diamond[\alpha_i\ \texttt{Cstit}]p_i \to \Diamond\left(\bigwedge_{1 \leq k \leq n}[\alpha_i\ \texttt{Cstit}]p_i\right) \tag{IA}$$

$$\odot_S[\alpha\ \texttt{Cstit}](p \to q) \to (\odot_S[\alpha\ \texttt{Cstit}]p \to \odot_S[\alpha\ \texttt{Cstit}]q) \tag{A5}$$
$$\Box \odot_S[\alpha\ \texttt{Cstit}]p \lor \Box \neg \odot_S[\alpha\ \texttt{Cstit}]p \tag{A6}$$
$$\odot_S[\alpha\ \texttt{Cstit}]p \to \odot_S[\alpha\ \texttt{Cstit}](K_\alpha p) \tag{A7}$$
$$K_\alpha p \to [\alpha\ \texttt{Cstit}]p \tag{KX}$$
$$\Diamond K_\alpha p \to K_\alpha \Diamond p \tag{US}$$
$$K_\alpha \Box p \to \odot_S[\alpha\ \texttt{Cstit}]p \tag{s.N}$$
$$\odot_S[\alpha\ \texttt{Cstit}]p \to \Diamond K_\alpha p \tag{s.Oic}$$
$$\odot_S[\alpha\ \texttt{Cstit}]p \to K_\alpha \Box \odot_S[\alpha\ \texttt{Cstit}]p \tag{Cl}$$

(Rules of inference)

- *Modus Ponens*, *Substitution*, *Necessitation* for $\Box$.

Axiom (IA) is the axiom for 'independence of agents'. Axiom (KX) encodes the 'practical knowledge' constraint. Axiom (US) encodes the 'uniformity of strategies'

---

[12] For Example 2.5, they get that for every $h \in H_{m_1}$ $\langle m_1, h \rangle \not\models \odot[\alpha\ \texttt{kstit}]G$. For For Example 2.6, they get that for every $h \in H_{m_1}$ $\langle m_1, h \rangle \models \odot[\alpha\ \texttt{kstit}]\neg G$. For Example 2.7, they get that for every $h \in H_{m_1}$ $\langle m_1, h \rangle \not\models \odot[\alpha\ \texttt{kstit}]W$.

constraint. Axioms (Oic) and $(s.Oic)$ concern the objective, resp. subjective, versions of Kant's directive of 'ought implies can'. Axiom $(s.N)$ (standing for 'subjective necessity') says that all that is historically necessary at epistemically indistinguishable situations must be a subjective obligation, and it results important in the proof of completeness. Axiom (Cl) (standing for 'closure') characterizes a desirable property of the logic, namely that if one subjectively ought to do something, then one knows that that is the case.

In the unpublished work [4], the axiom system Λ is shown to be sound and complete with respect to the class of epistemic utilitarian *bi-valued* KCO-models. Such models are more general than the ones we introduced in Definition 3.4. Instead of only one value function, there are two: one for the objective ought-to-do's, and the other for the subjective ones[13]. These models, then, are of the form $\langle T, \sqsubset$, **Choice**, $\{\sim_\alpha\}_{\alpha \in Ags}$, **Value**$_O$, **Value**$_S\rangle$. As such, the models in Definition 3.4 are particular instances of bi-valued models, in which both value functions assign the same value to each history of the tree.

As it turns out, if we construct an axiom system Λ_n by adding to Λ the axiom

$$\bigwedge_{1 \leq k \leq n} \diamond \left(\bigwedge_{1 \leq i \leq k-1} \neg p_i \wedge \diamond[\alpha \text{ Cstit}](p_k) \right) \rightarrow \bigvee_{0 \leq k \leq n-1} p_k \qquad (APC_n),$$

then we have that Λ_n is sound and complete with respect to the class of epistemic utilitarian bi-valued KCO-models where each agent has at most n possible choices at every moment. Therefore, there is a soundness and completeness result with respect to *finite-choice* models that extend those of Definition 3.4. Such a result presupposes a logic of ought-to-do that addresses the 'gambling problem', a common objection to the Meinong-Chisholm approach of ought-to-do in stit semantics (see [10] and [18]). The Meinong-Chisholm approach identifies what an agent ought to do with what it ought to be the case that the agent does, but this interpretation is vulnerable to a problem commonly known as the 'gambling problem', which Horty explains in detail in [10] (Chapter 3, subsection 3.4.2). We implement Horty's solution to this problem in the logic that we axiomatize, so that the semantics is in fact a generalization of the one introduced in Definitions 2.4 and 3.4. In order to address such a generalization and the soundness and completeness results, we need further definitions.

A function s on Ags is a selection function at m if it assigns to each α a member of **Choice**$_\alpha^m$. We denote by **Select**^{m} the set of all selection functions at m. For a fixed $m \in T$ and $\beta \in Ags$, we define **State**$_\beta^m = \{S \subseteq H_m \mid S = \bigcap_{\alpha \in Ags - \{\beta\}} s(\alpha)$, where $s \in$ **Select**$^m\}$. Next, we redefine the dominance orderings. For a fixed $\alpha \in Ags$ and $m_* \in T$, we first define an ordering $\leq$ on $\mathcal{P}(H_{m_*})$ such that for $X, Y \subseteq H_{m_*}$, $X \leq Y$ iff **Value**$_O(h) \leq$ **Value**$_O(h')$ for every $h \in X, h' \in Y$. Next we define the objective dominance ordering $\preceq$ on **Choice**$_\alpha^{m_*}$ such that for

---
[13]This extension is extremely useful for the proof of completeness, but its conceptual reach has not been a subject of our investigation as of yet.

$L, L' \in \mathbf{Choice}_\alpha^{m*}$, $L \preceq L'$ iff $\forall S \in \mathbf{State}_\alpha^{m*}, L \cap S \leq L' \cap S$. We write $L \prec L'$ iff $L \preceq L'$ and $L' \not\preceq L$. Finally, we define an objectively optimal set of actions $\mathbf{Optimal}_\alpha^{m*} =_{def} \{L \in \mathbf{Choice}_\alpha^{m*}; \text{there is no } L' \in \mathbf{Choice}_\alpha^{m*} \text{such that } L \prec L'\}$. As for the subjective ought-to-do's, we first define an ordering $\leq_s$ on $\mathcal{P}(H_{m*})$ such that for $X, Y \subseteq H_{m*}$, $X \leq_s Y$ iff $\mathbf{Value}_S(h) \leq_s \mathbf{Value}_S(h')$ for all $h \in X, h' \in Y$. Next we define a subjective ordering $\preceq_s$ on $\mathbf{Choice}_\alpha^{m*}$ such that for $L, L' \subseteq H_{m*}$, $L \preceq_s L'$ iff $\forall m$ such that $m_* \sim_\alpha m$, $\forall S \in \mathbf{State}_\alpha^m, [L]_\alpha^m \cap S \leq_s [L']_\alpha^m \cap S$. As expected, we write $L \prec_s L'$ iff $L \preceq_s L'$ and $L' \not\preceq_s L$. Finally, we define a subjectively optimal set of actions $\mathbf{S-optimal}_\alpha^{m*} =_{def} \{L \in \mathbf{Choice}_\alpha^{m*}; \text{there is no } L' \in \mathbf{Choice}_\alpha^{m*} \text{ such that } L \prec_s L'\}$. As expected, we write $L \prec_s L'$ iff $L \preceq_s L'$ and $L' \not\preceq_s L$.

With these definitions, we adapt the semantics of the formulas with ought-to-do operators, making the logic invulnerable to the 'gambling problem': we set $\langle m, h \rangle \models \odot[\alpha \ \mathtt{Cstit}]\varphi$ iff for every $L \in \mathbf{Choice}_\alpha^m$ such that $\langle m, h_L \rangle \not\models \varphi$ for some $h_L \in L$, there exists $L' \in \mathbf{Choice}_\alpha^m$ such that $L \prec L'$ and if $L'' = L'$ or $L' \preceq L''$, then $\langle m, h' \rangle \models \varphi$ for every $h' \in L''$. Similarly, we set $\langle m, h \rangle \models \odot_S[\alpha \ \mathtt{Cstit}]\varphi$ iff for every $L \in \mathbf{Choice}_\alpha^m$ such that $\langle m', h_L \rangle \not\models \varphi$ for some $h_L \in [L]_\alpha^{m'}(m \sim_\alpha m')$, there exists $L' \in \mathbf{Choice}_\alpha^m$ such that $L \prec_s L'$ and if $L'' = L'$ or $L' \preceq_s L''$, then for every m'' such that $m \sim_\alpha m''$, $\langle m'', h'' \rangle \models \varphi$ for every $h'' \in [L'']_\alpha^{m''}$.

Proposition 3.6. *The system Λ is sound with respect to the class of epistemic utilitarian bi-valued KCO-models.*

Proof. Let $\mathcal{M} = \langle T, \sqsubset, \mathbf{Choice}, \{\sim_\alpha\}_{\alpha \in Ags}, \mathbf{Value}_O, \mathbf{Value}_S \rangle$ be an epistemic utilitarian bi-valued KCO-frame. Let $\mathcal{V}$ be any pertinent valuation function.

- The **S5** axioms for $\Box, [\alpha \ \mathtt{Cstit}]$, K_α, and $\Box_K$, as well as axioms $(A1)$, $(A2)$, $(A3)$, $(A4)$, $(A6)$, (Oic), and (IA) are semantic validities, as established in [12].

- Axiom $(A7)$ is a semantic validity, straightforwardly from the semantics for $\odot_S[\alpha \ \mathtt{Cstit}]\varphi$.

- To see that $\mathcal{M} \models (KX)$, take $\langle m, h \rangle$ such that $\langle m, h \rangle \models K_\alpha \varphi$. Let $h' \in H_m$ such that $h' \in \mathbf{Choice}_\alpha^m(h)$. Reflexivity of $\sim_\alpha$ and semantic condition (**KX**) ensure that for every $h' \in H_m$ such that $h' \in \mathbf{Choice}_\alpha^m(h)$, $\langle m, h \rangle \sim_\alpha \langle m, h' \rangle$. Thus, for every $h' \in \mathbf{Choice}_\alpha^m(h)$, $\langle m, h' \rangle \models \varphi$, which means that $\langle m, h \rangle \models [\alpha \ \mathtt{Cstit}]\varphi$.

- To see that $\mathcal{M} \models (US)$, take $\langle m, h \rangle$ such that $\langle m, h \rangle \models \Diamond K_\alpha \varphi$ –which means that there exists $h' \in H_m$ such that $\langle m, h' \rangle \models K_\alpha \varphi$. Now, take an arbitrary situation $\langle m'', h'' \rangle$ such that $\langle m, h \rangle \sim_\alpha \langle m'', h'' \rangle$. The semantic condition (**US**) guarantees that there exists $h''' \in H_{m''}$ such that $\langle m, h' \rangle \sim_\alpha \langle m'', h''' \rangle$. Therefore, the fact that $\langle m, h' \rangle \models K_\alpha \varphi$ implies that $\langle m'', h''' \rangle \models \varphi$. With this, we have shown that $\langle m, h \rangle \models K_\alpha \Diamond \varphi$.

- To see that $\mathcal{M} \models (s.N)$, take $\langle m_*, h_* \rangle$ such that $\langle m_*, h_* \rangle \models K_\alpha \Box \varphi$. Let m be such that $m_* \sim_\alpha m$ (which means that there exist $j \in H_{m_*}$, $j' \in H_m$ such that $\langle m_*, j \rangle \sim_\alpha \langle m, j' \rangle$), and let $L \in \textbf{Choice}_\alpha^{m_*}$. Condition (US) ensures that there exists $h'_* \in H_m$ such that $\langle m_*, h_* \rangle \sim_\alpha \langle m, h'_* \rangle$. Our assumption that $\langle m_*, h_* \rangle \models K_\alpha \Box \varphi$ implies then that $\langle m, h'_* \rangle \models \Box \varphi$. Therefore, for any $h \in [L]_\alpha^m$, the fact that $h \in H_m$ yields that $\langle m, h \rangle \models \varphi$. Therefore, for every $L \in \textbf{Choice}_\alpha^{m_*}$ and every m such that $m_* \sim_\alpha m$, we have that $[L]_\alpha^m \subseteq |\varphi|^m$, which vacuously implies that $\langle m_*, h_* \rangle \models \odot_S[\alpha \ \texttt{Cstit}]\varphi$.

- To see that $\mathcal{M} \models (s.Oic)$, take $\langle m_*, h_* \rangle$ such that $\langle m_*, h_* \rangle \models \odot_S[\alpha \ \texttt{Cstit}]\varphi$. This implies that there exists $L_* \subseteq H_{m_*}$ such that $[L_*]_\alpha^{m'} \subseteq |\varphi|^{m'}$ for every $m' \in T$ such that $m_* \sim_\alpha m'$. Since $\sim_\alpha$ is reflexive, it is clear that $[L_*]_\alpha^{m_*} \subseteq |\varphi|^{m_*}$. Now, take $h_{*0} \in L_*$. Let $\langle m, h \rangle$ be such that $\langle m_*, h_{*0} \rangle \sim_\alpha \langle m, h \rangle$. From Definition 3.3, we have that $h \in [L_*]_\alpha^m$, so that $[L_*]_\alpha^m \subseteq |\varphi|^m$ implies that $\langle m, h \rangle \models \varphi$. Therefore, we have that there exists $h_{*0} \in H_{m_*}$ such that for every $\langle m, h \rangle$ such that $\langle m_*, h_{*0} \rangle \sim_\alpha \langle m, h \rangle$, $\langle m, h \rangle \models \varphi$. This means that $\langle m_*, h_{*0} \rangle \models K_\alpha \varphi$. Therefore, we have that $\langle m_*, h_* \rangle \models \Diamond K_\alpha \varphi$.

- To see that $\mathcal{M} \models (Cl)$, take $\langle m_*, h_* \rangle$ such that $\langle m_*, h_* \rangle \models \odot_S[\alpha \ \texttt{Cstit}]\varphi$. Let $\langle m, j \rangle$ be such that $\langle m_*, h_* \rangle \sim_\alpha \langle m, j \rangle$. Take $h \in H_m$. We want to show that for every $L \in \textbf{Choice}_\alpha^m$ such that $[L]^{m'} \not\subseteq |\varphi|^{m'}$ (for some m' such that $m \sim_\alpha m'$), there exists $L' \in \textbf{Choice}_\alpha^m$ such that $L \prec_s L'$ and if $L'' = L'$ or $L' \preceq_s L''$, then for every m'' such that $m \sim_\alpha m''$, we have that $[L'']_\alpha^{m''} \subseteq |\varphi|^{m''}$. Take $L \in \textbf{Choice}_\alpha^m$ such that there exists $m' \in T$ with $m \sim_\alpha m'$ and $[L]^{m'} \not\subseteq |\varphi|^{m'}$. Let N_L be an action in $\textbf{Choice}_\alpha^{m_*}$ such that $N_L \subseteq [L]_\alpha^{m_*}$. We know that such an action exists, in virtue of the (US) and (KX) constraints. Notice that transitivity of $\sim_\alpha$ entails that $[N_L]_\alpha^o = [L]_\alpha^o$ for any moment o, so we have that $[N_L]_\alpha^{m'} \not\subseteq |\varphi|^{m'}$. Since $\langle m_*, h_* \rangle \models \odot_S[\alpha \ \texttt{Cstit}]\varphi$, we get that there must exist $N \in \textbf{Choice}_\alpha^{m_*}$ such that $N_L \prec_s N$ and if $N' = N$ or $N \preceq_s N'$, then for every m'' such that $m_* \sim_\alpha m''$, $[N']_\alpha^{m''} \subseteq |\varphi|^{m''}$. Now, let L_N be an action in $\textbf{Choice}_\alpha^m$ such that $L_N \subseteq [N]_\alpha^m$ (which implies that $[L_N]_\alpha^o = [N]_\alpha^o$ for any moment o). We claim that $L \prec_s L_N$, and show our claim with the following argument. Let m'' be a moment such that $m \sim_\alpha m''$, and let $S \in \textbf{State}_\alpha^{m''}$. On one hand, we have that $[L]_\alpha^{m''} \cap S = [N_L]_\alpha^{m''} \cap S \leq_s [N]_\alpha^{m''} \cap S = [L_N]_\alpha^{m''} \cap S$ (a). On the other hand, we know that that there exists a moment m''' such that $m_* \sim_\alpha m'''$ and a state $S_0 \in \textbf{State}_\alpha^{m'''}$ such that $[N]_\alpha^{m'''} \cap S_0 \not\leq_s [N_L]_\alpha^{m'''} \cap S_0$: Therefore, we have that $[L_N]_\alpha^{m'''} \cap S_0 = [N]_\alpha^{m'''} \cap S_0 \not\leq_s [N_L]_\alpha^{m'''} \cap S_0 = [L]_\alpha^{m'''} \cap S_0$ (b). Together, (a) and (b) entail that $L \prec_s L_N$. Now, let $L'' \in \textbf{Choice}_\alpha^m$ such that $L'' = L_N$ or $L_N \preceq_s L''$. If $L'' = L_N$, then for every m'' such that $m \sim_\alpha m''$, $[L'']_\alpha^{m''} = [N]_\alpha^{m''} \subseteq |\varphi|^{m''}$. If $L_N \prec_s L''$, then an argument similar to the the one we used above to show that our claim was true renders that an action $N_{L''} \in \textbf{Choice}_\alpha^{m_*}$ such that $N_{L''} \subseteq [L'']_\alpha^{m_*}$ is such that $N \preceq_s N_{L''}$, so that $[L'']_\alpha^{m''} = [N_{L''}]_\alpha^{m''} \subseteq |\varphi|^{m''}$. With this, we have shown that $\langle m, h \rangle \models \odot_S[\alpha\ \texttt{Cstit}]\varphi$. Since this happens for every $h \in H_m$, we have that $\langle m, j \rangle \models \Box \odot_S [\alpha\ \texttt{Cstit}]\varphi$. Since this happens for any $\langle m, j \rangle$ such that $\langle m_*, h_* \rangle \sim_\alpha \langle m, j \rangle$, then we get that $\langle m_*, h_* \rangle \models K_\alpha \Box \odot_S [\alpha\ \texttt{Cstit}]\varphi$.

- Moreover, and as shown in [17], we have that $\mathcal{M} \models (APC)_n$ iff for every agent α and $m \in T$, $\textbf{Choice}_\alpha^m$ has at most n elements.

- *Modus Ponens*, Substitution, and Necessitation for $\Box$ preserve validity.

Therefore, we have shown that the system Λ is sound with respect to the class of epistemic bi-valued KCO-models. $\square$

We will only summarize the strategy used in [4] to prove the completeness of the axiom system Λ with respect to the class of epistemic utilitarian bi-valued KCO-models, and mention the relevant lemmas. Completeness of the axiom system for the fragment of $\mathcal{L}_{\textsf{KCO}}$ that includes only the stit operators and $\Box$ is standard, and a proof can be found in [17]. For the fragment extended with the $\odot[\alpha\ \texttt{Cstit}]$ operators, [12] provides a proof. The result in [4] details the proof of [12], and extends it to the full language –with the $\odot_S[\alpha\ \texttt{Cstit}]$ operators. Following the method used both by [17] and [12], [4] uses a canonical model for Λ, built from the syntax in the tradition of modal logic. The possible worlds in the domain of this canonical model are Λ-MCS's, and the relations among these worlds are defined so as to ensure the so-called *truth lemma*. The key element for getting the truth lemma for the $\odot_S[\alpha\ \texttt{Cstit}]$ operators lies in the definition of $\textbf{Value}_S$ in the canonical model. For such a definition, [12] provides some intuitions that [4] makes precise. If we denote by W^Λ the domain of the canonical model, we define $R_\Box$ as a relation over W^Λ such that for $w, v \in W^\Lambda$, $wR_\Box v$ iff for every φ, $\Box\varphi \in w \Rightarrow \varphi \in v$. For a given $w \in W^\Lambda$, the set $\{v \in W^\Lambda \mid wR_\Box v\}$ is denoted by $\overline{w}$. For a fixed $\alpha \in Ags$ and $w \in W^\Lambda$, let $\Gamma_\alpha^w = \{K_\alpha\varphi \mid \odot_S[\alpha\ \texttt{Cstit}]\varphi \in w\}$. If we take $\Gamma^w = \bigcup_{\alpha \in Ags} \Gamma_\alpha^w$, then $\textbf{Value}_S$ is defined by

$$\textbf{Value}_S(w) = \begin{cases} 1 \text{ iff } \Gamma^w \subseteq w \\ 0 \text{ otherwise.} \end{cases}$$

With this definition, [4] shows two important lemmas that enable the proof of the truth lemma for $\odot_S[\alpha\ \texttt{Cstit}]$:

Lemma 3.7. *For $\beta \in Ags$, we have that for every formula φ of $\mathcal{L}_{\textsf{KCO}}$ and every $w \in W^\Lambda$, $\odot_S[\beta\ \texttt{Cstit}]\varphi \in w$ iff $K_\beta\varphi \in v$ for every $v \in W^\Lambda$ such that $K_\beta\Box\varphi \in w \Rightarrow \varphi \in v$ and $\Gamma_\beta^v \subseteq v$.*[14]

Lemma 3.8. *For $\beta \in Ags$, for every $w \in W^\Lambda$, $\Gamma_\beta^w \subseteq w$ iff $\textbf{Choice}_\beta^{\overline{w}}(w) \in S-\textbf{Optimal}_\beta^{\overline{w}}$.*

---

[14] The proof of this lemma makes use of axioms $(s.N)$ and (Cl).

[4] uses these results to show completeness with respect to the canonical model. The canonical model is not exactly an epistemic utilitarian bi-valued KCO-model, since it is not based on a tree. However, one can show that there is an epistemic utilitarian bi-valued KCO-model that validates the same formulas as the canonical model. This is done in a two-step process. First, [4] introduces relational structures called Kripke-estit models, with a particular semantics for the formulas of the language $\mathcal{L}_{\mathsf{KCO}}$. Secondly, a correspondence theorem is drawn between Kripke-estit models and certain epistemic utilitarian bi-valued KCO-models. More precisely, this theorem states that every Kripke-esit model gives rise to an epistemic utilitarian bi-valued KCO-model that satisfies the same formulas. In order to prove completeness fully, then, one shows that the canonical structure built from the syntax is in fact a Kripke-estit model, proves the truth lemma for this last one, and gets completeness with respect to the class of Kripke-estit models. Then, the correspondence theorem allows us to prove completeness with respect to epistemic utilitarian bi-valued KCO-models.

This result is highly favorable in a framework of *responsibility*, and once again draws a desirable characteristic of our logic against the use of action types. The mixture of action types and knowledge poses difficulties for axiomatizations, as is demonstrated by the fact that variants of ATEL ([8]) have never been axiomatized.

To end on a less technical note, we address a few interesting properties concerning the interaction of different operators for the logic presented in Section 2:

$$\not\models \odot[\alpha\ \mathsf{Cstit}]\varphi \to \Diamond K_\alpha[\alpha\ \mathsf{Cstit}]\varphi.$$

Our solution to coin-flip problem #3 poses a counterexample, because $\langle m_1, h_1\rangle$ is such that $\langle m_1, h_1\rangle \models \odot[\alpha\ \mathsf{Cstit}]W$, but $\langle m_1, h_1\rangle \not\models \Diamond K_\alpha[\alpha\ \mathsf{Cstit}]W$, as witnessed by the facts that $\langle m_1, h_2\rangle \not\models [\alpha\ \mathsf{Cstit}]W$ and that $\langle m_2, h_3\rangle \not\models [\alpha\ \mathsf{Cstit}]W$.

$$\not\models \odot_{\mathcal{S}}[\alpha\ \mathsf{Cstit}]\varphi \to \odot[\alpha\ \mathsf{Cstit}]\varphi.$$

Our solution to coin-flip problem #2 poses a counterexample, because $\langle m_1, h_1\rangle$ is such that $\langle m_1, h_1\rangle \models \odot_{\mathcal{S}}[\alpha\ \mathsf{Cstit}]\neg G$, but $\langle m_1, h_1\rangle \not\models \odot[\alpha\ \mathsf{Cstit}]\neg G$.

$$\not\models \odot[\alpha\ \mathsf{Cstit}]\varphi \to \odot_{\mathcal{S}}[\alpha\ \mathsf{Cstit}]\varphi.$$

Our solution to coin-flip problem #3 poses a counterexample, because $\langle m_1, h_1\rangle$ is such that $\langle m_1, h_1\rangle \models \odot[\alpha\ \mathsf{Cstit}\ W]$, but $\langle m_1, h_1\rangle \not\models \odot_{\mathcal{S}}[\alpha\ \mathsf{Cstit}\ W]$.

4 Conclusion

This paper deals with important questions in the modeling of agency, practical knowledge, and responsibility. Central to the paper is the notion of 'ought-to-do'. We argue that to solve certain problems in the treatment of knowledge and obligations in stit logic, one possibility is to distinguish between objective and subjective versions of the ought-to-do. Through Kant's directive of 'ought implies can', it is seen that subjective ought-to-do's are relative to the practical knowledge of an agent. Then, contrary to conventional wisdom, we show that to model practical knowledge in stit logic, we do not need action types.

References

[1] R. Alur, T.A. Henzinger, and O. Kupferman. Alternating-time temporal logic. *Journal of the ACM*, 49(5):672–713, 2002.

[2] Can Baskent, Loes Olde Loohuis, and Rohit Parikh. On knowledge and obligation. *Episteme*, 9(2):171–188, 2012.

[3] N. Belnap, M. Perloff, and M. Xu. *Facing the future: agents and choices in our indeterminist world*. Oxford University Press, 2001.

[4] J. Broersen and A. Ramírez Abarca. Completeness and decidability of a logic of subjective oughts (work in progress). 2017.

[5] Fabrizio Cariani, Magdalena Kaufmann, and Stefan Kaufmann. Deliberative modality under epistemic uncertainty. *Linguistics and Philosophy*, 36(3):225–259, 2013.

[6] David Copp. 'ought' implies 'can' and the derivation of the principle of alternate possibilities. *Analysis*, 68(1):67–75, 2008.

[7] Andreas Herzig and Emiliano Lorini. A dynamic logic of agency I: Stit, capabilities and powers. *Journal of Logic, Language and Information*, 19(1):89–121, 2010.

[8] W. van der Hoek and M. Wooldridge. Cooperation, knowledge, and time: Alternating-time temporal epistemic logic and its applications. *Studia Logica*, 75(1):125–157, 2003.

[9] John Horty and Eric Pacuit. Action types in stit semantics. *Review of Symbolic Logic*, 2017. forthcoming.

[10] John F. Horty. *Agency and Deontic Logic*. Oxford University Press, 2001.

[11] N. Kolodny and J. MacFarlane. Ifs and oughts. *Journal of Philosophy*, 107(3):115–43, 2010.

[12] Yuko Murakami. Utilitarian deontic logic. *AiML-2004: Advances in Modal Logic*, 287, 2004.

[13] E. Pacuit, R. Parikh, and E. Cogan. The logic of knowledge based obligation. *Knowledge, Rationality and Action a subjournal of Synthese*, 149(2):311–341, 2006.

[14] D. Parfit. What we together do. Unpublished manuscript, 1988.

[15] Ralph Wedgwood. *Deontic modality*, chapter Objective and subjective 'ought', pages 142–168. Oxford Univeristy Press, 2016.

[16] Malte Willer. A remark on iffy oughts. *The Journal of Philosophy*, 109(7):449–461, 2012.

[17] Ming Xu. Decidability of deliberative stit theories with multiple agents. In *Temporal Logic*, pages 332–348. Springer, 1994.

[18] Ming Xu. Combinations of stit with ought and know. *Journal of Philosophical Logic*, 44(6):851–877, 2015.

Normative Conflicts in a Dynamic Logic of Norms and Codes

Ilaria Canavotto
ILLC, University of Amsterdam, The Netherlands
`i.canavotto@uva.nl`

Alessandro Giordani
Department of Philosophy, Catholic University of Milan, Italy
`alessandro.giordani@unicatt.it`

In this paper we introduce two conflict tolerant and dynamic deontic systems **DNC** and **DNC**$^+$ in which normative conflicts are analysed as conflicts between normative codes containing norms that prescribe the realisation of incompatible states of affairs. The systems present two crucial traits: first, norms and codes are explicitly represented in it; second, the connections between norms and codes and the way in which codes are updated by introducing new norms are properly defined. We will show how the systems can be used to fruitfully analyse paradigmatic cases of civil conflicts. Specifically, **DNC** and **DNC**$^+$ will allow us to model the genesis of a conflict by keeping track of which agent triggered it and, relatedly, to capture the basic distinction between civil disobedience and conscientious objection.

Keywords: Explicit Modal Logic; Dynamic Deontic Logic; Deontic Conflicts; Civil Disobedience.

1 Introduction

In the last decades, the study of normative conflicts in deontic logic has been guided by two main issues: *developing conflict tolerant deontic systems*, typically by changing the logical principles characterizing standard deontic logic (see [11, 12, 13]); given a conflict tolerant system, *developing solution procedures for deontic conflicts* mainly based on priority relations between obligations (see [14, 15, 16, 17, 18, 19]) or on selection functions in the tradition of input/output logics (see [21, 22, 23]).

In the present paper we supplement these lines of research by proposing a framework in which it is possible *to represent the dynamics giving rise to conflicts*. By

---

We would like to thank the referees of DEON 2018 for useful suggestions. Previous versions of this paper were presented in Amsterdam, Utrecht, and at Trends in Logic XVII. We would like to express our gratitude to the audience of these events for insightful comments and helpful feedback.

assuming this perspective, we will be able to draw key distinctions between different types of conflicts, and thus lay the basis for studying specific kinds of solution procedures, since conflicts of different types might require diverse solutions.

The basic idea is that, in order to model the origin of a conflict, we need to explicitly refer to the deontic codes of the agents involved, where a code can be thought of as the set of norms the agent accepts as binding in concrete situations. Conflicts can then arise either between codes of different agents or between the norms constituting the code of a single agent, as a result of an update of these codes by adding new norms. In this view, the interaction between different kinds of normative sources, specifically norms and codes, plays an essential role.

We account for this idea by designing a logic intended to explicitly represent

(*i*) the relations between norms;
(*ii*) the relations between norms and codes;
(*iii*) the dynamics induced by including a new norm in a certain code.

We do this by first developing a logic of norms **N** in section 2, which is then extended to a logic of norms and codes **NC** in section 3. **N** and **NC** are inspired by explicit modal logics [3, 10], as norms and codes are introduced in them as explicit sources of prescriptions. Unlike other logics of norms, such as input/output logics [21, 22, 23] or deontic systems based on default logics [18, 19], **N** and **NC** are not devised for representing logical relations between norms nor for deriving normative solutions given a conflict between norms in a code; rather, the two logics are specifically meant to be the basis for studying the *genesis* of such conflicts.

This becomes clear in section 4, where we make the system **NC** dynamic by defining update procedures which correspond, in the semantics, to model transformations. The dynamic logic of norms and codes **DNC** that we obtain is thus dynamic in the sense of public announcement logic and dynamic epistemic logic [4, 5, 25]. To the best of our knowledge, **DNC** is the first deontic system in which ideas from explicit and dynamic modal logics are merged. The update procedure encoded in this system corresponds to the event that an agent accepts a new norm in her code, and is different, both from a logical point of view and from a conceptual point of view, from updates following a public announcement and from deontic variants of preference update [6, 7, 20, 26].

In section 5, we illustrate how **DNC** can be used to analyse two paradigmatic cases of conflicts involving more than one agent, namely the cases of Antigone and Gandhi. In doing this, we show that our system has the resources to represent the origin of a conflict between codes by keeping track of which agent originates the conflict. In section 6, we then show that in **DNC** and one of its variants, which we call **DNC$^+$**, it is possible to model several key features that distinguish cases of civil disobedience from cases of conscientious objection. We take this to provide evidence that our framework is strong enough to distinguish important kinds of deontic conflicts. Section 7 then concludes by pointing to possible developments for future works.

2 Logic of Norms

The basic idea underlying the logic of norms **N** is that norms are abstract and consistent normative sources agents can adopt as directives for acting.[1] In what follows, we model this idea by formulating a semantic system in which: (i) each norm is characterized by what it explicitly prescribes, which is its *explicit* content; (ii) as each norm is *abstract*, its explicit content does not change across possible worlds; (iii) as each norm is *consistent*, its explicit content does not give rise to contradictions, so it corresponds to a non-empty set of possible worlds. This final assumption implies that inconsistencies can only be found between the prescriptions of different norms, and not within the content of an isolated norm.

Besides these conditions, which are specific of norms, two further minimal assumptions on normative sources in general will underlie both our logics **N** and **NC**. First, every normative source is required to explicitly prescribe its own satisfaction, since normative sources are directives and, as such, they at least direct us to situations where they are not violated. Second, whenever a normative source prescribes the satisfaction of a different norm, it is required to prescribe all explicit prescriptions of this norm. Thus, prescribing the satisfaction of a norm amounts to prescribing everything prescribed by that norm.

2.1 The Language $\mathcal{L}_\mathbf{N}$

The language $\mathcal{L}_\mathbf{N}$ of the logic of norms **N** is based on a countable set $\{n_i\}_{i\in\mathbb{N}}$ of names of basic norms and a countable set $\{p_i\}_{i\in\mathbb{N}}$ of propositional letters. The set of formulas $Fm(\mathcal{L}_\mathbf{N})$ of $\mathcal{L}_\mathbf{N}$ is then built according to the following rules:

$$\phi ::= p_i \mid sat(n_i) \mid \neg\phi \mid \phi \wedge \phi \mid \Box\phi \mid [1]\phi \mid n_i : \phi$$

The other logical connectives and the dual modalities $\Diamond$ and $\langle 1 \rangle$ are defined as usual.

The formulas of $\mathcal{L}_\mathbf{N}$ can be subdivided into ontic and deontic formulas. The ontic formulas are $\Box\phi$ and $[1]\phi$. Intuitively, the former says that ϕ holds at all possible worlds and the latter that ϕ holds at all worlds that are accessible from the current world. The two modalities are intended to capture the distinction between what is possible in general and what is possible in specific circumstances. The deontic formulas of $\mathcal{L}_\mathbf{N}$ include $sat(n_i)$ and $n_i : \phi$. The former says that norm n_i is satisfied and the latter that norm n_i explicitly prescribes ϕ.

The interaction between ontic and deontic modalities makes $\mathcal{L}_\mathbf{N}$ powerful enough to express the distinction between global and local conflicts and to define a modality to account for the implicit content of a norm.

---

[1] Hence, we are here assuming an *hyletic conception* of norms as proposed in [1, 2].

Definition 1 (Conflicts between norms).

(Global conflict) $n_i \perp n_j := \Box \neg(sat(n_i) \wedge sat(n_j))$
(Local conflict) $n_i \perp_1 n_j := [1]\neg(sat(n_i) \wedge sat(n_j))$

Intuitively, two norms are globally in conflict when there is no world in which they are both satisfied, i.e., when there is no way to satisfy both of them, while they are locally in conflict when there is no way to satisfy both of them in the current circumstances.

Definition 2 (ϕ is implicitly prescribed by n_i). $[n_i]\phi := \Box(sat(n_i) \to \phi)$.

Hence, what is implicitly prescribed by a norm is what is necessary for satisfying the norm. In what follows, we will say that what is implicitly prescribed by a norm constitutes its *implicit* content.

2.2 Semantics of $\mathcal{L}_\mathbf{N}$

In order to capture the basic conception of norms sketched at the beginning of this section, we interpret the language $\mathcal{L}_\mathbf{N}$ on models consisting of both ontic and deontic elements. At the ontic level, we assume a set W of possible worlds, a valuation V mapping propositional letters to the sets of worlds in which they are true, and a function R determining, for each possible world w, the set of worlds that are accessible from w. The function R is needed in order to model the distinction between what in possible in given circumstances and what is possible in general. At the deontic level, we assume a function S that determines, for every norm n_i, the set of worlds in which n_i is satisfied and a function P that determines, for each norm n_i, the set of explicit prescriptions of n_i, i.e., the explicit content of n_i. Importantly, the function P will assign an explicit content to each norm without reference to possible worlds: this will ensure that the content of a norm remains unchanged from a possible world to another, in line with the idea that norms are abstract normative sources characterized by their content.

Definition 3 (Model for $\mathcal{L}_\mathbf{N}$). *A model for $\mathcal{L}_\mathbf{N}$ is a tuple $M = \langle W, R, S, P, V \rangle$, where $W \neq \varnothing$ is a set of worlds, $R : W \to \wp(W)$ is an accessibility relation on W, $V : \{p_i\}_{i \in \mathbb{N}} \to \wp(W)$ is a valuation, and S and P are as follows:*

$S : \{n_i\}_{i \in \mathbb{N}} \to \wp(W)$

$P : \{n_i\}_{i \in \mathbb{N}} \to \wp(Fm(\mathcal{L}_\mathbf{N}))$

We require that the functions S and P satisfy the following conditions.

Conditions on norms. *For all n_i and n_j,*

N1 *(Explicit consistency)* $\quad \phi \in P(n_i) \Rightarrow \neg\phi \notin P(n_i)$

N2 *(Implicit consistency)* $\quad S(n_i) \neq \varnothing$

N3 *(Norm satisfaction)* $\quad sat(n_i) \in P(n_i)$

N4 *(Norm inclusion)* $\quad sat(n_i) \in P(n_j) \Rightarrow P(n_i) \subseteq P(n_j)$

In accordance with our assumptions on norms and normative sources, **N1** and **N2** ensure that every norm is consistent, both in the explicit sense that no norm explicitly prescribes a proposition and its negation and in the implicit sense that every norm is satisfied at at least one possible world. In addition, **N3** guarantees that every norm prescribes its own satisfaction and **N4** that, whenever a norm prescribes the satisfaction of another norm, the former also prescribes all propositions that are prescribed by the latter.

Definition 4 (Truth, $\mathcal{L}_\mathbf{N}$). *The notion of truth of a formula in a world of a model for $\mathcal{L}_\mathbf{N}$ is recursively defined as follows:*

$M, w \models p_i \Leftrightarrow w \in V(p_i)$
$M, w \models sat(n_i) \Leftrightarrow w \in S(n_i)$
$M, w \models \neg\phi \Leftrightarrow M, w \not\models \phi$
$M, w \models \phi \wedge \psi \Leftrightarrow M, w \models \phi \text{ and } M, w \models \psi$
$M, w \models \Box\phi \Leftrightarrow \forall v(v \in W \Rightarrow M, v \models \phi)$
$M, w \models [1]\phi \Leftrightarrow \forall v(v \in R(w) \Rightarrow M, v \models \phi)$
$M, w \models n_i : \phi \Leftrightarrow \phi \in P(n_i)$

In order to provide a sound interpretation of the formulas of $\mathcal{L}_\mathbf{N}$, models for $\mathcal{L}_\mathbf{N}$ should connect the explicit and implicit content of a norm in a suitable way.

Definition 5 (Suitability). *A model $M = \langle W, R, S, P, V \rangle$ for $\mathcal{L}_\mathbf{N}$ is suitable just in case it satisfies the following norm suitability condition:*

NS *(Norm suitability). For all $w \in W$, $\phi \in P(n_i)$ and $w \in S(n_i) \Rightarrow M, w \models \phi$.*

According to **NS**, a norm is satisfied only at those worlds where all the propositions it explicitly prescribes are true, which means, given definition 2, that what the norm explicitly prescribes is included in its implicit content. Since every norm prescribes its own satisfaction by **N3**, this ensures that every norm is satisfied precisely at those worlds where all the propositions it explicitly prescribes are true.

The notions of validity and logical consequence with respect to suitable models for $\mathcal{L}_\mathbf{N}$ are defined in the standard way.

2.3 The Axiom System N

The axiom system **N** is defined by three groups of axioms and rules:

Axioms and rules for propositional validities: any sufficient set.

Axioms and rules on the ontic part: Axioms on norms:
- **A1** $\Box(\phi \to \psi) \to (\Box\phi \to \Box\psi)$ **AN1** $n_i : \phi \to \Box(n_i : \phi)$
- **A2** $\Box\phi \to \phi$ **AN2** $n_i : \phi \to \neg(n_i : \neg\phi)$
- **A3** $\neg\Box\phi \to \Box\neg\Box\phi$ **AN3** $\neg\Box\neg sat(n_i)$
- **A4** $[1](\phi \to \psi) \to ([1]\phi \to [1]\psi)$ **AN4** $n_i : sat(n_i)$
- **A5** $\Box\phi \to [1]\phi$ **AN5** $n_j : sat(n_i) \wedge n_i : \phi \to n_j : \phi$
- **R1** If $\vdash_\mathbf{N} \phi$, then $\vdash_\mathbf{N} \Box\phi$ **AN6** $n_i : \phi \wedge sat(n_i) \to \phi$

The axioms and rules on the ontic modalities characterize $\Box$ as an **S5**-modality that is stronger than the **K**-modality [1]. The axioms on norms account for the basic traits of norms: **AN1** captures the fact that norms are *abstract*, and so that they are necessarily characterized by their content; **AN2** states that norms are *explicitly* consistent, so that no norm prescribes both ϕ and $\neg\phi$; **AN3** states that norms are *implicitly* consistent, so that every norm is in principle satisfiable; **AN4** captures the idea that norms prescribe at least their own satisfaction; finally, **AN5** states that a norm prescribing the satisfaction of another norm actually prescribes what is prescribed by this norm, while **AN6** states that a norm prescribing ϕ is satisfied only if ϕ is the case, so that satisfaction amounts to non-violation.

The following basic theorems are derivable.

- **TN1** $n_i : \phi \to [n_i]\phi$ by **AN6**, **AN1** and def. $[n_i]$
- **TN2** $n_i : \phi \wedge n_j : \neg\phi \to n_i \perp n_j$ by **AN6**, **AN1** and def. $\perp$

Due to space limitations we omit the proof of the following theorem.

Theorem 1 (Characterization). *The axiom system $\mathbf{N}$ is sound and complete with respect to the class of all suitable models for $\mathcal{L}_\mathbf{N}$.*

3 Logic of Norms and Codes

In accordance with the intuition that agents use norms to direct their conduct, the aim of the logic of norms and codes **NC** is to model the obligations induced by codes of agents, intended as sets of norms. To do this, we extend our conceptual apparatus with tools to refer to codes and to describe the relations between codes and norms. Specifically, we introduce codes as concrete normative sources determining what agents are bound to do in concrete situations.[2] This implies two main consequences.

First, codes are not essentially characterized by what they explicitly prescribe: since nothing prevents an agent from assuming different norms of conduct in diverse circumstances, the explicit prescriptions of a code can change across possible worlds. As a consequence, no normative source can prescribe the satisfaction of a code,

---

[2] Note that we take an agent to be any entity that has the ability to assume certain norms as binding. Hence, the set of agents can include, for instance, persons, communities, organizations, or states. Note also that codes are always intended as codes of agents.

except the code itself. Indeed, a code prescribes its own satisfaction because it is a directive (cf. section 2). Yet, the prescriptions of a code cannot refer to the satisfaction of another code, since an agent cannot bind herself to obey a normative source whose content can change in a way unknown to her. Similarly, a norm cannot prescribe the satisfaction of a code, because its content is fixed and, hence, it cannot contain references to prescriptions that might vary.

Second, codes can issue inconsistent prescriptions: since they are composite normative sources, they might consist of norms prescribing incompatible things. Hence, unlike norms, codes can be sources of internal deontic conflicts.

3.1 The Language $\mathcal{L}_{\mathbf{NC}}$

Let us fix a countable set Ag of names of agents and extend the alphabet of the language $\mathcal{L}_{\mathbf{N}}$ with a countable set $\{c_a\}_{a \in Ag}$ of names of agents' codes and with a modality O_a for every $a \in Ag$. The set of formulas $Fm(\mathcal{L}_{\mathbf{NC}})$ of $\mathcal{L}_{\mathbf{NC}}$ is then defined according to the following rules:

$$\phi ::= p_i \mid sat(n_i) \mid sat(c_a) \mid \neg\phi \mid \phi \wedge \phi \mid \Box\phi \mid [1]\phi \mid n_i : \phi \mid O_a\phi$$

where, in a formula like $n_i : \phi$, $\phi \in Fm(\mathcal{L}_{\mathbf{N}})$ and, in a formula like $O_a\phi$, $\phi \in Fm(\mathcal{L}_{\mathbf{N}}) \cup \{sat(c_a)\}$. As before, the other logical connectives and the modalities $\Diamond$ and $\langle 1 \rangle$ are defined as usual.

The intended reading of the new formulas is as follows: $sat(c_a)$ says that the code c_a of agent a is satisfied and $O_a\phi$ says that ϕ is obligatory for a in virtue of her own code. In accordance with our characterization of norms and codes, the conditions we impose on the construction of formulas of form $n_i : \phi$ and $O_a\phi$ guarantee that the prescriptions of a norm never concern a code and that the prescriptions of a code never concern the satisfaction of another code.

The new formulas that can be built in $\mathcal{L}_{\mathbf{NC}}$ allow us to account for the distinction between global and local conflicts not only between norms but also between a norm and a code and between two codes.

Definition 6 (Conflicts between a norm and a code and between two codes).

Global conflicts:
$n_i \perp c_a := \Box \neg (sat(n_i) \wedge sat(c_a))$
$c_a \perp c_b := \Box \neg (sat(c_a) \wedge sat(c_b))$

Local conflicts:
$n_i \perp_1 c_a := [1]\neg(sat(n_i) \wedge sat(c_a))$
$c_a \perp_1 c_b := [1]\neg(sat(c_a) \wedge sat(c_b))$

More importantly, the interaction between formulas on norms and formulas on codes allows us to express the fact that an agent accepts a norm as binding.

Definition 7 (Acceptance of a norm).

Agent a accepts norm n_i: $O_a sat(n_i)$.

Agent a accepts norm n_i as binding: $O_a sat(n_i) \wedge [1] O_a sat(n_i)$.

Accepting a norm means to be obliged to satisfy that norm in virtue of one's own code and, as we will see in a moment, in **NC** this corresponds in a precise sense to including a norm in one's own code. Accepting a norm as binding means to be obliged to satisfy that norm in virtue of one's own code *at all accessible worlds*. This kind of acceptance models the commitment of the agent to the norm, *viz.*, that she does not accept the norm merely for her convenience in the present circumstances.

3.2 Semantics of $\mathcal{L}_{\mathbf{NC}}$

Models for $\mathcal{L}_{\mathbf{NC}}$ differ from models for $\mathcal{L}_{\mathbf{N}}$ in two respects. First, the domain of the function S is extended to $\{n_i\}_{i\in\mathbb{N}} \cup \{c_a\}_{a\in Ag}$. In this way, S will determine, for every normative source named in $\mathcal{L}_{\mathbf{NC}}$, the set of worlds in which it is satisfied. Second, models for $\mathcal{L}_{\mathbf{NC}}$ include a new function C that determines, for every code c_a and world w, the set of explicit prescriptions of c_a at w. Hence, unlike the function P, C will assign an explicit content to each code relative to each world, thus capturing the idea that codes are concrete normative sources whose explicit content can vary from a world to another.

Definition 8 (Model for $\mathcal{L}_{\mathbf{NC}}$). *A model for $\mathcal{L}_{\mathbf{NC}}$ is a tuple $M = \langle W, R, S, P, C, V \rangle$, where W, R, V are as in definition 3, $S : \{n_i\}_{i\in\mathbb{N}} \cup \{c_a\}_{a\in Ag} \to \wp(W)$, and $C : \{c_a\}_{a\in Ag} \times W \to \wp(Fm(\mathcal{L}_{\mathbf{N}})) \cup \{sat(c_a)\}$. We require that models for $\mathcal{L}_{\mathbf{NC}}$ satisfy the conditions on norms **N1-N4** and the following conditions on codes.*

Conditions on codes: *for all n_i, c_a, and $w \in W$,*

C1 *(Code satisfaction)* $\quad sat(c_a) \in C(c_a, w)$

C2 *(Code inclusion)* $\quad sat(n_i) \in C(c_a, w) \Rightarrow P(n_i) \subseteq C(c_a, w)$

Thus, like norms, codes prescribe their own satisfaction and, whenever they prescribe the satisfaction of a norm, their explicit content includes all explicit prescriptions of that norm. Note that, unlike **N1-N4**, **C1** and **C2** allow for inconsistencies in the content of a code at some possible worlds. This is in line with the idea that, in some situations, there is nothing an agent can do to satisfy her code.

Definition 9 (Truth, $\mathcal{L}_{\mathbf{NC}}$). *The notion of truth of a formula in a world of a model for $\mathcal{L}_{\mathbf{NC}}$ is defined as in definition 4 with the addition of the following new cases:*

$M, w \models sat(c_a) \Leftrightarrow w \in S(c_a)$
$M, w \models O_a\phi \Leftrightarrow \phi \in C(c_a, w)$

In order to provide a sound interpretation of the formulas of $\mathcal{L}_{\mathbf{NC}}$, models for $\mathcal{L}_{\mathbf{NC}}$ should be suitable, in the sense that they satisfy the norm suitability condition **NS** (cf. definition 5) and a suitability condition for codes analogous to **NS**.

Definition 10 (Suitability). *A model $M = \langle W, R, S, P, C, V \rangle$ for $\mathcal{L}_{\mathbf{NC}}$ is suitable just in case it satisfies **NS** and the following code suitability condition:*

CS *(Code suitability)* For all $w \in W$, $\phi \in C(c_a, w)$ and $w \in S(c_a) \Rightarrow M, w \models \phi$.

CS guarantees that a code is satisfied at a possible world only if all the propositions it explicitly prescribes at that world are true there. Given that every code prescribes its own satisfaction by **C1**, this ensures that a code is satisfied at a possible world just in case all the propositions it prescribes at that world are true there. As before, validity and logical consequence are defined in the usual way.

3.3 The Axiom System NC

The axiom system **NC** is obtained by extending the axiom system **N** with the following axioms on codes.

Axioms on codes:
- **AC1** *(Code satisfaction)* $O_a sat(c_a)$
- **AC2** *(Code inclusion)* $O_a sat(n_i) \wedge n_i : \phi \to O_a \phi$
- **AC3** *(Code suitability)* $O_a \phi \wedge sat(c_a) \to \phi$

Axioms **AC1** to **AC3** parallel axioms **AN4** to **AN6** and characterize the prescriptions of codes as prescriptions of a normative source. Observe that, given definition 7, **AC2** states that whenever an agent accepts a norm, all prescriptions of that norm become obligatory for her in virtue of her code. Hence, the code of an agent includes all the norms she accepts by including all the explicit prescriptions of those norms. The following theorems will be important later on.

TC1 $[1]O_a\phi \wedge [1]O_b\neg\phi \to c_a \perp_1 c_b$ by **AC3**, logic of [1]
TC2 $[1]O_a sat(n_i) \wedge [1]O_b sat(n_j) \wedge n_i \perp_1 n_j \to c_a \perp_1 c_b$ by **AC3**, logic of [1]

Due to space limitations we omit the proof of the following theorem.

Theorem 2 (Characterization). *The axiom system **NC** is sound and complete with respect to the class of all suitable models for $\mathcal{L}_{\mathbf{NC}}$.*

4 Updating Codes: The System DNC

The logic **NC** is suitable to model the basic relations between norms and codes, but it is not suitable to model the fact that, when an agent finds herself in a new situation, she typically needs to specify her code by adding new norms. In fact, in most cases, a code is not specific enough to determine what the agent ought to do with respect to all decisions she has to take, and so that code is to be made more specific to direct the agent's conduct.

A natural way to improve our logic is to exploit the idea underlying dynamic epistemic logics: we can view the specification of a code as an update procedure that takes a model representing the initial deontic situation of the agent and returns the model representing the updated deontic situation in which the agent has included a new norm in her code. The updated situation will differ from the initial situation in two respects: (i) at every world, the explicit content of the code is extended so as to include all the explicit prescriptions of the new norm; (ii) the set of worlds in which the code is satisfied is restricted to the set of worlds in which both the original code and the new norm are satisfied. The next definition summarizes this procedure.

Definition 11 (Update model $M_{[c_a+n_i]}$). *Let M be a model for $\mathcal{L}_{\mathbf{NC}}$. For any agent a and norm n_i, the result of the inclusion of n_i in the code of a in M is represented by the model $M_{[c_a+n_i]} = \langle W, R, S_{[c_a+n_i]}, P, C_{[c_a+n_i]}, V \rangle$ where, for all $w \in W$*

$S_{[c_a+n_i]}$ *only differs from S in that* $S_{[c_a+n_i]}(c_a) = S(c_a) \cap S(n_i)$

$C_{[c_a+n_i]}$ *only differs from C in that*, $C_{[c_a+n_i]}(c_a, w) = C(c_a, w) \cup P(n_i)$

In order to represent the update of an agent's code within the language, we extend the alphabet of $\mathcal{L}_{\mathbf{NC}}$ with a modality $[c_a + n_i]$ for any c_a and n_i. We thus obtain a dynamic language $\mathcal{L}_{\mathbf{DNC}}$, whose formulas are build according to the following rules:

$$\phi ::= p_i \mid sat(n_i) \mid sat(c_a) \mid \neg\phi \mid \phi \wedge \phi \mid \Box\phi \mid [1]\phi \mid n_i : \phi \mid O_a\phi \mid [c_a + n_i]\phi$$

where no dynamic modality occurs in formulas like $n_i : \phi$ and $O_a\phi$.

Intuitively, $[c_a + n_i]\phi$ says that, after agent a includes the norm n_i in her code, ϕ is true. This intuitive reading is captured by the following semantic interpretation:

$$M, w \models [c_a + n_i]\phi \Leftrightarrow M_{[c_a+n_i]}, w \models \phi$$

The theorems below present the key results on updated models and allow us to use them in modelling the dynamics leading to the origin of a conflict.

Theorem 3. *Updating a model for $\mathcal{L}_{\mathbf{NC}}$ gives rise to a model for $\mathcal{L}_{\mathbf{NC}}$.*

Proof. Let M be a model for $\mathcal{L}_{\mathbf{NC}}$ and $M_{[c_a+n_i]}$ be the model obtained from M by updating c_a with n_i. Since the update procedure only affects the functions S and C when applied to c_a, we only need to check that $S_{[c_a+n_i]}$ and $C_{[c_a+n_i]}$ satisfy the requirements of definition 8 when applied to c_a. Given that $P(n_i) \subseteq Fm(\mathcal{L}_{\mathbf{N}})$, it is immediate to see that $S_{[c_a+n_i]}$ and $C_{[c_a+n_i]}$ are well defined. In addition, it can be verified that $C_{[c_a+n_i]}$ satisfies **C1** and **C2** by definition 11 and by the fact that M satisfies **C1**, **C2** and **N4** by hypothesis. □

According to the following lemma, the satisfaction of formulas of $\mathcal{L}_{\mathbf{N}}$ is invariant under model update. That is, formulas of $\mathcal{L}_{\mathbf{N}}$ cannot distinguish between a model and an update model obtained from it. This is due to the fact that update procedures do not affect the semantic components needed to evaluate formulas of $\mathcal{L}_{\mathbf{N}}$.

Lemma 1 (Restricted Monotonicity Lemma: *RML*). *Let $M_{[c_a+n_i]}$ be an update model. Then, for any $\phi \in Fm(\mathcal{L}_{\mathbf{N}})$ and $w \in W$, $M, w \models \phi \Leftrightarrow M_{[c_a+n_i]}, w \models \phi$.*

RML is crucial to prove that model updates preserve suitability.

Theorem 4 (Preservation of suitability). *Let $M_{[c_a+n_i]}$ be the update model obtained from a suitable model M for $\mathcal{L}_{\mathbf{NC}}$. Then $M_{[c_a+n_i]}$ is a suitable model for $\mathcal{L}_{\mathbf{NC}}$.*

Proof. By theorem 3, we already know that $M_{[c_a+n_i]}$ is a model for $\mathcal{L}_{\mathbf{NC}}$, so we only need to show that $M_{[c_a+n_i]}$ satisfies the suitability conditions **NS** and **CS**.

NS. $\phi \in P(n_j)$ and $w \in S(n_j) \Rightarrow M_{[c_a+n_i]}, w \models \phi$, for all $w \in W$.
Assume the antecedent. Since M is suitable by hypothesis, $M, w \models \phi$. But, since $\phi \in P(n_j) \subseteq \wp(Fm(\mathcal{L}_\mathbf{N}))$, $M_{[c_a+n_i]}, w \models \phi$ by RML.

CS $\phi \in C_{[c_a+n_i]}(c_b, w)$ and $w \in S_{[c_a+n_i]}(c_b) \Rightarrow M_{[c_a+n_i]}, w \models \phi$, for all $w \in W$.
The relevant case is when $b = a$. So, by applying definition 11, assume that $\phi \in C(c_a, w) \cup P(n_i)$ and $w \in S(c_a) \cap S(n_i)$. Then, either $\phi \in C(c_a, w)$ or $\phi \in P(n_i)$. In the former case, ϕ is either in $Fm(\mathcal{L}_\mathbf{N})$ or it is $sat(c_a)$. If $\phi \in Fm(\mathcal{L}_\mathbf{N})$, then, since $w \in S(c_a)$ and M is suitable, $M, w \models \phi$. The result then follows from RML. If ϕ is $sat(c_a)$, then the proof is immediate since $w \in S_{[c_a+n_i]}(c_a)$. Finally, if $\phi \in P(n_i)$, the proof proceeds as the proof of **NS**. □

At this point we can state the central theorem of this section. The proof, which is an easy application of the definition of truth and of definition 11, is omitted.

Theorem 5 (Reduction Laws). *The following reduction laws are valid in the class of suitable models for $\mathcal{L}_{\mathbf{NC}}$.*

Boolean cases and ontic operators
R1 $[c_a+n_i]p \leftrightarrow p$
R2 $[c_a+n_i]\neg\phi \leftrightarrow \neg[c_a+n_i]\phi$
R3 $[c_a+n_i](\phi \wedge \psi) \leftrightarrow [c_a+n_i]\phi \wedge [c_a+n_i]\psi$
R4 $[c_a+n_i]\Box\phi \leftrightarrow \Box[c_a+n_i]\phi$
R5 $[c_a+n_i][1]\phi \leftrightarrow [1][c_a+n_i]\phi$

Deontic operators: non-relevant cases
R6 $[c_a+n_i]n_j : \phi \leftrightarrow n_j : \phi$
R7 $[c_a+n_i]sat(n_j) \leftrightarrow sat(n_j)$
R8 $[c_a+n_i]O_b\phi \leftrightarrow O_b\phi$, if $b \neq a$
R9 $[c_a+n_i]sat(c_b) \leftrightarrow sat(c_b)$, if $b \neq a$

Deontic operators: relevant cases
R10 $[c_a+n_i]O_a\phi \leftrightarrow O_a\phi \vee n_i : \phi$
R11 $[c_a+n_i]sat(c_a) \leftrightarrow sat(c_a) \wedge sat(n_i)$

Given theorem 5, we can now define the axiom system **DNC** along the lines of the axiomatizations of dynamic epistemic logics.

Definition 12 (Axiom system **DNC**). *The axiom system **DNC** is obtained by extending the axiom system **NC** with the reduction laws stated in theorem 5 and the rule of necessitation for the dynamic modalities $[c_a+n_i]$.*

The following basic theorems are derivable.

TD1 $[c_a + n_i]O_a sat(n_i)$ by **AN4** and **R10**
TD2 $n_i : \phi \to [c_a + n_i]\Box O_a \phi$ by **AN1**, **AC2**, **TD1**
TD3 $n_i : \phi \to [c_a + n_i][1]O_a \phi$ by **TD1** and logic of [1]

Hence, after agent a specifies her code by including norm n_i, she in fact accepts n_i, and so she accepts everything that is prescribed by n_i, as it should be.

Theorem 6. *The system **DNC** is sound and complete with respect to the class of all suitable models for $\mathcal{L}_{\mathbf{NC}}$.*

Proof. The proof of soundness is immediate. The proof of completeness is standard and depends on the fact that theorem 5 ensures that every formula of $\mathcal{L}_{\mathbf{DNC}}$ is provably equivalent to a formula of $\mathcal{L}_{\mathbf{NC}}$, so the completeness result for **DNC** follows from the completeness result for **NC**.   □

5   DNC at Work: Two Paradigmatic Cases

In this and the next sections, we illustrate how the system **DNC** can be used to provide an insightful analysis of, and to draw important distinctions between, normative conflicts involving more than one agent. The sections aim both at clarifying how the framework introduced above works in practice and at providing evidence of its high expressive power. We do this, in the present section, by modeling two paradigmatic cases, namely those of Antigone and Gandhi, and, in the next section, by showing how **DNC** and a simple variant of it allow us to capture some of the main traits distinguishing conscientious objection from civil disobedience.

Antigone's case. *In Sophocles's tragedy, Antigone accepts the law of the gods, according to which her brother Polynices ought to be buried. Yet, the law of the state given by the ruler of Thebes, Creon, states that Polynices ought not to be buried.*

The case presents two normatively relevant agents, namely Antigone, whose code is denoted by c_a, and Creon, who is associated with the code of the state, denoted by c_{state}. The norms of interest are the norm of the gods n_1 prescribing that Polynices is buried ($n_1 : \phi$) and the conflicting norm n_2 promulgated by Creon and prescribing that Polynices is left unburied ($n_2 : \neg \phi$). Since the prescriptions of the two norms are jointly inconsistent, there is a global conflict between them ($n_1 \perp n_2$). We can use the dynamics of **DNC** to see how this conflict generates a local conflict between the concrete normative sources c_a and c_{state}. To be sure, according to the story, when Polynices dies, Antigone accepts the norm of the gods as binding ($O_a sat(n_1) \land [1]O_a sat(n_1)$), while Creon updates the code of the state with the conflicting norm n_2. As shown by the following inference, this generates a conflict between the codes c_a and c_{state}.

$n_1 : \phi \wedge n_2 : \neg\phi$	Assumption
$n_1 \perp n_2$	by **TN2**
$O_a sat(n_1) \wedge [1]O_a sat(n_1)$	Assumption
$O_a \phi \wedge [1]O_a \phi$	by **AC2**
$[c_{state} + n_2]O_a \phi \wedge [1]O_a \phi$	by logic of $[c_{state} + n_2]$
$[c_{state} + n_2]O_{state} \neg\phi \wedge [1]O_{state} \neg\phi$	by **TD1, TD3, R3**
$[c_{state} + n_2]c_{state} \perp_1 c_a$	by **TC1** and logic of $[c_{state} + n_2]$

What is particularly interesting is that the dynamic operator $[c_{state} + n_2]$ can be used to represent the origin of the conflict by keeping track of the fact that the clash between the two codes is due to Creon's decision to change the legal code rather than to Antigone's choice. Gandhi's case differs in this respect.

Gandhi's case. *Gandhi explicitly opposed the colonial rules imposed by the British Empire by employing a non-violent form of civil disobedience.*

As Antigone's case, Gandhi's case presents two normatively relevant agents, namely Gandhi, whose code is denoted by c_g, and the British Empire, whose code is denoted by c_{state}. We can think of the colonial rules opposed by Gandhi as the prescriptions of a norm n_1 establishing and regulating the colonial status of India. While the legal code of the British Empire includes the norm n_1 as binding ($O_{state} sat(n_1) \wedge [1]O_{state} sat(n_1)$), Gandhi's code commits him to violate n_1 ($O_g \neg sat(n_1) \wedge [1]O_g \neg sat(n_1)$). One can see that this gives rise to a local conflict between the two concrete normative sources c_{state} and c_g.

So far so good. But let us now analyse the conflict from a dynamic perspective. Suppose that Gandhi started opposing the colonial rules when he realized what it meant for India to be a British colony. We can then represent the origin of the conflict by introducing a norm n_2 prescribing to violate the colonial rules encoded by the norm n_1 ($n_2 : \neg sat(n_1)$). As shown by the following inference, the conflict between Gandhi and the British Empire thus arises when Gandhi specifies his code by accepting n_2.

$n_1 : sat(n_1) \wedge n_2 : \neg sat(n_1)$	**AN4** and assumption
$n_1 \perp n_2$	by **TN2**
$O_{state} sat(n_1) \wedge [1]O_{state} sat(n_1)$	Assumption
$[c_g + n_2]O_{state} sat(n_1) \wedge [1]O_{state} sat(n_1)$	by logic of $[c_g + n_2]$
$[c_g + n_2]O_g \neg sat(n_1) \wedge [1]O_g \neg sat(n_1)$	by **TD1, TD3, R3**
$[c_g + n_2]c_a \perp_1 c_{state}$	by **TC1** and logic of $[c_g + n_2]$

In this case the conflict can thus be represented as depending on an explicit decision of Gandhi to violate the laws of the state rather than on a change in the latter laws. This suggests that **DNC** is suitable to model a basic feature of civil disobedience as opposed to conscientious objection: while a civil disobedient, like

Gandhi, overtly opposes the current laws, a conscientious objector, like Antigone, at first opposes the current laws because the latter turn out to be wrong given her code – and not because she explicitly accepts a norm prescribing to oppose these laws. In other words, **DNC** seems to have the resources to account for the fact that the origin and motives of the opposition to the state are different in the two cases. We now explore a bit further how this distinction can be accounted for in **DNC**.

6 Further Applications and Extension

Despite being complex phenomena, cases of civil disobedience essentially involve three key elements [9, 24]. That is

(1) *conscientiousness*: a civil disobedient thinks that the laws of the state clash with the right conception of good or justice;

(2) *faithfulness to the law*: a civil disobedient is willing to accept the right laws and to communicate with the government;

(3) *constructive aim*: a civil disobedient aims at changing the laws of the state rather than overturning the entire legal system.

Cases of conscientious objection also involve conscientiousness, but might fail to involve faithfulness to the law and/or a constructive aim.

The analysis of these elements in **DNC** rests on the assumption that a code consists of the norms that better capture the agent's conception of justice or, more generally, the agent's deontic ideal. Under this assumption, $O_a \phi$ states that ϕ is obligatory for a in virtue of her deontic ideal. We can then analyse at least the first two elements as follows.

(1) *conscientiousness*: $O_{state}\phi_1 \wedge O_a\phi_2 \wedge \phi_1 \perp_1 \phi_2$.

(2) *faithfulness to the law*: $O_{state}\phi \wedge \langle 1 \rangle (\phi \wedge sat(c_a)) \to O_a\phi$, for all ϕ.

While (1) says that the ideal acknowledged by the state requires to do something that clashes with what is required by the ideal acknowledged by a, (2) says that the ideal acknowledged by a requires that a obeys all the prescriptions of the state that do not clash with the ideal itself.[3]

Can we also account for the presence of a constructive aim? Suppose that a is in a situation described by (1), i.e. $O_{state}\phi_1 \wedge O_a\phi_2 \wedge \phi_1 \perp_1 \phi_2$ is true. Then, there are two natural ways to express that a aims at changing the laws of the state, namely

(3a) *negative aim*: $O_a \neg O_{state}\phi_1$.

(3b) *positive aim*: $O_a O_{state}\phi_2$.

---

[3] Our analysis of faithfulness to the law thus represents faithfulness *pro tanto*: civil disobedients are willing to respect the current laws only to the extent that they do not clash with their conception of good or justice.

While (3a) says that in virtue of the deontic ideal she acknowledges, a ought to make the code of the state such that ϕ_1 ceases to be obligatory for the state, (3b) says that in virtue of the deontic ideal she acknowledges, a ought to make the code of the state such that ϕ_2 becomes obligatory for the state. If these formulas were available in $\mathcal{L}_{\mathbf{DNC}}$, then the presence of a constructive aim could be broken down into the presence of a negative aim and a positive aim. If we introduced a permission operator $P_a\phi := \neg O_a \neg \phi$, we could also express the presence of what we might call a *neutral aim* by means of the formula $O_a P_{state} \phi_2$. This would say that a ought to change c_{state} so that ϕ_2 at least becomes legally permitted.

Now, nesting of agent-relative obligation operators is not allowed in $\mathcal{L}_{\mathbf{DNC}}$, so the suggested representations of constructive (negative, positive, or neutral) aims are not available in **DNC**. Yet, it turns out that our dynamic system can be refined so as to allow the amount of nesting of obligation operators needed to model constructive aims and, at the same time, preserve the main results concerning update procedures. We devote the rest of this section to the presentation of this refinement. We start by introducing the static system $\mathbf{NC^+}$.

System $\mathbf{NC^+}$. The language $\mathcal{L}_{\mathbf{NC^+}}$ of $\mathbf{NC^+}$ is defined in such a way that more formulas are allowed in the scope of the deontic operators O_a. Specifically, let $\mathbf{O} = \{O_b\phi \mid b \in Ag \text{ and } \phi \in Fm(\mathcal{L}_{\mathbf{N}})\}$ and $\overline{\mathbf{O}} = \{\neg O_b\phi \mid b \in Ag \text{ and } \phi \in Fm(\mathcal{L}_{\mathbf{N}})\}$. Then, the set $Fm(\mathcal{L}_{\mathbf{NC^+}})$ of formulas of $\mathcal{L}_{\mathbf{NC^+}}$ is built as $Fm(\mathcal{L}_{\mathbf{NC}})$, except that in a formula like $O_a\phi$, $\phi \in Fm(\mathcal{L}_{\mathbf{N}}) \cup \{sat(c_a)\} \cup \mathbf{O} \cup \overline{\mathbf{O}}$. Hence, formulas of form $O_a O_b \phi$ and $O_a \neg O_b \phi$ are now allowed, even for $a \neq b$.

Intuitively, $O_a O_b \phi$ says that a ought to make the code of b such that ϕ becomes obligatory for b in virtue of her own code, while $O_a \neg O_b \phi$ says that a ought to make the code of b such that ϕ ceases to be obligatory for b in virtue of her own code.

In order to interpret formulas of $\mathcal{L}_{\mathbf{NC^+}}$, we adapt the notion of model for $\mathcal{L}_{\mathbf{NC}}$ by extending the set of formulas that can be included in a code.

Definition 13 (Model for $\mathcal{L}_{\mathbf{NC^+}}$). *A model for $\mathcal{L}_{\mathbf{NC^+}}$ is a tuple $M = \langle W, R, S, P, C, V \rangle$, where W, R, S, P, V are as in definition 8 and $C : \{c_a\}_{a \in Ag} \times W \to \wp(Fm(\mathcal{L}_{\mathbf{N}})) \cup \{sat(c_a)\} \cup \mathbf{O} \cup \overline{\mathbf{O}}$. We require that models for $\mathcal{L}_{\mathbf{NC^+}}$ satisfy all conditions satisfied by models of $\mathcal{L}_{\mathbf{NC}}$ plus the following:*

C3 $\neg O_b \phi \in C(c_a, w)$ *and* $\phi \in P(n_i) \Rightarrow \neg O_b sat(n_i) \in C(c_a, w)$

Condition **C3** can be read as a coherence requirement on codes: it says that, if a code c_a requires that an agent b is not obliged to realize ϕ in virtue of her code, then c_a also requires that b does not accept any norm prescribing ϕ. The definitions of truth and of suitable model for $\mathcal{L}_{\mathbf{NC^+}}$ are the same as definitions 9 and 10. In light of this, it is not difficult to see that the system $\mathbf{NC^+}$ obtained by extending the set of axioms and rules of $\mathbf{NC}$ with the axiom schema

AC4 $O_a \neg O_b \phi \land n_i : \phi \to O_a \neg O_b sat(n_i)$

is sound and complete with respect to the class of all suitable models for $\mathcal{L}_{\mathbf{NC}^+}$.

Dynamics. Our task is now to modify definition 11 in order to accommodate the fact that, by accepting a new norm n_i, an agent a might affect the satisfaction of codes of other agents, especially those prescribing that a should not accept n_i.

Definition 14 (Update model $M^+_{[c_a+n_i]}$). Let M be a model for $\mathcal{L}_{\mathbf{NC}^+}$. For any agent a and norm n_i, the result of the inclusion of n_i in c_a in M is represented by the model $M^+_{[c_a+n_i]} = \langle W, R, S^+_{[c_a+n_i]}, P, C^+_{[c_a+n_i]}, V \rangle$ where, for all $w \in W$

- $S^+_{[c_a+n_i]}$ only differs from S in that

$$S^+_{[c_a+n_i]}(c_a) = S(c_a) \cap S(n_i) \cap \{w \mid \neg O_a sat(n_i) \notin C(c_a, w)\}$$
$$S^+_{[c_a+n_i]}(c_b) = S(c_b) \cap \{w \mid \neg O_a sat(n_i) \notin C(c_b, w)\}, \text{ for all } b \neq a$$

- $C^+_{[c_a+n_i]}$ only differs from C in that $C^+_{[c_a+n_i]}(c_a, w) = C(c_a, w) \cup P(n_i)$

As we did in section 4, we can represent update procedures in the language by extending the set of formulas of $\mathcal{L}_{\mathbf{NC}^+}$ with dynamic formulas of form $[c_a + n_i]^+\phi$ for any c_a and n_i. These are given the following semantic interpretation:

$$M, w \models [c_a + n_i]^+\phi \Leftrightarrow M^+_{[c_a+n_i]}, w \models \phi$$

We can now prove results analogous to those obtained in section 4.

Theorem 7. *Updating a model for $\mathcal{L}_{\mathbf{NC}}$ gives rise to a model for $\mathcal{L}_{\mathbf{NC}}$.*

Proof. The proof is analogous to the proof of theorem 3, except that now we need to show that, besides **C1** and **C2**, update models satisfy **C3**. This follows immediately from definition 14 and the fact that static models satisfy **C3**. □

It is immediately verified that *RML* still holds and, hence, that model updates still preserve the norm suitability condition. The next theorem ensures that the code suitability condition is also preserved by model updates.

Theorem 8 (Preservation of code suitability). *Let M be a suitable model for $\mathcal{L}_{\mathbf{NC}^+}$. Then, for any c_b and n_i, $M^+_{[c_a+n_i]}$ satisfies the code suitability condition **CS**.*

Proof. We only sketch the proof for c_a. By applying definition 14, we need to show that the following is satisfied, for all $w \in W$:

If $\quad \phi \in C(c_a, w) \cup P(n_i)$ and $w \in S(c_a) \cap S(n_i) \cap \{w \mid \neg O_a sat(n_i) \notin C(c_a, w)\}$
then $\quad M^+_{[c_a+n_i]}, w \models \phi$.

There are two principal cases: either $\phi \in P(n_i)$ or $\phi \in C(c_a, w) \subseteq \wp(Fm(\mathcal{L}_\mathbf{N})) \cup \{sat(c_a)\} \cup \mathbf{O} \cup \overline{\mathbf{O}}$. The relevant case is when $\phi \in \overline{\mathbf{O}}$. For the sake of readability, let $X = \{w \mid \neg O_a sat(n_i) \notin C(c_a, w)\}$. Assume that $\neg O_b \psi \in C(c_a, w)$ and $w \in S(c_a) \cap S(n_i) \cap X \subseteq S(c_a)$. Then, $M, w \models \neg O_b \psi$, since M is suitable by hypothesis. Hence, $\psi \notin C(c_b, w)$. If $b \neq a$, this is sufficient to obtain the result. If $b = a$, suppose that $M^+_{[c_a+n_i]}, w \models O_a \psi$. Then, $\psi \in C(c_a, w) \cup P(n_i)$. Since $\psi \notin C(c_a, w)$, $\psi \in P(n_i)$. We thus have that $\neg O_a \psi \in C(c_a, w)$ and $\psi \in P(n_i)$. But then, by **C3**, $\neg O_a sat(n_i) \in C(c_a, w)$, against the hypothesis that $w \in X$. Therefore, $M^+_{[c_a+n_i]}, w \not\models O_a \psi$, as desired. □

It is now easily seen that the new semantics validates reduction laws for $\mathcal{L}_{\mathbf{DNC}^+}$.

Theorem 9 (Reduction laws for $\mathcal{L}_{\mathbf{DNC}^+}$). *The reduction laws obtained by replacing $[c_a+n_i]$ with $[c_a+n_i]^+$ in **R1-R8** and **R10** in theorem 5 plus the following reduction laws **R9$^+$** and **R11$^+$** are valid in the class of suitable models for $\mathcal{L}_{\mathbf{NC}^+}$.*

R9$^+$ $\quad [c_a + n_i]sat(c_b) \leftrightarrow sat(c_b) \land \neg O_b \neg O_a sat(n_i)$, *if* $b \neq a$

R11$^+$ $\quad [c_a + n_i]sat(c_a) \leftrightarrow sat(c_a) \land sat(n_i) \land \neg O_a \neg O_a sat(n_i)$

Theorem 9 ensures that the system $\mathbf{DNC}^+$ obtained by extending $\mathbf{NC}^+$ with the reduction laws for $\mathcal{L}_{\mathbf{DNC}^+}$ and the rule of necessitation for the dynamic modalities $[c_a + n_i]^+$ is sound and complete with respect to the class of all suitable models for $\mathcal{L}_{\mathbf{NC}^+}$.

7 Conclusion and Future Work

In this paper, we presented dynamic logics of norms and codes that explicitly represent abstract and concrete normative sources and the static and dynamic interactions between them. We have shown that the systems are highly expressive and that they can be used to account for important differences between conflicts involving two agents, especially those giving rise to episodes of civil disobedience as opposed to conscientious objection.

In ongoing and future works, we aim to develop the systems in different directions. First, we make explicit the connection between codes and deontic ideal suggested in section 5 by merging **NC** with a multi-agent version of standard deontic logic. Second, we generalize the structure of norms and codes in order to represent normative sources whose explicit content consists not only of categorical but also of conditional prescriptions. Finally, we investigate more complex procedures for updating and revising codes, so as to obtain a proper model of the dynamics of normative sources and normative systems. Besides considering procedures that can be compared to revision and contraction in *AGM* theory, as proposed for instance in [8, 14] in the framework of input/output and defeasible logics, we plan to enrich our framework with update procedures modelling normative acts where the issuer and the addressee of a norm are different agents, thus capturing the significant class of phenomena studied in [26]. The latter development would enable us to analyse how commands from an external authority generate obligations that are binding for an agent and, relatedly, to study the dynamics underlying Hohfeldian normative positions and complex deontic phenomena like contracts.

References

[1] Carlos E Alchourrón and Eugenio Bulygin. *Normative Systems*. Springer Verlag, New York and Wien, 1971.

[2] Carlos E Alchourrón and Eugenio Bulygin. The Expressive Conception of Norms. In Risto Hilpinen, editor, *New Studies in Deontic Logic: Norms, Actions, and the Foundations of Ethics*, pages 95–124. Springer Netherlands, Dordrecht, 1981.

[3] Sergei Artemov. The Logic of Justification. *The Review of Symbolic Logic*, 1(4):477–513, 2008.

[4] Alexandru Baltag, Lawrence S. Moss, and Slavomir Solecki. The Logic of Public Announcements, Common Knowledge, and Private Suspicious. In *Proceedings of the 7th Conference on Theoretical Aspects of Rationality and Knowledge (TARK 1998)*, pages 43–56. Morgan Kaufmann Publishers Inc., 1998.

[5] Johan van Benthem. *Logical Dynamics of Information and Interaction*. Cambridge University Press, Cambridge, 2011.

[6] Johan van Benthem, Davide Grossi, and Fenrong Liu. Priority Structures in Deontic Logic. *Theoria*, (2):116–152, 2014.

[7] Johan van Benthem and Fenrong Liu. Deontic Logic and Preference Change. *IfColog*, 1(2):1–46, 2014.

[8] Guido Boella, Gabriella Pigozzi, and Leendert Van der Torre. Normative Framework for Normative System Change. In Keith S. Decker, Jaime S. Sichman, Carles Sierra, and Cristiano Castelfranchi, editors, *Proceedings of the Eighth International Conference on Autonomous Agents and Multiagent Systems (AAMAS 2009)*, pages 169–176, 2009.

[9] Kimberley Brownlee. *Conscience and Conviction: The Case for Civil Disobedience*. Oxford University Press, Oxford, 2012.

[10] Melvin Fitting. The Logic of Proofs, Semantically. *Annals of Pure and Applied Logic*, 132(1):1–25, 2005.

[11] Lou Goble. A Logic for Deontic Dilemmas. *Journal of Applied Logic*, 3(3-4):461–483, 2005.

[12] Lou Goble. Normative Conflicts and The Logic of 'Ought'. *Noûs*, 43(3):450–489, 2009.

[13] Lou Goble. Prima Facie Norms, Normative Conflicts, and Dilemmas. In Dov Gabbay, John Horty, Xavier Parent, Ron van der Meyden, and Leendert van der Torre, editors, *Handbook of Deontic Logic and Normative Systems*, chapter 4, pages 241–352. College Publications, Milton Keynes, 2013.

[14] Guido Governatori and Antonino Rotolo. Changing Legal Systems: Legal Abrogations and Annulments in Defeasible Logic. *Logic Journal of the IGLP*, 18(1):157–194, 2010.

[15] Jörg Hansen. Deontic Logics for Prioritized Imperatives. *Artificial Intelligence and Law*, 14(1):1–34, 2006.

[16] Jörg Hansen. Reasoning About Permission and Obligation. In Sven Ove Hansson, editor, *David Makinson on Classical Methods for Non-Classical Problems*, pages 287–

333. Springer Netherlands, Dordrecht, 2014.

[17] John F. Horty. Reasoning with Moral Conflicts. *Noûs*, 37(4):557–605, 2003.

[18] John F. Horty. Defaults with Priorities. *Journal of Philosophical Logic*, 36(4):367–413, 2007.

[19] John F. Horty. *Reasons as Defaults*. Oxford University Press, 2012.

[20] Fenrong Liu. *Reasoning about Preference Dynamics*. Springer, Amsterdam, 2011.

[21] David Makinson and Leendert van der Torre. Input/Output Logics. *Journal of Philosophical Logic*, 29(4):383–408, 2000.

[22] David Makinson and Leendert van der Torre. Constraints for Input/Output Logics. *Journal of Philosophical Logic*, 30(2):155–185, 2001.

[23] Xavier Parent and Leendert van der Torre. Input/Output Logic. In Dov Gabbay, John Horty, Xavier Parent, Ron van der Meyden, and Leendert van der Torre, editors, *Handbook of Deontic Logic and Normative Systems*, chapter 8, pages 499–544. College Publications, Milton Keynes, 2013.

[24] William Smith. *Civil Disobedience and Deliberative Democracy*. Routledge, 2013.

[25] Hans P. van Ditmarsch, Wiebe van der Hoek, and Barteld Kooi. *Dynamic Epistemic Logic*. Springer, Berlin, 2008.

[26] Tomoyuki Yamada. Logical Dynamics of Some Speech Acts that Affect Obligations and Preferences. *Synthese*, 165(2):295–315, 2008.

Resolving Conflicting Obligations in Mīmāṃsā: A Sequent-based Approach

Agata Ciabattoni
TU Wien, Austria
agata@logic.at

Francesca Gulisano
Scuola Normale Superiore Pisa, Italy
francesca.gulisano@sns.it

Björn Lellmann
TU Wien, Austria
lellmann@logic.at

The Philosophical School of Mīmāṃsā provides a treasure trove of more than 2000 years worth of deontic investigations. In this paper we formalize the Mīmāṃsā approach of resolving conflicting obligations by giving preference to the more specific ones. From a technical point of view we provide a method to close a set of prima-facie obligations under a restricted form of monotonicity, using specificity to avoid conflicting obligations in a dyadic non-normal deontic logic. A sequent-based decision procedure for the resulting logic is also provided.

1 Introduction

The Mīmāṃsā is a philosophical school which originated in ancient India in the last centuries BCE and whose main focus was the exegesis of the prescriptive portions of the Indian Sacred Texts (the *Vedas*). To this aim over the course of more than two millennia, Mīmāṃsā authors have analyzed normative statements, resulting in theories considered early deontic logic [14]. Despite the undeniable importance of Mīmāṃsā in Indian philosophy, theology and law, and despite the rigorous structure of its texts lending themselves to formal analysis [6], virtually no logical formalization of the deontic concepts in Mīmāṃsā has been carried out so far. The main reason for this is that most Sanskritists are not trained in mathematical logic, and the untranslated or unanalyzed texts are inaccessible to logicians.

We thank E. Freschi for her invaluable help with the Indological aspects of this work, and the referees for their insightful comments concerning related literature. Funded by WWTF project MA16-28.

In order to enable readers to understand the Vedas independently of any authorial intention, and explain "what has to be done" in presence of seemingly[1] conflicting obligations, Mīmāṃsā authors have proposed a rich body of deontic, hermeneutical and linguistic principles (metarules), called *nyāyas*. Those principles are so modern, rational, scientific, and systematic [1] that they are still applied in Indian jurisprudence to decide court cases, e.g. [15].

To formalize Mīmāṃsā reasoning in a step-by-step bottom-up approach, we have transformed in [5] some of the deontic *nyāyas* into Hilbert axioms. This led to the introduction of the non-normal dyadic deontic logic bMDL, whose proof-calculus and semantics were successfully used there to analyze the seemingly conflicting obligations in the Vedas concerning the *Śyena* sacrifice. However, bMDL is only a first step towards the formalization of Mīmāṃsā reasoning. In particular, many *nyāyas* are still waiting to be found, translated from Sanskrit, and interpreted, which is the subject of ongoing work. Notice also that not all the *nyāyas* can be simply converted into Hilbert axioms. Some of these indeed offer more general interpretative principles to resolve apparent contradictions in the Vedas; prominent examples of such *nyāyas* are *Guṇapradhāna* and *Vikalpa*, which are investigated in this paper. The Vikalpa principle states that when there is a real conflict between obligations, any of the conflicting injunctions may be adopted as option: this principle is known in deontic logic as *disjunctive response* [10] and corresponds to the phenomenon of *floating conclusions* in nonmonotonic reasoning [17]. The Guṇapradhāna principle states that more specific rules override more generic ones. Already introduced by Śabara (3rd-5th c. CE), Guṇapradhāna is widely used, e.g., in Artificial Intelligence, where it was formulated much later and where it is known as *specificity principle*. These principles are also used to capture *defeasible* reasoning in the context of *nonmonotonic logics* [7, 18, 21, 23]. In particular, the specificity principle may lead to the loss of the monotonicity of the consequence relation: an obligation "α should be the case" could follow from a set of premises Γ, but it might be overruled by a more specific obligation β, so that it does not follow from the set $\Gamma \cup \{\beta\}$ anymore.

In this paper we further pursue the proof-theoretic approach to approximate Mīmāṃsā deontic reasoning by extending the deontic part of bMDL with a mechanism to capture the Guṇapradhāna principle. We provide a sequent calculus that derives what "has to be done" from the explicit prescriptions contained in the Vedas (*Śrauta* in Sanskrit), and a finite set of propositional facts by resolving conflicts using Guṇapradhāna (specificity), and which satisfies Vikalpa (disjunctive response).

Examples of sequent calculi for defeasible reasoning in normative contexts include [3, 12, 24] (the latter is applied in the context of an argument-based system).

As, e.g., in [10, 27] here we interpret the notion of a conditional obligation being more specific than another one as the conditions of the former implying those of

---

[1]The Vedas are assumed to be not contradictory and Mīmāṃsā authors invested all their efforts in creating a consistent deontic system.

the latter. Our calculus is built on the sequent calculus for the □-free fragment of bMDL, which turns out to be the dyadic version of non-normal deontic logic MD [4] (cf. Prop. 2.2 and [8]). Additional rules to derive all possible prescriptions are defined using limited monotonicity on the conditions of the (non-nested) prescriptions in the Vedas (prima-facie obligations) "up to conflicting obligations" relative to the given set of facts. These additional rules are motivated by the interpretation given by the Mīmāṃsā author Madhatīthi (9-10th c. CE) that more specific *Śrauta* provide exceptions to more general ones and that the latter apply to all circumstances but those indicated in the exceptions (or implied by them). Apart from this non-monotonic inference from prima facie to actual obligations, all inferences use the monotonic system bMDL. Thus we restrict nonmonotonic reasoning using the specificity principle to resolving possible conflicts between prima-facie obligations, but keep the inferences of the logic for arbitrary formulae deductive (i.e., monotone). This is inspired by [25] which states that Indian philosophers – in particular the Mīmāṃsā author Kumārila – tried to keep their arguments not defeasible "as much as possible". From a technical point of view the advantage is that the consequences of a set of prima-facie obligations can be constructed iteratively instead of by a fixed-point construction as e.g. in [13]. Moreover the system does not use key properties of non-monotonic logics (as, e.g., in [20]) which seem not to hold in Mīmāṃsā reasoning (e.g. cautious monotony, see Example 3.1). Finally, we show that the introduced system provides a decision procedure and satisfies the disjunctive response.

2 Basic Mīmāṃsā Deontic Logic

Basic Mīmāṃsā Deontic Logic bMDL was introduced in [5] as a first step towards the formalization of Mīmāṃsā reasoning. The idea was to define a logical system following a bottom-up approach of extracting deontic principles from the Mīmāṃsā texts. The resulting logic extends the alethic system S4 with the following axiom schemata for the deontic operator $\mathcal{O}(A/B)$, which intuitively reads as "A is obligatory under the condition B":

1. $(\Box(A \to B) \land \mathcal{O}(A/C)) \to \mathcal{O}(B/C)$
2. $\Box(B \to \neg A) \to \neg(\mathcal{O}(A/C) \land \mathcal{O}(B/C))$
3. $(\Box((B \to C) \land (C \to B)) \land \mathcal{O}(A/B)) \to \mathcal{O}(A/C)$

Axioms (1)-(3) arise by rewriting some of the Mīmāṃsā deontic interpretative principles (*nyāya*s) as logic formulas. E.g., (1) formalizes three different principles; among them the following reformulation of a Sanskrit *nyāya* in the *Tantrarahasya* (15th-17th c. CE) that can be abstracted as (See [6] for details)

> If the accomplishment of X presupposes the accomplishment of Y, the obligation to perform X prescribes also Y.

$$
\begin{array}{ll}
\text{(M)} \ \mathcal{O}(A \wedge B/C) \to \mathcal{O}(A/C) & \dfrac{A \leftrightarrow C \quad B \leftrightarrow D}{\mathcal{O}(A/B) \to \mathcal{O}(C/D)} \ \text{Cg} \\
\text{(D)} \ \neg(\mathcal{O}(A/B) \wedge \mathcal{O}(\neg A/B)) &
\end{array}
$$

Figure 1: The modal part of a Hilbert-style system for dyadic MD.

$$
\dfrac{}{p \Rightarrow p} \text{ init} \quad \dfrac{}{\bot \Rightarrow} \bot_L \quad \dfrac{\Gamma, B \Rightarrow \Delta \quad \Gamma \Rightarrow A, \Delta}{\Gamma, A \to B \Rightarrow \Delta} \to_L \quad \dfrac{\Gamma, A \Rightarrow B, \Delta}{\Gamma \Rightarrow A \to B, \Delta} \to_R
$$

$$
\dfrac{A \Rightarrow C \quad B \Rightarrow D \quad D \Rightarrow B}{\mathcal{O}(A/B) \Rightarrow \mathcal{O}(C/D)} \text{ Mon} \quad \dfrac{A, C \Rightarrow \quad B \Rightarrow D \quad D \Rightarrow B}{\mathcal{O}(A/B), \mathcal{O}(C/D) \Rightarrow} \text{ D} \quad \dfrac{A \Rightarrow}{\mathcal{O}(A/B) \Rightarrow} \text{ P}
$$

$$
\dfrac{\Gamma, A, A \Rightarrow \Delta}{\Gamma, A \Rightarrow \Delta} \text{Con}_L \quad \dfrac{\Gamma \Rightarrow A, A, \Delta}{\Gamma \Rightarrow A, \Delta} \text{Con}_R \quad \dfrac{\Gamma \Rightarrow \Delta}{\Gamma, A \Rightarrow \Delta} \text{W}_L \quad \dfrac{\Gamma \Rightarrow \Delta}{\Gamma \Rightarrow A, \Delta} \text{W}_R
$$

Figure 2: The sequent calculus G_{MD} for dyadic MD.

Remark 2.1. bMDL is weaker than most known deontic logics, e.g., those in [19]; in particular it has neither any deontic aggregation principles nor any form of factual or deontic detachment. In part this is due to our step-by-step methodology: so far indeed we have not found any mention of corresponding principles in the texts. However, the absence of (factual) detachment principles is also in line with the statement by one of the main authors of Mīmāṃsā, Prabhākara, that "*A prescription regards what has to be done. But it does not say that it has to be done*" (*Bṛhatī* I, 7th c. CE).

Here for simplicity we only consider the box-free fragment of bMDL, which coincides with the dyadic version of the logic MD [4] axiomatized as in Fig. 1 (Prop. 2.2). For space reasons we treat the propositional connectives $\wedge, \vee, \neg$ as defined by $\bot, \to$ in the usual way. In the following we will consider an extension of a *sequent calculus* for this logic, where a *sequent* is a tuple of multisets of formulas, written as $\Gamma \Rightarrow \Delta$. The rules of the sequent calculus G_{MD} are given in Fig. 2, those of the calculus $\mathsf{G}_{\mathsf{bMDL}}$ for bMDL from [5] in Fig. 3, where $\Gamma^{\square}$ denotes Γ in which all formulas not of the form $\square \varphi$ are deleted. Note that the usual sequent rules for $\wedge, \vee, \neg$ are derivable using the definitions in terms of $\bot, \to$. As usual, a *derivation* is a finite labelled tree where every node is labelled with a sequent such that the labels of a node follow from the labels of its children using the rules of the calculus. In particular, the leaves are labelled with conclusions of the zero-premise rules init or $\bot_L$, see also [26]. For G one of $\mathsf{G}_{\mathsf{MD}}, \mathsf{G}_{\mathsf{bMDL}}$ we write $\vdash_\mathsf{G} \Gamma \Rightarrow \Delta$ if there is a derivation of $\Gamma \Rightarrow \Delta$ in G. For the original semantic equivalent of the following proposition, see [8].

Proposition 2.2. *If* $\Gamma \Rightarrow \Delta$ *does not contain* $\square$, *then* $\vdash_{\mathsf{G}_{\mathsf{MD}}} \Gamma \Rightarrow \Delta$ *iff* $\vdash_{\mathsf{G}_{\mathsf{bMDL}}} \Gamma \Rightarrow \Delta$. *Hence the box-free fragment of* bMDL *is* MD.

$$\frac{\Gamma^\Box \Rightarrow \varphi}{\Gamma \Rightarrow \Box\varphi, \Delta} \, 4 \qquad \frac{\Gamma, \Box\varphi, \varphi \Rightarrow \Delta}{\Gamma, \Box\varphi \Rightarrow \Delta} \, T \qquad \frac{\Gamma^\Box, \varphi \Rightarrow \theta \quad \Gamma^\Box, \psi \Rightarrow \chi \quad \Gamma^\Box, \chi \Rightarrow \psi}{\Gamma, \mathcal{O}(\varphi/\psi) \Rightarrow \mathcal{O}(\theta/\chi), \Delta} \, \text{Mon}'$$

$$\frac{\Gamma^\Box, \varphi \Rightarrow}{\Gamma, \mathcal{O}(\varphi/\psi) \Rightarrow \Delta} \, D_1 \qquad \frac{\Gamma^\Box, \varphi, \theta \Rightarrow \quad \Gamma^\Box, \psi \Rightarrow \chi \quad \Gamma^\Box, \chi \Rightarrow \psi}{\Gamma, \mathcal{O}(\varphi/\psi), \mathcal{O}(\theta/\chi) \Rightarrow \Delta} \, D_2$$

Figure 3: The modal part of the sequent calculus G$_{bMDL}$ for bMDL from [5].

Proof. One direction of the equivalence follows from changing the rules of G$_{bMDL}$ into the corresponding rules of G$_{MD}$ possibly followed by the weakening rules W$_L$, W$_R$. The other direction follows since a derivation in G$_{MD}$ is a derivation in G$_{bMDL}$ with the addition of the structural rules of weakening W$_L$, W$_R$ and contraction Con$_L$, Con$_R$, which are admissible in G$_{bMDL}$ [5, Lem. 1]. Completeness and soundness of G$_{MD}$ for MD follow from general methods for constructing sequent calculi from axioms and proving cut elimination such as [16]. □

Remark 2.3. The mechanism for handling propositional facts employed in this paper differs from that in [5]: whereas there we encoded such assumptions as boxed formulas in the conclusion of a derivation, here we treat them as leaves. This has the welcome consequence that we can avoid the alethic modality □ including any question about its axiomatisation, in line with the view that Mīmāṃsā authors did not distinguish between necessity and epistemic certainty.

3 Defeasible reasoning in Mīmāṃsā

The specificity principle (*Guṇapradhāna*) is used in Mīmāṃsā to resolve apparent contradictions; these may occur in the set of Vedic (Śrauta) prescriptions or can be derived via the facts. For example, consider the Śrauta prescriptions: (a) A *Śūdra* (i.e., a member of the lower class) should not engage with the Veda, (b) Knowledge of the Vedas is a prerequisite for sacrificing and (c) A chariot maker should sacrifice. The additional fact (d) A chariot maker is a Śūdra, leads to (apparently) conflicting obligations, as extensively discussed by Mīmāṃsā author Jaimini (2nd c. BCE). The following example illustrates the kind of reasoning Mīmāṃsā authors employed to solve such kinds of conflicting obligations.

Example 3.1. Consider the obligations (a) $\mathcal{O}_{pf}(\text{agn}/\top)$ ("You ought to perform the ritual offering called Agnihotra") and (b) $\mathcal{O}_{pf}(\neg\text{agn}/\text{sdr})$ ("You ought not to perform the Agnihotra if you are a Śūdra"). By an implicit deduction from those two premises we could obtain two obligations: (c) $\mathcal{O}(\text{agn}/\gamma)$ and (d) $\mathcal{O}(\neg\text{agn}/\gamma \wedge \text{sdr})$. Now, let us interpret γ as being more specific than sdr, e.g. as "being a chariot maker" (chmk). Since chariot makers are Śūdra, the formulas chmk and chmk ∧ sdr are equivalent, and thus the obligations (c) and (d) give an apparent conflict. One of the solutions to this employed by Mīmāṃsā authors it to interpret (c) as an explicit Vedic (*Śrauta*) prescription, i.e. as $\mathcal{O}_{\sf pf}({\sf agn/chmk})$. In this case, using the specificity principle, the Mīmāṃsā authors derive the opposite of (d), (d') $\mathcal{O}({\sf agn/chmk} \wedge {\sf sdr})$.

However, also a state such that none of γ and ${\sf sdr}$ is more specific than the other is compatible with the Mīmāṃsā reasoning; for instance we can imagine a situation where you are asked to decide what to do if you are a Śūdra but you became a school teacher (${\sf sch}$). Also in this case, if (c) is interpreted as a *Śrauta* injunction $\mathcal{O}_{\sf pf}({\sf agn/sch})$, (d) should not follow anymore. In the latter case, writing $\hspace{0.1em}\sim\hspace{-0.9em}\mid\hspace{0.4em}$ for the consequence relation given by the implicit deduction from Śrauta to actual obligations, we have $\{\mathcal{O}_{\sf pf}({\sf agn}/\top), \mathcal{O}_{\sf pf}(\neg{\sf agn/sdr})\} \hspace{0.1em}\sim\hspace{-0.9em}\mid\hspace{0.4em} \mathcal{O}({\sf agn/sch})$ and also $\{\mathcal{O}_{\sf pf}({\sf agn}/\top), \mathcal{O}_{\sf pf}(\neg{\sf agn/sdr})\} \hspace{0.1em}\sim\hspace{-0.9em}\mid\hspace{0.4em} \mathcal{O}(\neg{\sf agn/sch} \wedge {\sf sdr})$, but in contrast also $\{\mathcal{O}_{\sf pf}({\sf agn}/\top), \mathcal{O}_{\sf pf}(\neg{\sf agn/sdr}), \mathcal{O}_{\sf pf}({\sf agn/sch})\} \not\hspace{0.1em}\sim\hspace{-0.9em}\mid\hspace{0.4em} \mathcal{O}(\neg{\sf agn/sch} \wedge {\sf sdr})$. Hence the Mīmāṃsākas' reasoning can provide a counterexample for *Cautious Monotony* – one of the classical principles of non-monotonic logics [9].

Here we continue the proof-theoretic approach initiated in [5] to reproduce Mīmāṃsā reasoning in a formal framework. We extend the sequent calculus G_{MD} for the logic MD with special rules $\mathsf{ga}_L, \mathsf{ga}_R$ to derive conditional obligations of the form $\mathcal{O}(A/B)$ from prima-facie obligations (i.e. Śrauta prescriptions) written as $\mathcal{O}_{\sf pf}(C/D)$, adopting limited forms of monotonicity (Sec. 3.1). The resulting calculus is shown to be decidable (Sec. 3.2), applies the specificity principle, and turns out to satisfy the disjunctive response (Sec. 3.3).

3.1 Sequent calculus for Specificity/Guṇapradhāna

In order to extend the sequent calculus for MD to capture the specificity principle, loosely following [10, p.281], we interpret the notion of *specificity* as entailment in the presence of (global) propositional assumptions. I.e., given a set $\mathfrak{F}$ of propositional *facts* about the world we say that proposition A is *at least as specific* as proposition B, if $\mathfrak{F}$ entails $A \to B$. Given this interpretation, the *specificity principle* can be understood as limiting monotonicity (the inference by ārtha above) of the operator $\mathcal{O}$ in the second argument in the following sense. Given a list $\mathfrak{L}$ of non-nested *prima facie obligations*, e.g., *Śrauta* prescriptions, and a proposition B, we should be licensed to infer the actual obligation $\mathcal{O}(A/B)$ if there is an injunction $\mathcal{O}_{\sf pf}(A/C)$ in $\mathfrak{L}$ such that B is at least as specific as C, i.e., we can infer using $\mathfrak{F}$ that $B \to C$, and there is no $\mathcal{O}_{\sf pf}(D/E)$ in $\mathfrak{L}$ such that B is at least as specific as E and E is at least as specific as C, and further the formulas A and E are inconsistent, i.e., we can infer $\neg(A \wedge E)$. However, while this implements the notion that more specific Śrauta obligations overrule less specific conflicting ones, this only resolves conflicts between propositions $\mathcal{O}_{\sf pf}(A_i/B_i)$ and $\mathcal{O}_{\sf pf}(A_j/B_j)$ in $\mathfrak{L}$ for which the conditions are comparable in the sense that either B_i implies B_j or B_j implies B_i. Hence, to make

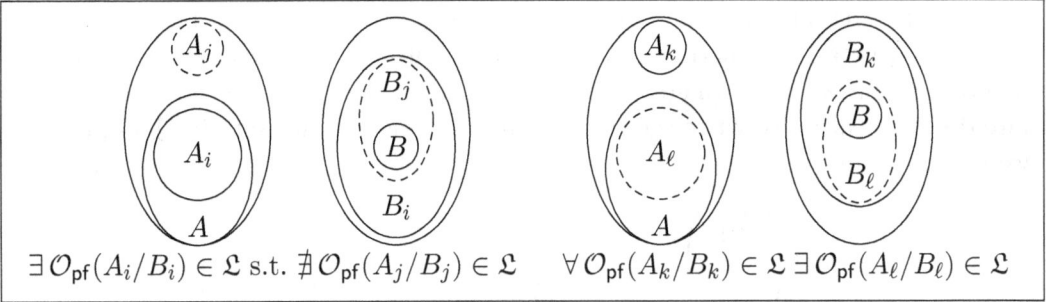

Figure 4: A graphical representation of the conditions for $\mathcal{O}(A/B)$ being derivable. Areas can be taken as formulas with containment representing entailment, i.e., more specific formulas are contained in less specific ones.

the resulting theory consistent with MD, following the Mīmāṃsā reasoning in Ex. 3.1 we add a further condition stating that there is no obligation $\mathcal{O}_{\mathsf{pf}}(A_k/B_k) \in \mathfrak{L}$ such that B is at least as specific as B_k, the enjoined A and A_k are inconsistent, and which is not overruled by a more specific obligation $\mathcal{O}_{\mathsf{pf}}(A_\ell/B_\ell)$ from $\mathfrak{L}$. Graphically, these two conditions can be visualised as in Fig. 4. In the following we make this formally precise, and prove a cut elimination theorem for the resulting system.

In the remainder of this paper we assume that $\mathfrak{F}$ is a finite set of sequents containing only propositional variables, which is *closed under cuts*, i.e., whenever $\Gamma \Rightarrow \Delta, p$ and $p, \Sigma \Rightarrow \Pi$ are in $\mathfrak{F}$, then so is $\Gamma, \Sigma \Rightarrow \Delta, \Pi$, and *closed under contractions*, i.e., whenever $\Gamma, p, p \Rightarrow \Delta$ or $\Gamma \Rightarrow p, p, \Delta$ are in $\mathfrak{F}$, then so are $\Gamma, p \Rightarrow \Delta$ and $\Gamma \Rightarrow p, \Delta$ respectively. We call $\mathfrak{F}$ the set of *(propositional) facts*. Note that, since every propositional formula is equivalent to a formula in conjunctive normal form, using this definition we can stipulate arbitrary propositional formulas as facts. We further assume a finite set $\mathfrak{L}$ of formulas of the form $\mathcal{O}_{\mathsf{pf}}(A_m/B_m)$ where A_m and B_m do not contain the $\mathcal{O}$-operator. We call these formulas *prima facie obligations*.

To capture the intuition for the specificity principle given above in a well-behaved sequent system, we first need to make the notion of implication used there formally precise. In particular, we would like to define a notion of inference $\vdash$ from the facts in $\mathfrak{F}$ depending on the set $\mathfrak{L}$, such that we can derive a formula $\mathcal{O}(A/B)$ if and only if both of the following hold:

- there is $\mathcal{O}_{\mathsf{pf}}(A_i/B_i) \in \mathfrak{L}$ such that $\mathfrak{F} \vdash B \Rightarrow B_i$ and $\mathfrak{F} \vdash A_i \Rightarrow A$ and for all $\mathcal{O}_{\mathsf{pf}}(A_j/B_j) \in \mathfrak{L}$ we have: ($\mathfrak{F} \nvdash B \Rightarrow B_j$ or $\mathfrak{F} \nvdash B_j \Rightarrow B_i$ or $\mathfrak{F} \nvdash A_j, A \Rightarrow$)

- for all $\mathcal{O}_{\mathsf{pf}}(A_k/B_k) \in \mathfrak{L}$ we have: $\mathfrak{F} \nvdash B \Rightarrow B_k$ or $\mathfrak{F} \nvdash A_k, A \Rightarrow$ or there is a $\mathcal{O}_{\mathsf{pf}}(A_\ell/B_\ell) \in \mathfrak{L}$ such that: ($\mathfrak{F} \vdash B \Rightarrow B_\ell$ and $\mathfrak{F} \vdash B_\ell \Rightarrow B_k$ and $\mathfrak{F} \vdash A_\ell \Rightarrow A$).

Remark 3.2. The formulas we want to infer might have nested deontic operators. Indeed, they should capture key prescriptions like "under the condition of having to perform sacrifice α under the conditions β, you ought to do γ".

To turn this into sequent rules (the rules $\mathsf{ga}_L, \mathsf{ga}_R$ in Def. 3.3 below), we convert every (meta-)conjunction and universal quantifier in this characterization into different premises, while (meta-)disjunctions and existential quantifiers yield a split into different rules. To write the rules in an economic way, for sets $\mathcal{P}, \mathcal{Q}_i$ of premises we use the notation

$$\dfrac{\mathcal{P} \cup \begin{pmatrix} \mathcal{Q}_1 \\ \vdots \\ \mathcal{Q}_n \end{pmatrix}}{\Gamma \Rightarrow \Delta} \quad \text{for} \quad \left\{ \dfrac{\mathcal{P} \cup \mathcal{Q}_1}{\Gamma \Rightarrow \Delta}, \ldots, \dfrac{\mathcal{P} \cup \mathcal{Q}_n}{\Gamma \Rightarrow \Delta} \right\}$$

In case $\mathcal{Q}_i$ is a singleton we also omit the braces.

Since the rules now also will mention *underivability*, we further need to add a judgment for this to some of the sequents, written as $\mathfrak{F} \nvdash_{\mathsf{G_{MD}ga}_{\mathfrak{L}}\mathsf{cut}}$, with the intended meaning that the sequent is not derivable from the facts $\mathfrak{F}$ in the system $\mathsf{G_{MD}ga}_{\mathfrak{L}}\mathsf{cut}$ in the sense defined below (Def. 3.4). Thus we will obtain a set of rules ga_R introducing a formula of the form $\mathcal{O}(A/B)$ on the right hand side of the sequent. For technical reasons we will also add rules ga_L introducing such a formula on the left hand side – these essentially follow from absorbing inferences using the axiom D into the previous rule, and we will show below (Lem. 3.7) that they do not change the set of derivable sequents.

Definition 3.3. Let $\mathfrak{L} = \{\mathcal{O}_{\mathsf{pf}}(A_1/B_1), \ldots, \mathcal{O}_{\mathsf{pf}}(A_n/B_n)\}$ be a finite set of non-nested prima facie obligation formulas and let $\mathfrak{F}$ be a set of propositional sequents. The rules of $\mathsf{ga}_{\mathfrak{L}}$ are given in Fig. 5. A *proto-derivation* with conclusion $\Gamma \Rightarrow \Delta$ in the system $\mathsf{G_{MD}ga}_{\mathfrak{L}}$ from assumptions $\mathfrak{F}$ is a finite labelled tree, where each internal node is labelled with a sequent, each leaf is labelled with an initial sequent, a sequent from $\mathfrak{F}$, or an *underivability statement* $\mathfrak{F} \nvdash_{\mathsf{G_{MD}ga}_{\mathfrak{L}}\mathsf{cut}} \Sigma \Rightarrow \Pi$, such that the label of every internal node is obtained from the labels of its children using the rules of $\mathsf{G_{MD}}$ or $\mathsf{ga}_{\mathfrak{L}}$. The notion of a *proto-derivation in the system* $\mathsf{G_{MD}ga}_{\mathfrak{L}}\mathsf{cut}$ is defined analogously, but also permitting applications of the *cut rule*

$$\dfrac{\Gamma \Rightarrow \Delta, A \quad A, \Sigma \Rightarrow \Pi}{\Gamma, \Sigma \Rightarrow \Delta, \Pi} \; \mathsf{cut}\;.$$

The *depth* of a proto-derivation is the depth of the underlying tree, i.e., the maximal length of a branch in the tree plus one.

Definition 3.4. A proto-derivation in $\mathsf{G_{MD}ga}_{\mathfrak{L}}$ (in $\mathsf{G_{MD}ga}_{\mathfrak{L}}\mathsf{cut}$) from $\mathfrak{F}$ is *valid* if for each of the underivability statements $\mathfrak{F} \nvdash_{\mathsf{G_{MD}ga}_{\mathfrak{L}}\mathsf{cut}} \Sigma \Rightarrow \Pi$ occurring as one of the leafs of that derivation there is no valid proto-derivation of $\Sigma \Rightarrow \Pi$ in $\mathsf{G_{MD}ga}_{\mathfrak{L}}\mathsf{cut}$ from $\mathfrak{F}$. In case there is such a valid proto-derivation we also write $\mathfrak{F} \vdash_{\mathsf{G_{MD}ga}_{\mathfrak{L}}} \Gamma \Rightarrow \Delta$ and $\mathfrak{F} \vdash_{\mathsf{G_{MD}ga}_{\mathfrak{L}}\mathsf{cut}} \Gamma \Rightarrow \Delta$ respectively.

$$\dfrac{\begin{array}{l}\{B \Rightarrow B_i\} \quad \cup \quad \{A_i \Rightarrow A\} \\ \cup \quad \left\{\left(\begin{array}{l}\{\mathfrak{F} \nvdash_{\mathsf{GMDga}_{\mathfrak{L}}\mathsf{cut}} B \Rightarrow B_j\} \\ \{\mathfrak{F} \nvdash_{\mathsf{GMDga}_{\mathfrak{L}}\mathsf{cut}} B_j \Rightarrow B_i\} \\ \{\mathfrak{F} \nvdash_{\mathsf{GMDga}_{\mathfrak{L}}\mathsf{cut}} A_j, A \Rightarrow \}\end{array}\right) \;\bigg|\; \mathcal{O}_{\mathsf{pf}}(A_j/B_j) \in \mathfrak{L}\right\} \\ \cup \quad \left\{\left(\begin{array}{l}\{\mathfrak{F} \nvdash_{\mathsf{GMDga}_{\mathfrak{L}}\mathsf{cut}} B \Rightarrow B_k\} \\ \{\mathfrak{F} \nvdash_{\mathsf{GMDga}_{\mathfrak{L}}\mathsf{cut}} A_k, A \Rightarrow \} \\ \{B \Rightarrow B_1\} \cup \{B_1 \Rightarrow B_k\} \cup \{A_1 \Rightarrow A\} \\ \vdots \\ \{B \Rightarrow B_n\} \cup \{B_n \Rightarrow B_k\} \cup \{A_n \Rightarrow A\}\end{array}\right) \;\bigg|\; \mathcal{O}_{\mathsf{pf}}(A_k/B_k) \in \mathfrak{L}\right\}\end{array}}{\Rightarrow \mathcal{O}(A/B)}\; \mathsf{ga}_R$$

$$\dfrac{\begin{array}{l}\{D \Rightarrow B_i\} \quad \cup \quad \{A_i, C \Rightarrow \} \\ \cup \quad \left\{\left(\begin{array}{l}\mathfrak{F} \nvdash_{\mathsf{GMDga}_{\mathfrak{L}}\mathsf{cut}} D \Rightarrow B_j \\ \mathfrak{F} \nvdash_{\mathsf{GMDga}_{\mathfrak{L}}\mathsf{cut}} B_j \Rightarrow B_i \\ \mathfrak{F} \nvdash_{\mathsf{GMDga}_{\mathfrak{L}}\mathsf{cut}} A_j \Rightarrow C)\end{array}\right) \;\bigg|\; \mathcal{O}_{\mathsf{pf}}(A_j/B_j) \in \mathfrak{L}\right\} \\ \cup \quad \left\{\left(\begin{array}{l}\mathfrak{F} \nvdash_{\mathsf{GMDga}_{\mathfrak{L}}\mathsf{cut}} D \Rightarrow B_j \\ \mathfrak{F} \nvdash_{\mathsf{GMDga}_{\mathfrak{L}}\mathsf{cut}} A_j \Rightarrow C \\ \{D \Rightarrow B_1\} \cup \{B_1 \Rightarrow B_j\} \cup \{A_1, C \Rightarrow \} \\ \vdots \\ \{D \Rightarrow B_n\} \cup \{B_n \Rightarrow B_j\} \cup \{A_n, C \Rightarrow \}\end{array}\right) \;\bigg|\; \mathcal{O}_{\mathsf{pf}}(A_j/B_j) \in \mathfrak{L}\right\}\end{array}}{\mathcal{O}(C/D) \Rightarrow}\; \mathsf{ga}_L$$

Figure 5: The rules of $\mathsf{ga}_{\mathfrak{L}}$ for $\mathfrak{L} = \{\mathcal{O}_{\mathsf{pf}}(A_1/B_1), \ldots, \mathcal{O}_{\mathsf{pf}}(A_n/B_n)\}$, with $i = 1, \ldots, n$.

Note that underivability statements are always evaluated in the system with the cut rule. Since the definition of a valid proto-derivation involves the notion of a valid proto-derivation itself, it is not immediately clear that this notion is well-defined. We will show in Thm. 3.10 below that this is indeed the case.

Example 3.5. Consider the prima-facie obligations given by $\mathfrak{L} = \{\mathcal{O}_{\mathsf{pf}}(\mathsf{agn}/\top), \mathcal{O}_{\mathsf{pf}}(\neg\mathsf{agn}/\mathsf{sdr})\}$ (cf. Ex. 3.1) and the set $\mathfrak{F} = \emptyset$ of facts. Taking the formula $\mathcal{O}_{\mathsf{pf}}(\mathsf{agn}/\top)$ as the formula $\mathcal{O}_{\mathsf{pf}}(A_i/B_i)$ in the general scheme of Fig. 5, we obtain the rules in Fig. 6. In particular, the sequent $\Rightarrow \mathcal{O}(\mathsf{agn}/\mathsf{sch})$ would be derivable using, e.g., an instance of the rule

$$\dfrac{B \Rightarrow \top \quad \mathsf{agn} \Rightarrow A \quad \mathfrak{F} \nvdash_{\mathsf{GMDga}_{\mathfrak{L}}\mathsf{cut}} \mathsf{agn}, A \Rightarrow \quad \mathfrak{F} \nvdash_{\mathsf{GMDga}_{\mathfrak{L}}\mathsf{cut}} B \Rightarrow \mathsf{sdr} \quad B \Rightarrow \top \quad \top \Rightarrow \top \quad \mathsf{agn} \Rightarrow A \quad \mathfrak{F} \nvdash_{\mathsf{GMDga}_{\mathfrak{L}}\mathsf{cut}} B \Rightarrow \mathsf{sdr}}{\Rightarrow \mathcal{O}(A/B)}$$

$$
\begin{array}{c}
\{B \Rightarrow \mathsf{T}\} \\
\cup \{\mathsf{agn} \Rightarrow A\}
\end{array}
\cup
\left(\begin{array}{l}
\mathfrak{F} \nvdash_{\mathsf{GMDga}_\mathcal{L}\mathsf{cut}} B \Rightarrow \mathsf{T} \\
\mathfrak{F} \nvdash_{\mathsf{GMDga}_\mathcal{L}\mathsf{cut}} \mathsf{T} \Rightarrow \mathsf{T} \\
\mathfrak{F} \nvdash_{\mathsf{GMDga}_\mathcal{L}\mathsf{cut}} \mathsf{agn}, A \Rightarrow
\end{array}\right)
\cup
\left(\begin{array}{l}
\mathfrak{F} \nvdash_{\mathsf{GMDga}_\mathcal{L}\mathsf{cut}} B \Rightarrow \mathsf{sdr} \\
\mathfrak{F} \nvdash_{\mathsf{GMDga}_\mathcal{L}\mathsf{cut}} \mathsf{sdr} \Rightarrow \mathsf{T} \\
\mathfrak{F} \nvdash_{\mathsf{GMDga}_\mathcal{L}\mathsf{cut}} \neg\mathsf{agn}, A \Rightarrow
\end{array}\right)
$$

$$
\cup
\left(\begin{array}{l}
\mathfrak{F} \nvdash_{\mathsf{GMDga}_\mathcal{L}\mathsf{cut}} B \Rightarrow \mathsf{T} \\
\mathfrak{F} \nvdash_{\mathsf{GMDga}_\mathcal{L}\mathsf{cut}} \mathsf{agn}, A \Rightarrow \\
\{B \Rightarrow \mathsf{T}\} \cup \{\mathsf{T} \Rightarrow \mathsf{T}\} \cup \{\mathsf{agn} \Rightarrow A\} \\
\{B \Rightarrow \mathsf{sdr}\} \cup \{\mathsf{sdr} \Rightarrow \mathsf{T}\} \cup \{\neg\mathsf{agn} \Rightarrow A\}
\end{array}\right)
$$

$$
\cup
\left(\begin{array}{l}
\mathfrak{F} \nvdash_{\mathsf{GMDga}_\mathcal{L}\mathsf{cut}} B \Rightarrow \mathsf{sdr} \\
\mathfrak{F} \nvdash_{\mathsf{GMDga}_\mathcal{L}\mathsf{cut}} \neg\mathsf{agn}, A \Rightarrow \\
\{B \Rightarrow \mathsf{T}\} \cup \{\mathsf{T} \Rightarrow \mathsf{sdr}\} \cup \{\mathsf{agn} \Rightarrow A\} \\
\{B \Rightarrow \mathsf{sdr}\} \cup \{\mathsf{sdr} \Rightarrow \mathsf{sdr}\} \cup \{\neg\mathsf{agn} \Rightarrow A\}
\end{array}\right)
$$

$$
\Rightarrow \mathcal{O}(A/B)
$$

Figure 6: The rules from Ex. 3.5.

Similarly, taking the formula $\mathcal{O}_{\mathsf{pf}}(A_i/B_i)$ to be $\mathcal{O}_{\mathsf{pf}}(\neg\mathsf{agn}/\mathsf{sdr})$ we obtain, e.g.

$$
\frac{B \Rightarrow \mathsf{sdr} \quad \neg\mathsf{agn} \Rightarrow A \quad \mathfrak{F} \nvdash_{\mathsf{GMDga}_\mathcal{L}\mathsf{cut}} \mathsf{T} \Rightarrow \mathsf{sdr} \quad \mathfrak{F} \nvdash_{\mathsf{GMDga}_\mathcal{L}\mathsf{cut}} \neg\mathsf{agn}, A \Rightarrow}{\Rightarrow \mathcal{O}(A/B)}
$$
$$
B \Rightarrow \mathsf{sdr} \quad \mathsf{sdr} \Rightarrow \mathsf{T} \quad \neg\mathsf{agn} \Rightarrow A \quad B \Rightarrow \mathsf{sdr} \quad \mathsf{sdr} \Rightarrow \mathsf{sdr} \quad \neg\mathsf{agn} \Rightarrow A
$$

which serves to derive the sequent $\Rightarrow \mathcal{O}(\neg\mathsf{agn}/\mathsf{sch} \wedge \mathsf{sdr})$. Finally, using ga_L with $\mathcal{O}_{\mathsf{pf}}(\neg\mathsf{agn}/\mathsf{sdr})$ for the formula $\mathcal{O}_{\mathsf{pf}}(A_i/B_i)$ yields a derivation of $\mathcal{O}(\mathsf{agn}/\mathsf{sch} \wedge \mathsf{sdr}) \Rightarrow$ and thus $\Rightarrow \neg\mathcal{O}(\mathsf{agn}/\mathsf{sch} \wedge \mathsf{sdr})$. Note that even for just two prima-facie obligations we obtain many (often redundant) rules.

The following lemma is useful to shorten derivations.

Lemma 3.6. *For every formula A we have $\mathfrak{F} \vdash_{\mathsf{GMDga}_\mathcal{L}\mathsf{cut}} \Gamma, A \Rightarrow A, \Delta$.*

Proof. By straightforward induction on the complexity of A, using the rule **Mon** in the modal case. □

We now show that the rule ga_L indeed is a mere technical convenience.

Lemma 3.7. *If there is a valid proto-derivation of $\Gamma \Rightarrow \Delta$ in $\mathsf{GMDga}_\mathcal{L}\mathsf{cut}$ from $\mathfrak{F}$, then there is a valid proto-derivation of $\Gamma \Rightarrow \Delta$ from $\mathfrak{F}$ in the same system without the rule ga_L.*

Proof. We show how to replace every application a rule ga_L by an application of ga_R and **cut**. Suppose we have an application of ga_L as given in Fig. 5. From the premises $A_j, C \Rightarrow$ (if any) using weakening and the $\rightarrow_R$ rule we obtain $A_j \Rightarrow C \rightarrow \bot$. Further, from every underivability statement $\mathfrak{F} \nvdash_{\mathsf{G_{MD}ga_{\mathfrak{L}}cut}} A_j \Rightarrow C$ we obtain $\mathfrak{F} \nvdash_{\mathsf{G_{MD}ga_{\mathfrak{L}}cut}} A_j, C \rightarrow \bot \Rightarrow$, since, if for the latter there were a valid proto-derivation, we could extend it to one of the former via

$$\dfrac{\dfrac{\overline{C \Rightarrow C, \bot}\ \text{Lem. 3.6}}{\Rightarrow C, C \rightarrow \bot}\ \rightarrow_R \qquad A_j, C \rightarrow \bot \Rightarrow}{A_j \Rightarrow C}\ \text{cut}$$

But then we have all the premises necessary to apply the rule ga_R with conclusion $\Rightarrow \mathcal{O}(C \rightarrow \bot/D)$. From this we obtain the conclusion of the application of ga_L as follows:

$$\dfrac{\Rightarrow \mathcal{O}(C \rightarrow \bot/D) \qquad \dfrac{\dfrac{\overline{\bot, C \Rightarrow}\ \bot_L \qquad \overline{C \Rightarrow C}\ \text{Lem. 3.6}}{C \rightarrow \bot, C \Rightarrow}\ \rightarrow_L \qquad \dfrac{\overline{D \Rightarrow D}\ \text{Lem. 3.6}}{\mathcal{O}(C \rightarrow \bot/D), \mathcal{O}(C/D) \Rightarrow}\ \text{cut}}{\mathcal{O}(C/D) \Rightarrow}}$$

□

In order to unravel the definition of valid proto-derivations and to be able to provide a decision procedure, we show the redundancy (actually the eliminability) of the cut rule in valid proto-derivations. First we obtain:

Lemma 3.8. *If* $\mathfrak{F} \vdash_{\mathsf{G_{MD}ga_{\mathfrak{L}}}} \Gamma \Rightarrow \Delta$, *then* $\mathfrak{F} \vdash_{\mathsf{G_{MD}ga_{\mathfrak{L}}cut}} \Gamma \Rightarrow \Delta$.

Proof. Straightforward since every rule in $\mathsf{G_{MD}}$ is a rule in $\mathsf{G_{MD}cut}$, and since the underivability statements range over the same system for valid proto-derivations in both $\mathsf{G_{MD}ga_{\mathfrak{L}}}$ and $\mathsf{G_{MD}ga_{\mathfrak{L}}cut}$ □

In order to fully control the underivability statements involved in the notion of a valid derivation, we further need to show the converse of this statement.

Theorem 3.9 (Partial cut elimination). *If* $\mathfrak{F} \vdash_{\mathsf{G_{MD}ga_{\mathfrak{L}}cut}} \Gamma \Rightarrow \Delta$, *then* $\mathfrak{F} \vdash_{\mathsf{G_{MD}ga_{\mathfrak{L}}}} \Gamma \Rightarrow \Delta$.

Proof. We show how to eliminate topmost applications of the *multicut rule*

$$\dfrac{\Gamma \Rightarrow \Delta, A^n \qquad A^m, \Sigma \Rightarrow \Pi}{\Gamma, \Sigma \Rightarrow \Delta, \Pi}\ \mathsf{mcut}$$

from a proto-derivation, preserving validity (here A^n is the multiset containing n copies of A). Since cut is a case of mcut and mcut is derivable using $\mathsf{Con}_L, \mathsf{Con}_R$ and cut, this suffices. The proof is by double induction on the complexity of the cut formula A and the sum of the depths of the derivations of the two premises of the application of mcut (see [26, Sec. 4.1.9] for the classical case without underivability statements).

If the complexity of the cut formula is 0, then it is a propositional variable, and hence not principal in a modal or propositional rule or a rule from $\mathsf{ga}_\mathcal{L}$. Thus, as usual, we permute mcut into the premises of the last applied rules using the inner induction on the depths of the derivations, until it is absorbed by an application of weakening, or reaches the leaves of the proto-derivation. In this case the premises of the multicut are initial sequents or elements of $\mathfrak{F}$. If at least one of these is an initial sequent, the multicut is eliminated as usual, if both sequents are elements of $\mathfrak{F}$ we use that $\mathfrak{F}$ is closed under contraction and cuts and replace the multicut with the corresponding element of $\mathfrak{F}$.

So assume that the complexity of the cut formula is $n+1$. Again, using the inner induction on the depth of the proto-derivation we permute the multicut into the premise(s) of the last applied rules, until it is principal in the last rules of the derivations of both premises of the multicut. In case the cut formula is propositional we use the standard transformation, see [26].

The only interesting case is where the cut formula is a deontic formula. If the last applied rules both are among P, D, Mon, then the transformation is essentially as for the system G_{MD}. E.g., if the last applied rules were Mon and D, the multicut has the following form:

$$\dfrac{\dfrac{C \Rightarrow A \quad D \Rightarrow B \quad B \Rightarrow D}{\mathcal{O}(C/D) \Rightarrow \mathcal{O}(A/B)} \text{ Mon} \quad \dfrac{A, E \Rightarrow \quad B \Rightarrow F \quad F \Rightarrow B}{\mathcal{O}(A/B), \mathcal{O}(E/F) \Rightarrow} \text{ D}}{\mathcal{O}(C/D), \mathcal{O}(E/F) \Rightarrow} \text{ mcut}$$

Using the induction hypothesis on the complexity of the cut formula we obtain valid proto-derivations of the conclusions of

$$\dfrac{C \Rightarrow A \quad A, E \Rightarrow}{C, E \Rightarrow} \text{ mcut} \qquad \dfrac{D \Rightarrow B \quad B \Rightarrow F}{D \Rightarrow F} \text{ mcut} \qquad \dfrac{F \Rightarrow B \quad B \Rightarrow D}{F \Rightarrow D} \text{ mcut}$$

Now an application of the rule D yields the sequent $\Gamma, \mathcal{O}(C/D), \Sigma, \mathcal{O}(E/F) \Rightarrow \Delta, \Pi$. In case both principal formulas of the application of D are cut formulas, we proceed similarly, only using the rule P in the last step. The other cases of the modal rules are similar.

In the most interesting cases at least one of the premises of the cut was derived using a rule from $\mathsf{ga}_\mathcal{L}$. We consider all the different cases.

Suppose that the two last applied rules were ga_R and Mon. Then the two derivations end in an instance of a rule from

$$
\cfrac{
\begin{array}{l}
\{B \Rightarrow B_i\} \quad \cup \quad \{A_i \Rightarrow A\} \\
\cup \left\{ \left(\begin{array}{l} \mathfrak{F} \not\vdash_{\mathsf{G_{MD}ga_{\mathfrak{L}}cut}} B \Rightarrow B_j \\ \mathfrak{F} \not\vdash_{\mathsf{G_{MD}ga_{\mathfrak{L}}cut}} B_j \Rightarrow B_i \\ \mathfrak{F} \not\vdash_{\mathsf{G_{MD}ga_{\mathfrak{L}}cut}} A_j, A \Rightarrow) \end{array} \right) \;\middle|\; \mathcal{O}_{\mathsf{pf}}(A_j/B_j) \in \mathfrak{L} \right\} \\
\cup \left\{ \left(\begin{array}{l} \mathfrak{F} \not\vdash_{\mathsf{G_{MD}ga_{\mathfrak{L}}cut}} B \Rightarrow B_k \\ \mathfrak{F} \not\vdash_{\mathsf{G_{MD}ga_{\mathfrak{L}}cut}} A_k, A \Rightarrow \\ \{B \Rightarrow B_1\} \cup \{B_1 \Rightarrow B_k\} \cup \{A_1 \Rightarrow A\} \\ \vdots \\ \{B \Rightarrow B_n\} \cup \{B_n \Rightarrow B_k\} \cup \{A_n \Rightarrow A\} \end{array} \right) \;\middle|\; \mathcal{O}_{\mathsf{pf}}(A_k/B_k) \in \mathfrak{L} \right\}
\end{array}
}{\Rightarrow \mathcal{O}(A/B)} \; \mathsf{ga}_R \quad (1)
$$

and

$$\cfrac{A \Rightarrow C \quad B \Rightarrow D \quad D \Rightarrow B}{\mathcal{O}(A/B) \Rightarrow \mathcal{O}(C/D)} \; \mathsf{Mon}$$

respectively. By induction hypothesis on the complexity of the cut formula we obtain valid proto-derivations of $D \Rightarrow B_i$ and $A_i \Rightarrow C$, as well as for $1 \leq \ell \leq n$ the sequents $D \Rightarrow B_\ell$ and $B_\ell \Rightarrow B_k$ and $A_\ell \Rightarrow C$ whenever the corresponding sequents occur in the application of ga_R. Further, for every underivability statement $\mathfrak{F} \not\vdash_{\mathsf{G_{MD}ga_{\mathfrak{L}}cut}} B \Rightarrow B_j$ together with derivability of $B \Rightarrow D$ we obtain the underivability statement $\mathfrak{F} \not\vdash_{\mathsf{G_{MD}ga_{\mathfrak{L}}cut}} D \Rightarrow B_j$ by contradiction: assuming there is a valid proto-derivation of $D \Rightarrow B_j$ in $\mathsf{G_{MD}ga_{\mathfrak{L}}cut}$ from $\mathfrak{F}$ we could apply cut to this and $B \Rightarrow D$ to obtain $\mathfrak{F} \vdash_{\mathsf{G_{MD}ga_{\mathfrak{L}}cut}} B \Rightarrow B_j$, in contradiction to $\mathfrak{F} \not\vdash_{\mathsf{G_{MD}ga_{\mathfrak{L}}cut}} B \Rightarrow B_j$. Similarly, for every underivability statement $\mathfrak{F} \not\vdash_{\mathsf{G_{MD}ga_{\mathfrak{L}}cut}} A_j, A \Rightarrow$ using derivability of $A \Rightarrow C$ we obtain the underivability statement $\mathfrak{F} \not\vdash_{\mathsf{G_{MD}ga_{\mathfrak{L}}cut}} A_j, C \Rightarrow$. Hence we can apply the rule ga_R to obtain a proto-derivation of $\Rightarrow \mathcal{O}(C/D)$. By the reasoning above, all the underivability statements hold, hence the proto-derivation is valid.

The cases where the two last applied rules were ga_R and D with only one of the principal formulas a cut formula or Mon and ga_L are similar, in each case finishing with an application of ga_L.

For the case where the last rules were ga_R and P, we claim that it actually cannot occur. For otherwise the derivations end in an instance of (1) and

$$\cfrac{A \Rightarrow}{\mathcal{O}(A/B) \Rightarrow} \; \mathsf{P} \; .$$

However, then for $i = j$ we have valid proto-derivations for all three of $B \Rightarrow B_j$ and $B_j \Rightarrow B_i$ and $A_j, A \Rightarrow$. The first one is the first premise of the application of ga_R, the second one follows from Lem. 3.6 since $i = j$, and the last one follows from the premise of P using W_L. But then the proto-derivation of $\Rightarrow \mathcal{O}(A/B)$ cannot have been valid since for some of the underivability statements in the premises of the rule ga_R there is a valid proto-derivation.

The case where the last rules were ga_R and D with both principal formulas of the latter cut formulas is analogous to the previous case.

This leaves the case where the last rules were ga_R and ga_L, which likewise cannot happen. For suppose it did, then the derivations would end in (1) and

$$\cfrac{\{B \Rightarrow B_k\} \ \cup \ \{A_k, A \Rightarrow \} \ \cup \ \left\{\begin{pmatrix} \mathfrak{F} \nvdash_{\mathsf{GMDga}_{\mathfrak{L}}\mathsf{cut}} B \Rightarrow B_\ell \\ \mathfrak{F} \nvdash_{\mathsf{GMDga}_{\mathfrak{L}}\mathsf{cut}} B_\ell \Rightarrow B_k \\ \mathfrak{F} \nvdash_{\mathsf{GMDga}_{\mathfrak{L}}\mathsf{cut}} A_\ell \Rightarrow A \end{pmatrix} \Big| \ \mathcal{O}_{\mathsf{pf}}(A_\ell/B_\ell) \in \mathfrak{L}\right\} \ \cup \ \left\{\begin{pmatrix} \mathfrak{F} \nvdash_{\mathsf{GMDga}_{\mathfrak{L}}\mathsf{cut}} B \Rightarrow B_\ell \\ \mathfrak{F} \nvdash_{\mathsf{GMDga}_{\mathfrak{L}}\mathsf{cut}} A_\ell \Rightarrow A \\ \{B \Rightarrow B_1\} \cup \{B_1 \Rightarrow B_\ell\} \cup \{A_1, A \Rightarrow \} \\ \vdots \\ \{B \Rightarrow B_n\} \cup \{B_n \Rightarrow B_\ell\} \cup \{A_n, A \Rightarrow \} \end{pmatrix} \Big| \ \mathcal{O}_{\mathsf{pf}}(A_\ell/B_\ell) \in \mathfrak{L}\right\}}{\mathcal{O}(A/B) \Rightarrow} \ \mathsf{ga}_L$$

But then in particular the application of the rule ga_R has one of the premises $\mathfrak{F} \nvdash_{\mathsf{GMDga}_{\mathfrak{L}}\mathsf{cut}} B \Rightarrow B_k$ and $\mathfrak{F} \nvdash_{\mathsf{GMDga}_{\mathfrak{L}}\mathsf{cut}} A_k, A \Rightarrow$ or all of the three premises

$$B \Rightarrow B_m \qquad B_m \Rightarrow B_k \qquad A_m \Rightarrow A$$

for some $m \leq n$. However, the first case gives a contradiction with the premise $B \Rightarrow B_k$ of the application of ga_L using validity of the proto-derivation. The second case gives a contradiction with the premise $A_k, A \Rightarrow$ of ga_L, again using validity of the proto-derivation. Finally, the third case gives a contradiction because the application of ga_L contains one of the premises

$$\mathfrak{F} \nvdash_{\mathsf{GMDga}_{\mathfrak{L}}\mathsf{cut}} B \Rightarrow B_m \qquad \mathfrak{F} \nvdash_{\mathsf{GMDga}_{\mathfrak{L}}\mathsf{cut}} B_m \Rightarrow B_k \qquad \mathfrak{F} \nvdash_{\mathsf{GMDga}_{\mathfrak{L}}\mathsf{cut}} A_m \Rightarrow A$$

and the proto-derivation is valid. Hence this case also cannot occur. □

3.2 Applications of cut elimination

Thm. 3.9 is the basis for a number of important results. First and foremost we obtain that the notion of valid proto-derivations (Def. 3.4) actually makes sense.

Theorem 3.10. *The notion of a valid proto-derivation is well-defined.*

Proof. Cut elimination together with Lem. 3.8 shows that we can replace every underivability statement in the rules $\mathsf{ga}_L, \mathsf{ga}_R$ by a statement of the form $\mathfrak{F} \nvdash_{\mathsf{GMDga}_{\mathfrak{L}}}$ $\Sigma \Rightarrow \Pi$ and define valid proto-derivations in terms of cut-free derivations. Since the modal nesting depth properly decreases in the modal rules, the definition of a valid proto-derivation is hence equivalent to a *stratified definition*, where we define the notion of a valid proto-derivation of *rank n* such that every sequent occurring in the derivation has modal nesting depth at most n, and all the underivability statements refer to underivability using valid proto-derivations of rank smaller than n. □

The proof of the previous theorem serves to illustrate one of the main differences between the approach followed here and approaches which model conditional obligations using non-monotone inference or defeasible rules such as [13, 18]: since the underivability statements in the premises of a rule can be restricted to sequents of smaller modal nesting depth, we can avoid having to perform a fixed-point computation. In particular, for checking whether a non-nested conditional obligation formula is derivable we only need to check (classical) derivability for purely propositional sequents. This is possible because the underivability statements in the rules ga_L, ga_R only depend on the list of prima-facie obligations, and not on obligations which themselves are derivable. Consequently, we obtain decidability of the logic:

Theorem 3.11 (Decidability). *The set of all sequents for which there is a valid proto-derivation in* $\mathsf{G}_{\mathsf{MD}}\text{ga}_{\mathfrak{L}}\text{cut}$ *from* $\mathfrak{F}$ *is decidable.*

Proof. First we show that a sequent has a valid proto-derivation in $\mathsf{G}_{\mathsf{MD}}\text{ga}_{\mathfrak{L}}$ if and only if it has a valid proto-derivation using rules of the system $\mathsf{G}^*_{\mathsf{MD}}\text{ga}_{\mathfrak{L}}$, which is obtained from $\mathsf{G}_{\mathsf{MD}}\text{ga}_{\mathfrak{L}}$ by dropping the contraction rules $\text{Con}_L, \text{Con}_R$ and replacing the propositional rules $\rightarrow_L, \rightarrow_R$ with their invertible versions where the principal formula is copied into the premises:

$$\frac{\Gamma, A \rightarrow B, B \Rightarrow \Delta \quad \Gamma, A \rightarrow B \Rightarrow A, \Delta}{\Gamma, A \rightarrow B \Rightarrow \Delta} \rightarrow^*_L \qquad \frac{\Gamma, A \Rightarrow B, A \rightarrow B, \Delta}{\Gamma \Rightarrow A \rightarrow B, \Delta} \rightarrow^*_R$$

Equivalence of the systems is obtained by first showing equivalence of the propositional rules and their versions above in the presence of weakening and contraction, and then showing that the contraction rules are admissible in $\mathsf{G}^*_{\mathsf{MD}}\text{ga}_{\mathfrak{L}}$.

The proof of the latter is, as usual, by induction on the depth of the proto-derivation, using that the set $\mathfrak{F}$ of facts is closed under contraction: If the depth is 1, then, the proto-derivation consists of an initial sequent or a sequent from $\mathfrak{F}$. Since $\mathfrak{F}$ is assumed to be closed under contractions, the admissibility follows. If the depth is $n > 1$ we distinguish cases according to the last applied rule in the proto-derivation. The only non-trivial case is when that rule was one of the modal rules, i.e. the rule D, since none of the other rules has two formulae on the same side of the conclusion. In this case, the proto-derivation ends in

$$\frac{A, A \Rightarrow \quad B \Rightarrow B \quad B \Rightarrow B}{\mathcal{O}(A/B), \mathcal{O}(A/B) \Rightarrow} \text{D}$$

Applying the induction hypothesis on the proto-derivation of the first premise we obtain a proto-derivation of the sequent $A \Rightarrow$, and an application of the rule P yields the desired sequent $\mathcal{O}(A/B) \Rightarrow$.

To check whether for a given sequent there is a valid proto-derivation in the system $\mathsf{G}_{\mathsf{MD}}\text{ga}_{\mathfrak{L}}\text{cut}$ from $\mathfrak{F}$, by Thm. 3.9 it is enough to search through the possible proto-derivations in $\mathsf{G}_{\mathsf{MD}}\text{ga}_{\mathfrak{L}}$, which by the previous considerations is equivalent to searching through proto-derivations in $G^*_{\mathsf{MD}}\mathsf{ga}_\mathfrak{L}$. For the latter we perform (depth-first) *backwards proof search*, following a *local loop checking* strategy to prevent rule applications where every formula of a premise already occurs in the conclusion. Upon encountering an underivability statement the decision procedure calls itself recursively and simply flips the answer. The procedure terminates, because the modal nesting depth of the underivability statements in the premises of $\mathsf{ga}_L, \mathsf{ga}_R$ is lower than that of the conclusion, the rules have bounded branching factor and the local loop checking strategy implies that the proto-derivations themselves have bounded depth. □

Furthermore, we obtain that the rules $\mathsf{ga}_\mathfrak{L}$ are compatible with deontic logic MD in the sense that they do not yield any conflicting obligations:

Theorem 3.12 (Consistency). *For any $\mathfrak{L}$ and $\mathfrak{F}$ not containing the empty sequent, the consequences of $\mathfrak{L}$ under $\mathfrak{F}$ are consistent over* MD, *i.e.,* $\mathfrak{F} \nvdash_{\mathsf{G_{MD}ga_\mathfrak{L}cut}} \Rightarrow \bot$. *Hence there is no $\mathcal{O}(A/B)$ with $\mathfrak{F} \vdash_{\mathsf{G_{MD}ga_\mathfrak{L}cut}} \Rightarrow \mathcal{O}(A/B) \wedge \mathcal{O}(\neg A/B)$.*

Proof. By inspection it is clear that all the rules in the calculus $\mathsf{G_{MD}ga_\mathfrak{L}}$ have the *subformula property relative to* $\mathfrak{L}$ in the sense that every formula occurring in a premise of a rule, including the underivability statements, is a subformula of a formula occurring in its conclusion or in $\mathfrak{L}$. Since the empty sequent is not in $\mathfrak{F}$, and apart from W_R there is no rule introducing $\bot$ on the right hand side of a sequent, we cannot derive $\Rightarrow \bot$. The second statement follows from derivability of $\mathcal{O}(A/B) \wedge \mathcal{O}(\neg A/B) \Rightarrow$ and cut. □

3.3 The disjunctive response/Vikalpa

The described system rejects any inferences which would result in conflicting obligations. In particular, for a set $\mathfrak{L} = \{\mathcal{O}_{\mathsf{pf}}(a/b), \mathcal{O}_{\mathsf{pf}}(c/d)\}$ and a set $\mathfrak{F} = \{a, c \Rightarrow\}$ establishing that a and c are not jointly possible, neither of the formulas $\mathcal{O}(a/b \wedge d)$ and $\mathcal{O}(c/b \wedge d)$ will be derivable. While this is as intended, intuitively still the disjunction of a and c should be obligatory, i.e., the formula $\mathcal{O}(a \vee c/b \wedge d)$ should be derivable, and similarly for sets of formulas $\{\mathcal{O}_{\mathsf{pf}}(a_1/b_1), \ldots, \mathcal{O}_{\mathsf{pf}}(a_n/b_n)\}$ where all the a_i are not jointly possible. Amazingly, this principle, which is called the *disjunctive response* in [10], was formulated already more than two millennia ago in one of the founding texts of the Mīmāṃsā school, the *Pūrva Mīmāṃsā Sūtras* of Jaimini under the name of *vikalpa*, whose English translation (and reformulation) is

> If a prescription enjoins X and a prohibition forbids one to perform the same act X, and no other interpretation is possible, the act X should be considered optional (*vikalpa*), although this leads to the problematic situation that either the one or the other is transgressed.

Thus, checking whether our rendering of the specificity principle satisfies this principle is a good test for checking suitability to capture both the intuitive and the Mīmāṃsā notion of obligation. Indeed, generalizing the above to sets of obligations, and adding that all the enjoined acts should be possible we have:

Theorem 3.13. *Let* $X = \{\mathcal{O}_{\mathsf{pf}}(A_1/B_1), \ldots, \mathcal{O}_{\mathsf{pf}}(A_n/B_n)\} \subseteq \mathfrak{L}$ *be a set such that* $\mathfrak{F} \nvdash_{\mathsf{GMDga}_{\mathfrak{L}}\mathsf{cut}} A_i \Rightarrow$ *for every* $i \leq n$, *and for every* $\mathcal{O}_{\mathsf{pf}}(C/D) \in \mathfrak{L} \smallsetminus X$ *with* $\mathfrak{F} \vdash_{\mathsf{GMDga}_{\mathfrak{L}}\mathsf{cut}} \bigwedge_{i \leq n} B_i \Rightarrow D$ *we have* $\mathfrak{F} \nvdash_{\mathsf{GMDga}_{\mathfrak{L}}\mathsf{cut}} \bigvee_{i \leq n} A_i, C \Rightarrow$. *Then* $\mathfrak{F} \vdash_{\mathsf{GMDga}_{\mathfrak{L}}\mathsf{cut}} \Rightarrow \mathcal{O}(\bigvee_{i \leq n} A_i / \bigwedge_{i \leq n} B_i)$.

Proof. We show that we have all the premises to apply the rule ga_R. From the propositional rules we obtain $\mathfrak{F} \vdash_{\mathsf{GMDga}_{\mathfrak{L}}\mathsf{cut}} A_1 \Rightarrow \bigvee_{i \leq n} A_i$ and $\mathfrak{F} \vdash_{\mathsf{GMDga}_{\mathfrak{L}}\mathsf{cut}} \bigwedge_{i \leq n} B_i \Rightarrow B_1$. Moreover, for every $j \leq n$ we obtain $\mathfrak{F} \nvdash_{\mathsf{GMDga}_{\mathfrak{L}}\mathsf{cut}} A_j, \bigvee_{i \leq n} A_i \Rightarrow$, since otherwise in particular we would have $\mathfrak{F} \vdash_{\mathsf{GMDga}_{\mathfrak{L}}\mathsf{cut}} A_j, A_j \Rightarrow$, and hence $\mathfrak{F} \vdash_{\mathsf{GMDga}_{\mathfrak{L}}\mathsf{cut}} A_j \Rightarrow$. Moreover, by assumption, for every $\mathcal{O}_{\mathsf{pf}}(C/D) \in \mathfrak{L} \smallsetminus X$ we have either $\mathfrak{F} \nvdash_{\mathsf{GMDga}_{\mathfrak{L}}\mathsf{cut}} \bigwedge_{i \leq n} B_i \Rightarrow D$ or $\mathfrak{F} \nvdash_{\mathsf{GMDga}_{\mathfrak{L}}\mathsf{cut}} C, \bigvee_{i \leq n} A_i \Rightarrow$. Now applying ga_R yields $\Rightarrow \mathcal{O}(\bigvee_{i \leq n} A_i / \bigwedge_{i \leq n} B_i)$. □

It should be noted that for the statement of the theorem it is not relevant whether the A_i from the set X are jointly possible or not, only that their disjunction $\bigvee_{i \leq m} A_i$ is not blocked by any C from outside that set. In particular, it also applies to the case where the A_i are not jointly possible. Thus, our system as described indeed satisfies the disjunctive response resp. vikalpa.

4 Conclusion

We have explored connections between the Mīmāṃsā school of Indian philosophy and symbolic deontic logic concerning the *specificity principle*. We investigated a notion of specificity based on a sequent calculus for MD and explored some of its properties. Apart from the technical content, this paper illustrates some of the vast potential for cross-fertilisation between Mīmāṃsā and deontic logic. Of course, many aspects both of the Mīmāṃsā philosophy and of the proposed formal system are still waiting to be unearthed; among them, how to modify our rules $\mathsf{ga}_{\mathfrak{L}}$ to accommodate for new deontic axioms extracted from *nyāyas* that might be added to bMDL. We further conjecture that it is possible to show a completeness result with respect to the class of neighbourhood models under a certain set of global assumptions along the lines of [11]; due to the infinite nature of the set of global assumptions this is not entirely straightforward. In view of Remark 2.3 it would also be interesting to see whether our approach can be generalized to handle nested prima-facie obligations. We also plan to implement the introduced calculus and try to use it to prove Mīmāṃsā conjectures about Vedic sacrifices; e.g., whether the so-called Full and New Moon sacrifice is the archetype of all vegetable sacrifices.

From a more philosophical point of view, we recall the extensive discussion in the deontic literature about the difference between Contrary-To-Duty obligations and instances of the specificity principle [28] and about the connections of this topic with the problem of factual detachment [2, 22]. It would be interesting to see whether our proposed methods and calculi can be used to analyze more closely where the Mīmāṃsā authors would be situated in this debate.

References

[1] K. L. Bathia. *Legal Language and Legal Writing*. Universal Law Publishing Co, 2010.

[2] M. Beirlaen, J. Heyninck, and C. Straßer. Structured argumentation with prioritized conditional obligations and permissions. *Journal of Logic and Computation*, 2018.

[3] P. A. Bonatti and N. Olivetti. Sequent calculi for propositional nonmonotonic logics. *ACM Transactions on Computational Logic*, 3(2):226–278, 2002.

[4] B. F. Chellas. *Modal Logic*. Cambridge University Press, 1980.

[5] A. Ciabattoni, E. Freschi, F. A. Genco, and B. Lellmann. Mīmāṃsā deontic logic: proof theory and applications. In *TABLEAUX 2015*, volume 9323 of *LNCS*, pages 323–338. Springer, 2015.

[6] A. Ciabattoni, E. Freschi, F. A. Genco, and B. Lellmann. Understanding prescriptive texts: Rules and logic as elaborated by the Mīmāṃsā school. *Online Journal of World Philosophies*, 2:47–66, 2017.

[7] J. P. Delgrande and T. H. Schaub. Compiling specificity into approaches to nonmonotonic reasoning. *Artificial Intelligence*, 90(1):301 – 348, 1997.

[8] E. Freschi, A. Ollett, and M. Pascucci. Duty and sacrifice. a logical analysis of the Mīmāṃsā theory of Vedic injunctions. Submitted, 2018.

[9] D. M. Gabbay. Theoretical foundations for non-monotonic reasoning in expert systems. In *Logics and models of concurrent systems*, pages 439–457. Springer, 1985.

[10] L. Goble. Prima facie norms, normative conflicts, and dilemmas. In *Handbook of Deontic Logic and Normative Systems*, pages 241–351. College Publications, 2013.

[11] R. Goré, C. Kupke, and D. Pattinson. Optimal tableau algorithms for coalgebraic logics. In *TACAS 2010*, LNCS, pages 114–128. Springer, 2010.

[12] G. Governatori and A. Rotolo. Logic of violations: A Gentzen system for reasoning with contrary-to-duty obligations. *The Australasian Journal of Logic*, 4:193–215, 2006.

[13] J. F. Horty. *Reasons as Defaults*. Oxford University Press, 2012.

[14] C. H. Huisjes. *Norms and logic*. Thesis, University of Groningen, 1981.

[15] D. Kapoor. www.patheos.com/blogs/drishtikone/2008/03/indian-supreme-court-uses-mimamsa-school-legal-reasoning-decide-case/, 2008. Accessed 2018-03-12.

[16] B. Lellmann and D. Pattinson. Constructing cut free sequent systems with context restrictions based on classical or intuitionistic logic. In *ICLA 2013*, volume 7750 of *LNCS*, pages 148–160. Springer, 2013.

[17] D. Makinson and K. Schlechta. Floating conclusions and zombie paths: Two deep difficulties in the 'directly skeptical' approach to inheritance nets. *Artificial Intelligence*,

48:199–209, 1991.
[18] D. Nute. Defeasible logic. In *INAP 2001*, volume 2543 of *LNCS*, pages 151–169. Springer, 2003.
[19] E. Orlandelli. Proof analysis in deontic logics. In *DEON 2014*, volume 8554 of *LNCS*, pages 139–148. Springer, 2014.
[20] X. Parent. Maximality vs. optimality in dyadic deontic logic. *Journal of Philosophical Logic*, 43:1101–1128, 2013.
[21] H. Prakken and G. Sartor. A system for defeasible argumentation, with defeasible priorities. In *FAPR'96*, volume 1085 of *LNCS*, pages 510–524. Springer, 1996.
[22] C. Straßer. A deontic logic framework allowing for factual detachment. *Journal of Applied Logic*, 9(1):61–80, 2011.
[23] C. Straßer and A. Antonelli. Non-monotonic logic. In *The Stanford Encyclopedia of Philosophy*. Stanford University, 2016.
[24] C. Straßer and O. Arieli. Sequent-based argumentation for normative reasoning. In *DEON 2014*, volume 8554 of *LNCS*, pages 224–240. Springer, 2014.
[25] J. Taber. Is Indian logic nonmonotonic? *Philosophy East and West*, 54:143–170, 2004.
[26] A. S. Troelstra and H. Schwichtenberg. *Basic Proof Theory*. Cambridge University Press, 2nd edition, 2000.
[27] L. W. N. van der Torre. Violated obligations in a defeasible deontic logic. In *ECAI 94*, pages 371–375. Wiley, 1994.
[28] L. W. N. van der Torre and Y. Tan. The many faces of defeasibility in defeasible deontic logic. In *Defeasible Deontic Logic*, pages 79–121. Kluwer, 1997.

A Look at Chisholm's Ethics of Requirement

MARIAN J. R. GILTON
University of California, Irvine

The aim of this paper is to informally study Chisholm's Ethics of Requirement (EOR) by considering what resources it has to offer for thinking about several of the paradoxes of Standard Deontic Logic It is shown that EOR evades Chisholm's paradox in a novel way, and that it escapes additional problems concerning moral conflict given in Sartre's dilemma and Plato's dilemma. EOR is also able to avoid Prior's paradox of the good Samaritan.

1 Introduction

Chisholm's work in deontic logic has been profoundly influential. There is a vast literature surrounding the paradox that bears his name, and the basic concepts of his formal system, called 'The Ethics of Requirement' (EOR), are a well-known locus of the notion of defeat in moral reasoning.[1] And yet, EOR has received little study as its own formal system. Given its historical importance and conceptual influence, it deserves further attention. Moreover, Chisholm's notion of defeat has also been deeply influential in epistemology. In his famous *Theory of Knowledge* ([5]), he cites [3] and [4], his two main works on EOR, as giving a thorough and more general account of what he means by 'defeat.' It is therefore of wider interest, too, to study EOR as a formalization of this philosophically significant notion of defeat.

The aim of this paper is to informally study EOR by considering what resources it has to offer for thinking about several of the paradoxes of Standard Deontic Logic (SDL). In section 2 we consider how EOR addresses various kinds of moral conflict. It is shown that EOR evades Chisholm's paradox in a novel way, and that it escapes additional problems concerning moral conflict given in Sartre's dilemma and Plato's dilemma. EOR is also able to avoid Prior's paradox of the Good Samaritan. While there are other paradoxes to consider, EOR's ability to evade those considered here is a promising start. The remainder of this introductory section presents the basic definitions and axioms of EOR.

I am grateful to three anonymous reviewers for insightful comments and suggestions, and to Thomas Gilton for careful reading and editing.

[1] Chisholm first put forward EOR in [3] and further developed it in [4] and [6]. Its influence can be seen especially in [10] and in [9].

The foundational difference between SDL and EOR is that the former takes *obligation* as its primitive starting point, whereas EOR defines obligation in terms of a relation of *requirement*. Requirement is a relation between states of affairs. That the state p requires the state q is symbolized as $p\text{R}q$. The intended meaning of Chisholm's idea of requirement is quite broad; that $p\text{R}q$ is intended to capture the idea that p calls for q, or that q is fitting to p. For example, promise-making requires promise-keeping, and wrongdoing requires repentance and making amends. Chisholm gives other helpful examples from aesthetic considerations: "the dominant seventh requires the chord of the tonic; one color in the lower left calls for a complementary color in the upper right" ([3] p. 147). To my mind, the most intuitive examples are culinary ones: a piece of sweet, rich chocolate cake really *calls for* an accompanying cup of hot, bitter coffee. Many words ending in "worthy" or "able" suggest other instances of requirement: a praise*worthy* action is one that calls for it being praised, while a despic*able* one requires that it be despised.[2]

In EOR, the definition of obligation in terms of this requirement relation relies upon the fact that requirements may be overridden:

Definition 1. There is a requirement for q which is *overridden* iff $\exists p$, $\exists r$, such that $[(p \wedge p\text{R}q) \wedge (r \wedge \neg((p \wedge r)\text{R}q))]$.

A state of affairs is obligatory if there is a requirement for it which has not been overridden:

Definition 2. Oq iff $\exists p \, \neg \exists r \, [(p \wedge p\text{R}q) \wedge (r \wedge \neg((p \wedge r)\text{R}q))]$.

However, rewriting this definition as $\exists p \, \forall r \, \neg[(p \wedge p\text{R}q) \wedge (r \wedge \neg((p \wedge r)\text{R}q))]$ reveals that it suffers from a couple of shortcomings.[3] First, if p requires q but p does not obtain, it follows that q is obligatory. Similarly, if p is anything at all that obtains but does not require q, that is, if $p \wedge \neg(p\text{R}q)$, then q is again obligatory. These are clearly consequences that Chisholm would not have wanted.

We can avoid these issues by amending the definition slightly:

Definition 2'. Oq iff $[\exists p \, (p \wedge p\text{R}q) \wedge \neg \exists r \, (r \wedge \neg((p \wedge r)\text{R}q))]$.

We note here a few key differences between SDL and EOR. First, whereas obligations in SDL are fixed, in EOR they are contingent upon what is required by the conjunction of all states of affairs which happen to obtain at any given point in time.[4] Second, it is a well-known theorem of SDL that every action is either obligatory, impermissible, or optional. But in EOR, one can distinguish further between states

---

[2] See [1] chapter 7 for a thorough discussion of various more specific notions of requirement.

[3] I am grateful to a reviewer for pointing out the full extent of the trouble here.

[4] And yet, a notion of time is not built into the system. As we will see in section 2, redeveloping EOR by integrating a temporal logic may be necessary in order to allow for formalizing instances of unmet obligations.

of affairs that are *optional* and those that are *indifferent*; within the category of the optional we can further define *supererogatory* and *infravetatory* states of affairs. As first raised by [16], SDL cannot accommodate these distinctions.

Absent from the original EOR paper, the following axioms for the requirement relation first appeared in [4].

(AR1) $pRq \to \exists x, \exists y\ xRy$.

(AR2) $pRq \to \Box(pRq)$.

(AR3) $pRq \to \neg\Box[\neg(p \wedge q)]$, (or equivalently, $pRq \to \Diamond(p \wedge q)$).

(AR4) $\exists x, \exists y, \exists z\ [\neg\Box\neg(x \wedge y) \wedge (xRz) \wedge (yR\neg z)]$.

(AR5) $(pRs \wedge qRs) \to (p \vee q)Rs$.

(AR6) $(pRq \wedge pRs) \to pR(q \wedge s)$.

(AR7) $(p \vee q)Rs \to (pRs \vee qRs)$.

Most of these axioms are straightforward, but we first note a peculiar consequence of AR3. Suppose that some q is obligatory, as determined by definition 2′. Then there is some p that obtains and that requires q, and this requirement is not overridden. From this and AR3, it follows that q must obtain.[5] To see this, suppose that $\neg q$ is true. Then it must be that $(p \wedge \neg q)Rq$, since p's requirement for q is undefeated. But then AR3 would have it that $\Diamond(p \wedge q \wedge \neg q)$, which is not true.

This may be good reason to reject AR3. The axiom seems to have been intended to capture the idea that it should be possible for a requirement to be met. But AR3 as it stands may not be the best way to capture this idea, if we grant that there are things that could require their own negation: each uncompleted piece of the jigsaw puzzle calls for its own completion, and a grave injustice calls for being put right.

We note also that AR4 is intended to capture a notion of genuine moral conflict while allowing for the possibility that all such conflicts may be resolved. The key is that the conflict occurs at the level of requirements but not at the level of obligations. In SDL, one is forced to consider moral conflict in terms of it being obligatory that p and it being obligatory that $\neg p$. This leads to a number of well-known problems, some of which will be discussed below. EOR instead says that, in an instance of moral conflict, there is a requirement for p and there is *some other* requirement for $\neg p$. If one of these requirements overrides the other, then there is no conflict of obligations. Note, also, the 'agglomeration' or 'aggregation' principle contained in AR6: again, we are at the level of requirements rather than obligations, so it need not be the case that $(Op \wedge Oq) \to O(p \wedge q)$. Agglomeration of obligations is known to cause other problems in SDL.[6]

---

[5]I'm grateful to a reviewer for pointing this out.
[6]See especially [17].

2 Three puzzles

In this section, we consider three prominent problems for SDL and what conceptual resources EOR has for addressing these problems. We consider first the contrary-to-duty paradox, or Chisholm's paradox, and the issues it raises concerning conditional obligation. We then consider various forms of moral conflict, followed by a look at the Good Samaritan paradox.

2.1 The contrary-to-duty paradox

Chisholm's famous example is given by the following four statements:[7]

1. It ought to be that a certain man go to the assistance of his neighbors.
2. It ought to be that if he does go he tell them he is coming.
3. If he does not go then he ought not to tell them he is coming.
4. He does not go.

The paradox arises because these four statements seem to be mutually consistent, and yet when formalized in SDL, they lead to a contradiction:

1. Og
2. $O(t/g)$
3. $O(\neg t/\neg g)$
4. $\neg g$

where $O(t/g)$ stands for the conditional obligation, that is, 'Ot provided that g.' From 1. and 2. we derive Ot, and from 3. and 4. we derive $O\neg t$. But it is an axiom of SDL that $Op \rightarrow \neg O \neg p$, which in this case gives us $\neg O \neg t$. So we have both $O\neg t$ and $\neg O \neg t$. Since the situation described in 1.-4. seems plausible—perhaps even familiar to our own experience—the challenge is to give a consistent formalization of these contrary-to-duty imperatives.

Much, though not all, of the literature on Chisholm's paradox concerns what sort of detachment rule should be employed for the conditional obligations given in 2. and 3.[8] At least part of the problem seems to be that the conclusion drawn from 2. and 3. arise in different ways. We have the following two detachment rules:

---

[7]First given in [2] p. 34-35.
[8]See especially [15] for an interesting rendition of a Chisholm-type paradox which, at the heart of the matter, is not at all concerned with conditional obligation.

Factive Detachment $(p \wedge O(q/p)) \to Oq$.

Deontic Detachment $(Op \wedge O(q/p)) \to Oq$.

It is factive detachment that allows one to conclude $O\neg t$ from 3. and 4., while it is deontic detachment that provides the conclusion of Ot for from 1. and 2. If we have good reason to accept one detachment rule while rejecting the other, then we escape the paradox.[9]

EOR provides a different sort of solution to the puzzle since the notion of conditional obligation is replaced with the requirement relation. First, we formalize the Chisholm quartet in EOR using the technical language of requirements rather than obligations. The first line is most naturally thought of as asserting that there is some state of affairs p such that pRg and p obtains, and the second line is to be formalized similarly.[10] The statements themselves give no indication as to what state of affairs p might be the one that is doing the requiring. Presumably, for 1. it is some general fact about the relationship of neighbors that makes it fitting for one neighbor to go to the assistance of another. And for 2., it might be some general fact about the act of making a visit that makes it fitting for the visitor to give the other person the courtesy of advanced notice. The third line is given by $\neg gR\neg t$. So we have

1.′ pRg

2.′ $qR(g \to t)$

3.′ $\neg gR\neg t$

4.′ $\neg g$

What can we conclude from these premises? Presumably, we should also include as premises that both p and q obtain. So the final question to ask is: what is required by $(p \wedge q \wedge \neg g)$? Since Jones has in fact decided not to go, it seems most fitting that he not tell his neighbor that he is coming. Thus we avoid any conflict between a requirement for t and one for $\neg t$. This is all true despite the fact that p's requirement for Jones to go is still in the background: it is still true that there is a requirement for Jones to go. In this sense, we can still say that it would be better if Jones changed his mind and instead actually went to the assistance of his neighbors. But it is precisely his failure to meet p's requirement for g which has generated the requirement for $\neg t$.

---

[9] See [7]. See also [8] p. 113 ff. for discussion and a review of the literature.

[10] The reader will note that, while Chisholm's original formulation in natural language uses the word "ought", this does not mean that we must formalize these statements using the notion of obligation given in definition 2. We instead first formalize things in terms of requirement relations, which hold necessarily, and then consider all of the contingent facts (crucially, here, that $\neg g$ obtains) to determine which, if any, of these requirements result in an actual obligation.

2.2 Dilemmas from Sartre and Plato

Conflicting obligations can be either explicit and direct, or implicit and indirect. The former case is known as Sartre's dilemma, where the moral agent in question finds herself obligated both to bring about p and to not bring about p. [11] gives the following example:

1. It is obligatory that I now meet Jones (say, as promised to Jones, my friend).

2. It is obligatory that I now do not meet Jones (say, as promised to Smith, another friend).

If such a situation seems implausible, imagine a moral agent who is particularly susceptible to peer pressure and people pleasing. Thus, we can imagine that she frequently finds herself expressing the same likes and dislikes as those she happens to be speaking with at the moment. She considers whether or not to attend a party that Jones is hosting. In speaking about it with Jones, she promises to attend. But in speaking about it with Smith, who thinks ill of Jones and all his parties, she finds herself agreeing with Smith and says she will not go.

We formalize this as:

1. Oj

2. $O \neg j$

In SDL, we can derive from 1 that $\neg O \neg j$, which then explicitly contradicts 2. The puzzle is that this situation seems to be quite possible in ordinary life, so we want to be able to formalize it consistently.

On the other hand, there are many scenarios which lead to implicit and indirect conflicts of obligations, and these are known as instances of Plato's dilemma. Suppose that, instead of explicitly promising contradictory actions, you have simply double-booked yourself, having promised to be at one place at noon and to be across town for a different appointment, also at noon. Because, like all mere mortals you lack the ability to bi-locate, you cannot possibly fulfill both promises, and so fulfilling one will result in your failing to fulfill the other. Thus, you have implicitly conflicting obligations. And this resulted from your oversight in managing your calendar, and so your predicament is your own fault.

Or suppose, by no fault of your own, you are similarly pulled in two different directions. In this vein, [11] offers the following scenario:

1. I'm obligated to meet you for a light lunch at noon.

2. I'm obligated to rush my choking child to the hospital at noon.

Here you could not have foreseen the conflicting commitments at noon, and so you are not at fault. Moreover, in this case the obligation to rush to the care of your child is clearly much stronger than the obligation to keep a lunch date.

Using the language of EOR, the issue for both dilemmas seems to be of the form:

1. $p \wedge p\text{R}q$,

2. $r \wedge r\text{R}s$,

3. $\neg \Diamond (q \wedge s)$.

In the case where s is $\neg q$, we recover Sartre's dilemma.

There are a few different issues at work in situations of this form. First, the conflict may be a strict logical impossibility, as in the case of Sartre's dilemma, or it may be that the real conflict is a matter of practical impossibility given the circumstances. In Plato's dilemma, one could easily fulfill both obligations if only the times were different.

Second, there are the relative strengths of the conflicting obligations. It may be that one obligation is stronger than the other, in which case the resolution is simply for the moral agent to follow through on the overriding obligation while failing to follow through on the lesser obligation. EOR is straightforwardly designed to handle these types of situations.

But suppose instead that the two conflicting obligations are equally strong. What then? Thinking first of Sartre's dilemma, we might say that the promise to Jones and the promise to Smith are of equal weight. This may be the case if, say, our moral agent's level of friendship to each is sufficiently similar that her promises to each 'pull' her equally in opposite directions. Such equal and opposite forces could also be at work in the implicit conflict of an instance of Plato's dilemma, supposing that the appointments in your calendar are of equal importance. In this case, the two requirements must be mutually defeating. It cannot be that $(p \wedge r)\text{R}q$, and it cannot be that $(p \wedge r)\text{R}s$. If both of these requirements were true, then by A6 we could conclude that $(p \wedge r)\text{R}(q \wedge s)$, which, together with AR3, contradicts line 3.

So what *do* the combined circumstances of $(p \wedge r)$ require? It would be nice to have a formal method for determining the requirement relations between states of affairs in order to always have a principled answer to the question. But in the absence of such a method at present, we can still sketch some rough answers to the question, What is most fitting, given that $(p \wedge r)$? Or better yet, What is most fitting given that, not just $(p \wedge r)$, but also given that, separately, these two states of affairs would require conflicting actions? By using the framework of requirements, we find that the best moral advice to give our agent might not actually be among the initial options of q or s. Where at first it seemed like she would have to choose between conflicting obligations, now it seems that she has to ask herself a broader question: given that I am under equal and opposite requirements, what is the most fitting course of action? In some cases it may be most fitting to arbitrarily choose between q and s. But in general the answer may be something entirely new. In the case of our peer pressured agent, the most important thing for her to do might be to reflect on how she came to be in such a bind in the first place, and thereby come to see her need for moral growth. The situation calls for her to exercise virtuous autonomy in order to no longer make promises based on what she thinks her friends want to hear. Whether or not she actually goes to the party is not her most pressing concern.

Chisholm's analysis of the contrary-to-duty paradox provides some insight for these cases of moral conflict as well:

> [W]e are required to consider the familiar duties associated with blame, confession, restoration, reparation, punishment, repentance, and remedial justice, in order to be able to answer the question: 'I have done something I should not have done–so what should I do now?'... For most of us need a way of deciding, not only what we ought to do, but also what we ought to do after we fail to do some of the things we ought to do. ([2] p. 35-36)

When a moral conflict arises, either as a result of a moral agent's genuine failure or from circumstances beyond her control, the way forward may be one of repentance and confession, or of simply back-tracking and charting a new course. If I have double-booked myself, or made promises which I should not have made, then I should let the relevant parties know of my mistake, and ask for their forgiveness and perhaps their help in sorting out the situation. The resolution of a conflict between performing one action or another may best be given by a third option.

2.3 The Good Samaritan paradox

Consider the following two statements:

1. It ought to be the case that Jones helps Smith who has been robbed.

2. It ought to be the case that Smith has been robbed.

Statement 1 seems morally correct, while 2 seems obviously objectionable[11]. But from formalizing 1 in SDL we can derive 2. This is because the only way to formalize 1 seems to be $O(h \wedge r)$. Moreover, $(h \wedge r) \to r$, and in SDL we can conclude from this that $O(h \wedge r) \to Or$. This is the good Samaritan paradox: an obligation to help the robbed implies an obligation for there to be a robbery.[12]

---

[11]Excepting, perhaps, situations in which Smith is the evil Prince John and it ought to be that he is robbed by our hero Robin Hood.

[12]This was first given by [13] and subsequently discussed by [12] as well as [14].

The root of the trouble seems to be that, in order for it to be obligatory that the robbed person receives assistance, the robbery must have in fact taken place. Jones is not obligated to help Smith if Smith is not in fact in need of help. How do we account for this sense in which the robbery is a necessary part of the situation which leads to the obligation? In EOR, we can put the fact of the robbery r on the left hand side of the requirement relations: given the fact of r (and other relevant facts about, perhaps, Jone's ability to help) it is fitting that h. So we have $r \wedge rRh$, and this is why—assuming nothing else obtains that overrides this requirement—Jones ought to help Smith. Formalizing the situation in EOR thus escapes the problem of concluding that it ought to be that Smith is robbed. The robbery is part of the state of affairs that *does* the requiring. As such it is no part of that which is *required*, and so it is not obligated.[13]

3 Conclusion

We have seen that EOR is able to avoid several of the well known paradoxes of SDL. The two levels of moral commitment given by requirements and obligations allow for the simultaneous recognition of a genuine moral conflict and a principled resolution to the conflict. While there may be conflicting requirements, this does not result in conflicting obligations. Either one requirement will decidedly override the other and become an obligation, or the two requirements will be mutually defeating and may generate an obligation for something entirely new.

There are of course many other paradoxes of SDL, to which it may turn out EOR is also susceptible. In addition, it may be that EOR has new and interesting paradoxes of its own. But the provisional success of EOR in evading the paradoxes and puzzles considered here shows that it is a viable alternative to SDL deserving of further study. Such further study could begin with a demonstration of the consistency of the axioms AR1-AR6. In addition, it would be worth considering what further revisions to the definition of defeat would allow for Op to obtain while p does not obtain, perhaps by incorporating a temporal logic to indicate not only what states of affairs obtain right now, but which might obtain in the future. Further work could consider the extent to which Chisholm's notion of a requirement coincides with various notions of a reason, such as that given in [9].

References

[1] John Broome. *Rationality through reasoning*. John Wiley & Sons, 2013.

---

[13] While this account is, I think, a reason to prefer EOR to SDL, both systems are in need of augmentation in order to properly express the idea that it is one and the same Smith who is in need of the help formalized by h and subject to the robbery given by r. And this shortcoming shared by the two formalisms seems to be at the heart of the matter with this puzzle.

[2] Roderick Chisholm. Contrary-to-duty imperatives and deontic logic. *Analysis*, 24(2):33–36, 1963.
[3] Roderick M. Chisholm. The ethics of requirement. *American Philosophical Quarterly*, 1(2):147–153, 1964.
[4] Roderick M. Chisholm. Practical reason and the logic of requirement. In S. Korner, editor, *Practical Reason*, pages 1–17. Basil Blackwell, Oxford, 1974.
[5] Roderick M. Chisholm. *Theory of Knowledge*. Prentice Hall, Englewood Cliffs, third edition, 1989.
[6] Roderick M. Chisholm. Knowldge and the challenge of the skeptic. In John. R. White, editor, *Ethics and Intrinsic Value*, pages 3–23. Universitatsverlag C. Winter, Heidelberg, 2001.
[7] P. S. Greenspan. Conditional Oughts and Hypothetical Imperatives. *Journal of Philosophy*, 72(10):259–276, 1975.
[8] Risto Hilpinen and Paul McNamara. Deontic Logic: A Historical Survey and Introduction. In Dov Gabbay, John Horty, Xavier Parent, Ron van der Meyden, and Leendert van der Torre, editors, *Handbook of Deontic Logic and Normative Systems*, pages 3–136. College Publications, London, 2013.
[9] John F Horty. *Reasons as Defaults*. Oxford University Press, 2012.
[10] Barry Loewer and Marvin Belzer. Dyadic deontic detachment. *Synthese*, 54(2):295–318, 1983.
[11] Paul McNamara. Deontic logic. In Edward N. Zalta, editor, *The Stanford Encyclopedia of Philosophy*. Metaphysics Research Lab, Stanford University, winter 2014 edition, 2014.
[12] P. H. Nowell-Smith and E. J. Lemmon. Escapism: The logical basis of ethics. *Mind*, 69(275):289–300, 1960.
[13] A. N. Prior. Escapism: The logical basis of ethics. In A. I. Melden, editor, *Journal of Symbolic Logic*, pages 610–611. University of Washington Press, 1958.
[14] Lennart Åqvist. Good samaritans, contrary-to-duty imperatives, and epistemic obligations. *Noûs*, 1(4):361–379, 1967.
[15] Catharine Saint-Croix and Richmond H. Thomason. Chisholm's paradox and conditional oughts. *Journal of Logic and Computation*, page exw003, 2016.
[16] J. O. Urmson. Saints and Heroes. In A. I. Melden, editor, *Essays in Moral Philosophy*. University of Washington Press, 1958.
[17] Bas C. van Fraassen. Values and the Heart's Command. *The Journal of Philosophy*, 70(1):5–19, 1973.

Ability and Responsibility in General Action Logic

ALESSANDRO GIORDANI
Department of Philosophy, Catholic University of Milan
alessandro.giordani@unicatt.it

In the last two decades logicians from different backgrounds have focused on the characterization of the notions of ability and responsibility. In the present paper I build on these lines of research and develop a system of modal logic of action which involves elements from both dynamic action logic and *stit* logic. The main advantage of the present system lies in the possibility of analysing the fact that an agent brings about a certain state of affairs in two distinct components: the fact that the agent performs a specific basic action and the fact that a state of affairs is a consequence of the performed action. This kind of analysis allows us to introduce a novel account of the notions of epistemic ability and knowingly doing and a comprehensive conceptual framework for classifying different levels of responsibility.

Keywords: action logic; responsibility; intentionality; knowingly doing; ability.

1 Opening

Responsibility is a central concept in both moral and legal theory, since our judgements concerning the moral or legal consequences of the conduct of an agent crucially depend on the fact that she can be held responsible for having performed an action or having produced a state of affairs. So, when is an agent responsible for her actions? Suppose you are doing a multiple choice test. You look at the questions, identify the problems to be solved, and select the answers you consider to be correct. In a case like this, in standard circumstances, you are able to select any of the possible answers, you know that you are selecting an answer you consider to be correct, and you check the corresponding box intentionally, with the aim of providing a correct answer. Thus, in accordance with the purpose of the examiner, you can be considered fully responsible for the outcome. Generalizing, we may assume that an agent can be considered fully responsible for ϕ when she has produced ϕ by having intentionally activated her ability to do an action producing ϕ, i.e. by having chosen to perform an action α that she is able to perform and that has ϕ as

---

I would like to thank the referees of DEON 2018 for helpful comments and for pushing me to further develop the analysis of the notion of causal connection.

one of its consequences, knowing that ϕ is one of its consequences, and aiming at achieving ϕ by doing that action type.[1]

The elements of responsibility According to the previous analysis, a basic characterization of the notion of full responsibility involves the following elements.

ELEMENTS	relative to action α	relative to state of affairs that ϕ
actuality	a is doing α	a is producing ϕ via α
ability	a is able to do α	a is able to produce ϕ via α
avoidability	a could avoid doing α	a could avoid producing ϕ via α
intentionality	a is intentionally doing α	a is intentionally producing ϕ
awareness	a is aware of doing α	a is aware of producing ϕ

Before going on, let me motivate the choice of including action types and distinguishing between actions and consequences of actions within the present framework. As to the first point, considering action types is essential for assessing a responsible conduct, since the agent's conduct can be assessed only when it is identified as an instance of a general type. Thus, in criminal law, offenses consisting in actions are defined by reference to action types, the crimes, so that a certain conduct can be punished only if it is recognized as a crime in the first place (see [9] ch.5). As to the second point, the distinction between actions and consequences of actions is central for a general theory of responsibility (see [18], ch.3). For instance, in order to understand the notion of negligence, it is necessary to introduce a distinction between crimes of harmful consequences and crimes of harmful actions: crimes of harmful actions cannot occur negligently, since, due to the logical dependence of harm on the action, there is no causal process that can go awry, while a negligent conduct is possible only where there is a conceptual gap between the action and the harmful consequence (see [9] ch.7).

The logic of responsibility In the light of this, what I want to develop is a logical framework rich enough to give us the possibility:

1. to distinguish actions and consequences
2. to identify actual actions and actual consequences
3. to identify possible actions and possible consequences
4. to identify intentional actions and intentional consequences
5. to identify the ability to perform actions and to produce consequences

In this paper, I will sharply distinguish the agent's conduct from the actions the agent performs. This distinction is to be intended as follows: the agent's conduct is

---

[1]See [1], §2.01: (1) A person is not guilty of an offense unless his liability is based on conduct that includes a voluntary act or the omission to perform an act of which he is physically capable.

what the agent is actually doing in accordance with her choice, i.e., an actual event, while the actions the agent performs are the action types under which her conduct can be classified. Hence, we can think of the agent's conduct as a bunch of action tokens corresponding to the types under which the conduct can be classified. Thus, your actual conduct during a multiple choice test can be classified in different ways, and so one can say that you are crossing a certain box, that you are answering a certain question or, in general, that you are doing a test. In addition, I will focus on the analysis of ability and responsibility with respect to a single agent, the extension of the framework to a multi-agent system being left for future work.

The paper is structured as follows. In section 2, I introduce and discuss a novel system of general action logic. In section 3, I propose a specific account of the notions of epistemic ability and knowingly doing and provide a comprehensive conceptual framework for classifying different levels of responsibility. In doing that I will briefly compare the present system with some of the most interesting systems of action logic currently available. In section 4, I conclude and point to some developments.

2 Introducing General Action Logic

The language $\mathcal{L}_{GAL}$ of the system GAL of *general action logic* is based on a countable set $\{p_i\}_{i \in \mathbb{N}}$ of variables for propositions and a countable set $\{x_i\}_{i \in \mathbb{N}}$ of variables for basic actions. The only agent involved is a.

The sets $Tm(\mathcal{L}_{GAL})$ of terms of $\mathcal{L}_{GAL}$ is defined as follows:

$$\alpha ::= x_i \mid \overline{\alpha} \mid \alpha \sqcup \alpha \mid \alpha \sqcap \alpha \mid 1$$

The sets $Fm(\mathcal{L}_{GAL})$ formulas of $\mathcal{L}_{GAL}$ is defined as follows:

$$\phi ::= p_i \mid \neg \phi \mid \phi \wedge \phi \mid done_a(\alpha) \mid \Box_a \phi \mid [a]\phi \mid \mathbb{I}_a \phi \mid \mathbb{J}_a \phi, \text{ where } \alpha \in Tm(\mathcal{L}_{GAL})$$

The intended interpretation of the modal formulas is as follows: (i) $done_a(\alpha)$ states that an action of type α has just be performed by a; (ii) $\Box_a \phi$ states that ϕ is a state of affairs that is fixed when a is acting, i.e. that all the actions a is able to perform can only occur in states where ϕ is the case; (iii) $[a]\phi$ states that ϕ is a consequence of the current conduct of a, i.e. that all the actions a is performing can only result in states where ϕ is the case; (iv) $\mathbb{I}_a \phi$ states that the agent has an intentional conduct aiming at ϕ, i.e. that what a intends to perform can only result in states where ϕ is the case; (v) $\mathbb{J}_a \phi$ states that the agent is justified in believing that ϕ is the case. The difference between current conduct, as represented by the operator $[a]$, and intentional conduct, as represented by the operator $\mathbb{I}_a$, is subtle but significant in the present framework: first, it is possible for a to be intentionally performing α even if her current conduct will not result in a state where α is performed, so that the current conduct of a is properly described as an *attempt* to perform α; conversely, it is possible for a to be performing α even if a has no intention to produce α, so that the current conduct of a is properly described as an *accidental* performance of α. This rich set of operators allows us to characterize in an intuitive way the elements of responsibility.

Definition 1. *Consequence in a state.*

Consequence in a state: $\Box \phi := \Box_a[a]\phi$

A consequence in a state is a state of affairs fixed by the current circumstances: it is something that necessarily follows from the way the world is at a certain stage, independently of the possible conducts of a.

Definition 2. *Consequence of an action.*

Consequence of an action: $[\alpha]_a \phi := \Box_a[a](done_a(\alpha) \to \phi)$

A consequence of an action α is a state of affairs fixed by the current circumstances together with the action: it is something that necessarily follows from the fact that α has been performed by a at a certain stage.

Definition 3. *Actual action and actual consequence.*

1. *Actual action:* $[a]done_a(\alpha)$
2. *Actual consequence via* α: $[a]done_a(\alpha) \land [\alpha]_a \phi$

An action is occurring if the agent's conduct can only result in states where the action is performed, while an actual consequence via α is simply a consequence of an action α that the agent is currently performing.

Definition 4. *Possible action and possible consequence.*

1. *Possible action:* $\Diamond_a[a]done_a(\alpha)$
2. *Consequence of a possible* α: $\Diamond_a[a]done_a(\alpha) \land [\alpha]_a \phi$

An action is possible if the agent is able to act in such a way that her conduct can only result in states where the action is performed, while a consequence of a possible action α is simply a consequence of an α which the agent is able to perform.

Definition 5. *Intentional action and intentional consequence.*

1. *Intentional action:* $\mathbb{I}_a done_a(\alpha)$
2. *Consequence of an intentional* α: $\mathbb{I}_a done_a(\alpha) \land [\alpha]_a \phi$

An action is intentional if the agent's intentional conduct can only result in states where the action is performed, while a consequence via α is intentional if it is a consequence of an intentional action α.

Definition 6. *Ability to act and ability to produce.*

1. *Ability to do* α: $\Diamond_a[a]done_a(\alpha)$
2. *Ability to produce* ϕ *via* α: $\Diamond_a[a]done_a(\alpha) \land [\alpha]_a \phi$

In the present framework, a possible action is an action the agent is able to perform and a possible consequence via α is simply a consequence of an action α which the agent is able to perform. Hence, we can define the ability to perform actions and to produce consequences via actions in terms of possible actions and possible consequences. In [15] the notion of causal ability, which is the notion defined here, is contrasted with the more interesting notion of epistemic ability. I will come back to such distinction in section 3.

Definition 7. *Knowledge.*

$\mathbb{K}_a \phi := \mathbb{J}_a \phi \wedge \phi$

The standard definition is adopted according to which a knows that ϕ precisely when a is justified in believing that ϕ and it is true that ϕ. Finally, note that the fact that a is aware of doing α can be identified with the fact that a knows that she is doing α. All the elements of responsibility come then to be definable in the present framework.

ELEMENTS	relative to an action α	relative to a state of affairs ϕ via α
actuality	$[a]done_a(\alpha)$	$[a]done_a(\alpha) \wedge [\alpha]_a \phi$
ability	$\Diamond_a [a]done_a(\alpha)$	$\Diamond_a [a]done_a(\alpha) \wedge [\alpha]_a \phi$
avoidability	$\Diamond_a [a]done_a(\overline{\alpha})$	$\Diamond_a [a]done_a(\alpha) \wedge [\alpha]_a \neg \phi$
intentionality	$\mathbb{I}_a done_a(\alpha)$	$\mathbb{I}_a done_a(\alpha) \wedge \mathbb{I}_a \phi \wedge \mathbb{J}_a [\alpha]_a \phi$
awareness	$\mathbb{J}_a [a]done_a(\alpha)$	$\mathbb{J}_a [a]done_a(\alpha) \wedge \mathbb{J}_a [\alpha]_a \phi$

The definitions of intentionality and awareness will be further discussed in section 3 in connection with the notion of responsibility.

2.1 Semantics

The system GAL is a development of action logic (see [6, 7, 11, 16, 17, 18]), which incorporates some key elements of *stit* logic (see [4, 5, 13, 14, 15, 20]). Thus, the agent's conduct is related to a set of transitions between states, which in turn are intended as stages of possible histories within a given transition system. In particular, GAL is based on three main ideas.

Firstly, in systems of action logic based on dynamic logic, action types are represented by sets of transitions, according to the idea that a set of α-transitions represents all the ways in which action type α can be dynamically accomplished in the current state. By contrast, in the present system action types are represented by sets of states, according to the idea that a set of α-states represents all the ways in which action type α is accomplished (see [6]). In this setting, α-transitions can be defined in terms of transitions landing to α-states, so that the descriptive power of standard action logic is preserved. Figure 1 depicts α and β transitions both as represented in standard action logic and as represented in terms of generic transitions ending in states where α and β have just been accomplished.

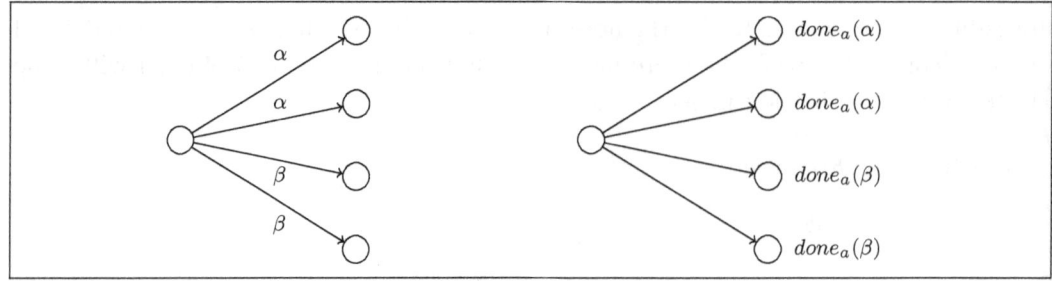

Figure 1: Transitions

Secondly, in *stit* logic the ability of an agent is represented in terms of the actions that the agent can choose, according to the idea that the agent's possible choices can be represented by a partition of the histories flowing out from a certain state. By contrast, in the present system the ability of an agent is represented in terms of the actions that the agent performs in alternative states, differing from the current one in terms of what the agent chooses to perform in them. The idea is then that an agent is able to perform a certain action precisely when there is an alternative state, consistent with the ability of the agent, where the agent is actually performing that action. In this setting choices can be explicitly defined in terms of chosen actions, so that the descriptive power of *stit* logic is not only preserved, but incremented in as much as, besides the consequences, also the chosen actions are explicitly taken into consideration. Figure 2 depicts an agent doing α, and able to do α and β, both as represented in *stit* and as represented in terms of alternative states. Here, the alternative states are the states that are accessible, given the agent's ability, and that can access, given the agent's conduct, only to β states.

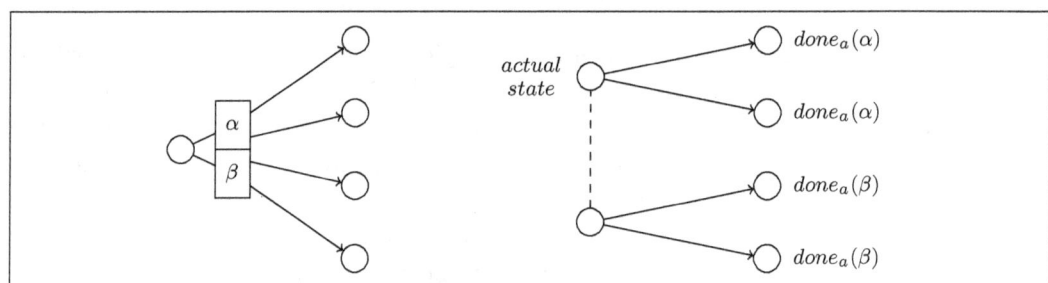

Figure 2: Actual and potential actions

Finally, in action logic what the agent is intentionally doing is not represented, while in *stit* logic what the agent is intentionally doing is determined at a state only relative to a certain history. By contrast, in the present system, for each state, the agent's intentional conduct is completely specified in terms of a function that returns the possible outcomes of such conduct. Thus, in the present setting, we are able to specify both what the agent is intentionally doing and the alternative courses of action compatible with her intentional conduct, together with the alternative courses of action she could have chosen and performed in virtue of her ability. Figure 3 represents an agent who is able to do both α and β, and is intentionally doing α instead of β, even if it is not necessary for the agent's intentional conduct to result in a state where α is done, since there are states accessible relative to the agent's conduct that are not α states.

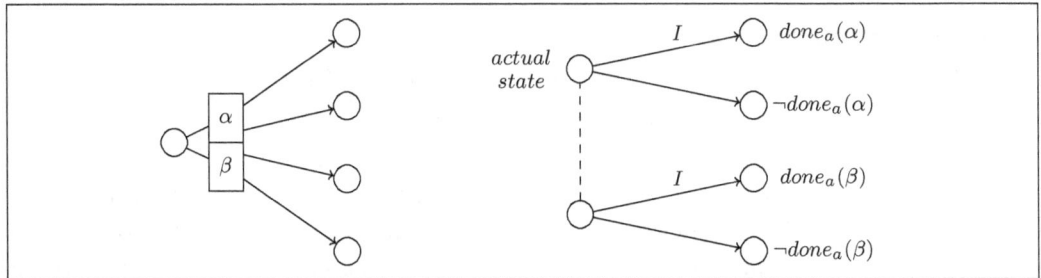

Figure 3: Actual and intentional actions

The framework here adopted is based on the following notion of frame, which is a development of the ontic frame proposed in [6]. The crucial novelty differentiating the present approach from the one adopted there is that the relation of one step accessibility is divided in two components:

(i) a relation A_a of possibility given the *current ability* of the agent, whose instances are the ones represented by dashed lines in the previous figures;

(ii) a relation R_a of one step accessibility given the *current conduct* of the agent, whose instances are the ones represented by arrows in the previous figures.

An important point to take into account in interpreting the present framework is that R_a is intended to model one step transitions, in conformity to the *xstit* approach developed in [4] and [5], so that the states R_a points to are to be considered as successive with respect to the states from which R_a starts. By contrast, A_a is intended to model connections between states that are accessible relative to the agent's ability, so that the states A_a points to are to be considered as simultaneous with respect to the state from which A_a starts. In fact, a state A_a points to is just a possible state in which the agent is performing a different conduct.

Definition 8. *Frame for $\mathcal{L}_{GAL}$.*

A frame for $\mathcal{L}_{GAL}$ is a tuple $\langle W, D_a, A_a, R_a, I_a, \mathcal{I}_a, J_a \rangle$, which can be viewed as the result of combining three specific frames:

1. a frame $\langle W, D_a, A_a, R_a \rangle$ for a logic of ability and actions;

2. a frame $\langle W, I_a, \mathcal{I}_a \rangle$ for a logic of intentional actions;
3. a frame $\langle W, J_a \rangle$ for a logic of justification.

$W \neq \emptyset$ is our set of states. The other elements occurring in a frame for $\mathcal{L}_{GAL}$ are assumed to satisfy the following conditions.

1. $D_a : Tm(\mathcal{L}_{GAL}) \to \wp(W)$ is the function that determines which actions have just been performed at a state. Thus, for each $\alpha \in Tm(\mathcal{L}_{GAL})$, $D_a(\alpha)$ is the set that contains a state precisely when α is an action that has just been performed at that state. It is assumed that D_a provides an intuitive connection between action operators and operators on sets of worlds.

Conditions on D_a:
(1) $D_a(1) = W$
(2) $D_a(\overline{\alpha}) = W - D_a(\alpha)$
(3) $D_a(\alpha \sqcup \beta) = D_a(\alpha) \cup D_a(\beta)$
(4) $D_a(\alpha \sqcap \beta) = D_a(\alpha) \cap D_a(\beta)$

As expected: (1) some action has been just done at any state, since the agent is assumed not to be inert; (2) performing the negation of α coincides with omitting α; (3) performing the disjunction of two actions coincides with doing one of the two actions; (4) performing the conjunction of two actions coincides with performing both the actions in parallel.

2. $A_a : W \to \wp(W)$ is the function returning the set of states that are accessible given the ability of a, and so $A_a(w)$ is the set of states where the conduct of a is one of the possible conducts a is able to select at w. It is assumed that a is able to select the current conduct and that a's ability is invariant across $A_a(w)$.

Conditions on A_a:
(1) for each $w \in W$, $w \in A_a(w)$
(2) for each $w \in W$, $v \in A_a(w) \Rightarrow A_a(v) = A_a(w)$

3. $R_a : W \to \wp(W)$ is the function returning the set of states that are accessible given the agent's conduct, and so $R_a(w)$ is the set of states that can results from the conduct of a at w. It is assumed that a is always acting in a certain way and that any consequence of the agent's conduct is settled at the next state.

Conditions on R_a:
for each $w \in W$, $R_a(w) \neq \emptyset$
for each $w, v \in W$, $v \in R_a(w) \Rightarrow A_a(v) \subseteq R_a(w)$

4. $I_a : W \to \wp(W)$ is the function returning the set of states that are accessible given the intentional aims of a, and so $I_a(w)$ is the set of states consistent with what a intends to achieve at w. It is assumed that what a is intentionally doing is consistent. In addition, what a intends to do at w is explicitly represented by the function $\mathcal{I}_a : W \to \wp(Fm(\mathcal{L}_{GAL}))$, so that $\mathcal{I}_a(w)$ is the set of formulas that encodes the intentions of a at w.

Conditions on I_a:
for each $w \in W$, $I_a(w) \neq \emptyset$

5. $J_a : W \to \wp(W)$ is the function returning the set of states that are accessible given the body of evidence of a, and so $J_a(w)$ is the set of states consistent with what a is justified to believe at w. It is assumed that what a is justified to believe is consistent, so that there is always at least a state in $J_a(w)$.

Conditions on J_a:
for each $w \in W$, $J_a(w) \neq \varnothing$

In the present context, justification is not assumed to be factive. An important consequence of this fact is that it is possible for an agent to be justified in believing that she is doing something even if she will fail to achieve what she intends to do.

Definition 9. *Model for $\mathcal{L}_{GAL}$.*

A model for $\mathcal{L}_{GAL}$ is a tuple $M = \langle W, D_a, A_a, R_a, I_a, \mathcal{I}_a, J_a, V \rangle$, where

1. $\langle W, R_a, A_a, D_a, I_a, \mathcal{I}_a, J_a \rangle$ is a frame for $\mathcal{L}_{GAL}$;
2. $V : \{p_i\}_{i \in \mathbb{N}} \to \wp(W)$ is a modal valuation.

As usual, V determines which elementary propositions are true at a state in W.

Definition 10. *Truth in a model for $\mathcal{L}_{GAL}$.*

The notion of truth is recursively defined as follows:

$M, w \models p_i \Leftrightarrow w \in V(p_i)$
$M, w \models \neg \phi \Leftrightarrow M, w \not\models \phi$
$M, w \models \phi \wedge \psi \Leftrightarrow M, w \models \phi$ and $M, w \models \psi$
$M, w \models done_a(\alpha) \Leftrightarrow w \in D_a(\alpha)$
$M, w \models \Box_a \phi \Leftrightarrow \forall v (v \in A_a(w) \Rightarrow M, v \models \phi)$
$M, w \models [a] \phi \Leftrightarrow \forall v (v \in R_a(w) \Rightarrow M, v \models \phi)$
$M, w \models \mathbb{I}_a \phi \Leftrightarrow \forall v (v \in I_a(w) \Rightarrow M, v \models \phi)$ and $\phi \in \mathcal{I}_a(w)$
$M, w \models \mathbb{J}_a \phi \Leftrightarrow \forall v (v \in J_a(w) \Rightarrow M, v \models \phi)$

It can be observed that not all models are appropriate for $\mathcal{L}_{GAL}$. In fact, in an appropriate model for $\mathcal{L}_{GAL}$, there should be a correspondence between I_a, which is the function representing the agent's intentions from the semantic point of view, and $\mathcal{I}_a$, which is the function that explicitly represents the agent's intentions. In particular, it seems intuitive to require that what is explicitly intended according to $\mathcal{I}_a$ be also implicitly intended according to I_a. It is also intuitive to require that the epistemic agent is justified in believing that she is actually performing what she is intending to perform. The following definition captures these requirements.

Definition 11. *Appropriate model for $\mathcal{L}_{GAL}$.*

A model for $\mathcal{L}_{GAL}$ is *appropriate* provided it satisfies:

A1: $\phi \in \mathcal{I}_a(w)$ and $v \in I_a(w) \Rightarrow M, v \models \phi$

A2: $\phi \in \mathcal{I}_a(w)$ and $v \in J_a(w) \Rightarrow M, v \models [a]\phi$

Finally, the notion of logical consequence in GAL is defined as usual, but with respect to the class of appropriate models. Hence, $X \Vdash_{GAL} \phi$ if and only if, for every *appropriate* model M and $w \in W$, $M, w \models X \Rightarrow M, w \models \phi$.

Hyperintensionality It is worth noting that, in contrast with most of the systems of logic of action currently available, GAL is hyperintensional with respect to the intentional modality. This trait of the system turns out to be extremely important in representing what an agent is responsible for. In fact, it is certainly impossible for ordinary agents to be aware of all the consequences of their actions or to intend to produce all the consequences of what they are intentionally producing. In addition, it even happens that agents do not intend to produce all the *known* consequences of what they intend to produce. In our setting these intuitions can be respected, since the intentional modality is explicit, in the sense of [2], and it can be proved that what an agent intend to do is not closed under known implications.

2.2 Axiomatization

The axiom system of GAL consists of the following groups of axioms and rules.

Group 1 (axioms for $\Box_a$):
$\Box$1: $\Box_a(\phi \to \psi) \to (\Box_a\phi \to \Box_a\psi)$
$\Box$2: $\Box_a\phi \to \phi$
$\Box$3: $\neg\Box_a\phi \to \Box_a\neg\Box_a\phi$
$\Box$R: $\phi/\Box_a\phi$

Group 2 (axioms for $done_a$):
D1: $done_a(1)$
D2: $done_a(\alpha) \leftrightarrow \neg done_a(\overline{\alpha})$
D3: $done_a(\alpha \sqcup \beta) \leftrightarrow done_a(\alpha) \vee done_a(\beta)$
D4: $done_a(\alpha \sqcap \beta) \leftrightarrow done_a(\alpha) \wedge done_a(\beta)$

Group 3 (axioms for $[a]$):
A1: $[a](\phi \to \psi) \to ([a]\phi \to [a]\psi)$
A2: $[a]\phi \to \neg[a]\neg\phi$
A3: $[a]\phi \to [a]\Box_a\phi$
AR: $\phi/[a]\phi$

Group 4 (axioms for $\mathbb{J}_a$ and $\mathbb{I}_a$):
J1: $\mathbb{J}_a(\phi \to \psi) \to (\mathbb{J}_a\phi \to \mathbb{J}_a\psi)$
J2: $\mathbb{J}_a\phi \to \neg\mathbb{J}_a\neg\phi$
J3: $\mathbb{I}_a\phi \to \mathbb{J}_a[a]\phi$
JR: $\phi/\mathbb{J}_a\phi$

The axioms reflects the semantic conditions on D_a, A_a, R_a, I_a, $\mathcal{I}_a$, J_a. The proof of the following theorem is omitted due to limitations of space.

Theorem 1. *The axiom system of GAL is sound and strongly complete with respect of the class of all appropriate models for $\mathcal{L}_{GAL}$.*

3 Interpreting Ability and Responsibility

As said above, GAL is a system of action logic incorporating fundamental intuitions codified in *stit* logic. In particular, being based on a distinction between actual conduct, actions and consequences of actions, GAL presents similarities with the systems of dynamic action logic proposed in [11].[2] In addition, being based on a transition semantics in a framework including action types, GAL is strictly connected with both the systems of $xstit$ logic proposed in [4, 5] and the systems of $stit$ logic with types proposed in [15]. In fact, as in $xstit$ logic, actions take effect in successive states and, as in $stit$ logic with types, actions are classified with reference to a set of corresponding action types.

3.1 Ability in Comparison with the $xstit$ Tradition

The idea, underlying $xstit$ logic, that an agent sees to it that ϕ when her choice is sufficient for achieving ϕ at the next state is here captured by $[a]\phi$, and so the notion of $xstit$ knowingly doing is definable as $\mathbb{K}_a[a]\phi$. Furthermore, a first notion of causal $xstit$-ability, intended as possibility of choosing a conduct leading to a state where ϕ is the case, is definable as $\Diamond_a[a]\phi$, while a notion of epistemic $xstit$-ability, intended as knowing how to produce a certain consequence via a certain action, can be defined as $\Diamond_a\mathbb{K}_a[a]\phi$.

Definition 12. *The xstit modalities.*

1. Knowingly doing: $\mathbb{K}_a[a]\phi$
2. Having the causal $xstit$-ability to do: $\Diamond_a[a]\phi$
3. Having the epistemic $xstit$-ability to do: $\Diamond_a\mathbb{K}_a[a]\phi$

However, as shown in [15], the concepts of causal $xstit$-ability and epistemic $xstit$-ability seem to be insufficient to model some interesting notions concerning what the agent is able to do. In a paradigmatic case proposed in [15], the agent lives in a world where a coin is on a table.

The agent has to bet whether the coin is heads up or tails up and wins if she guesses it right. Two scenarios are then presented as in figure 4.

These scenarios are such that:

w_1	and	w_2	are states where the coin is heads up;
w_3	and	w_4	are states where the coin is tails up;
v_1	and	v_3	are states where the agent has called heads;
v_2	and	v_4	are states where the agent has called tails.

---

[2]Unfortunately, it is not possible here to carry out an in-depth analysis of the similarities and the differences, as far as the expressive power is concerned, between GAL and the systems in [11]. A summary description of the peculiar traits of the present system is the following: the agent is not assumed to be performing a unique action at each state; the agent is not assumed to be successful in performing her intentional actions; a distinction is made between intentional and non-intentional conduct, intentional and non-intentional actions, and intentional and non-intentional consequences; the intentional operator is hyperintentional; a distinction is made between justification and knowledge.

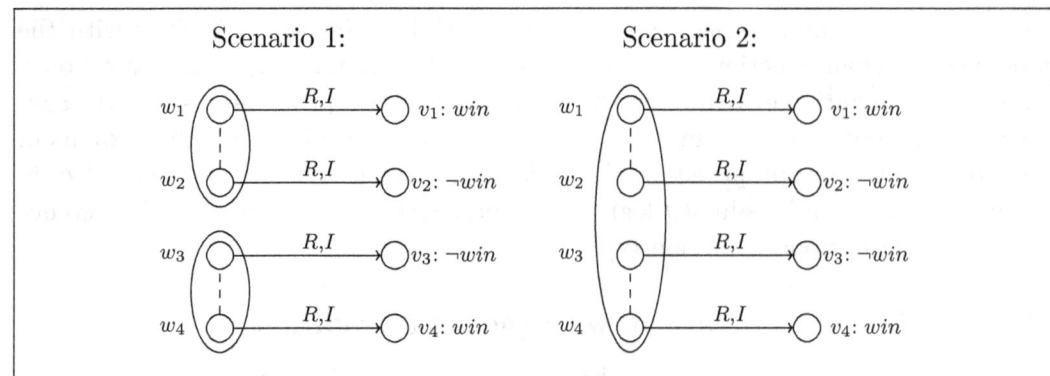

Figure 4: Scenarios

The ellipses enclose the states that are indistinguishable relative to the agent's knowledge. Hence, in the first scenario, the agent knows whether the coin has been placed heads up or tails up, while in the second one she lacks this piece of information. In this case, it seems intuitive to say that, in the first scenario, the agent knows how to win, while, in the second one, she lacks this ability. Still, it is not difficult to see that $\Diamond_a \mathbb{K}_a[a]win$ fails in both scenarios, so that we cannot use this formula to capture the notion of epistemic ability in the sense of knowing how to do something.

In the present framework this case presents no problem, since the notion of knowingly doing can be further analyzed and the notion of epistemic ability can be properly defined. In particular, due to the possibility of distinguishing actions and action consequences and to the availability of action types, we can discern the fact that the agent performs a specific action from the fact that a state of affairs is a consequence of the performed action, and so introduce the following refinement of the previous definitions.

Definition 13. *The analyzed modalities.*

1. Knowingly doing: $\mathbb{K}_a[a]done_a(\alpha) \wedge \mathbb{K}_a[\alpha]_a\phi$
2. Having the causal ability to do: $\Diamond_a[a]done_a(\alpha) \wedge [\alpha]_a\phi$
3. Having the epistemic ability to do: $\Diamond_a[a]done_a(\alpha) \wedge \mathbb{K}_a[\alpha]\phi$

As we can see, having the epistemic ability to produce ϕ coincides with having the ability to perform an action which is known to have ϕ as a consequence. This definition is a natural improvement of the *xstit*-definition and has many interesting connections with the solution proposed in [15], but for reasons of space a throughout study of these connections is left to another work.

3.2 The Notion of Responsibility

In our language, the basic concepts involved in the definition of the notion of responsibility are appropriately expressible. According to [9], three different inquiries are to be made before attributing responsibility.

1. The first inquiry is whether a certain action can be attributed to a particular agent. This inquiry aims at answering the question relative to whether $[a]done_a(\alpha)$ is the case.

2. The second inquiry is whether a certain harmful consequence can be attributed to a particular agent. This inquiry aims at answering the question relative to $[a]done_a(\alpha)$ and $[\alpha]\phi$.

3. The final inquiry is whether the action producing harm be attributed to the agent as a culpable action. This inquiry aims at answering the question relative to $[a]done_a(\alpha)$ and $[\alpha]\phi$ and to the levels of *mens rea*.

Hence, the first two inquiries aim to assess the objective elements of an action, while the final one is focused on the subjective elements characterizing the agent. In particular, since obtaining ϕ via α requires doing α and producing ϕ as a consequence of having done α, the subjective elements of an action are related to the performance of α, the causal connection between α and ϕ, and ϕ itself.

Definition 14. *Basic concepts involved in the notion of responsibility.*

Objective elements		Subjective elements	
successful action:	$[a]done_a(\alpha)$	intention to act:	$\mathbb{I}_a done_a(\alpha)$
causal connection:	$[\alpha]_a\phi$	causal awareness:	$\mathbb{J}_a[\alpha]_a\phi$
final consequence:	ϕ	intention to produce:	$\mathbb{I}_a\phi$

Thus, a first asset of the present framework is that all the objective and subjective concepts involved in attributing responsibility are representable. In particular, in terms of the basic elements described in section 2, *the conjunction of all the subjective elements characterizes intentionality*: $\mathbb{I}_a done_a(\alpha) \wedge \mathbb{I}_a\phi \wedge \mathbb{J}_a[\alpha]_a\phi$.

A second asset is that, using the previous concepts, it is possible to distinguish attempts, successful actions, and unsuccessful actions.

Definition 15. *Basic concepts involved in the notion of intentional action.*

1. attempt: $\mathbb{I}_a done_a(\alpha) \wedge \mathbb{I}_a\phi \wedge \mathbb{K}_a[\alpha]_a\phi$
2. successful attempt: $\mathbb{I}_a done_a(\alpha) \wedge \mathbb{I}_a\phi \wedge [a]done_a(\alpha) \wedge \mathbb{K}_a[\alpha]_a\phi$
3. unsuccessful attempt: $\mathbb{I}_a done_a(\alpha) \wedge \mathbb{I}_a\phi \wedge [a]\neg done_a(\alpha) \wedge \mathbb{K}_a[\alpha]_a\phi$

The notion of attempt to obtain a state of affairs is the generic one. In our setting, an attempt to obtain ϕ is given by the intention of performing an action that leads to ϕ, that is $\mathbb{I}_a done_a(\alpha) \wedge \mathbb{I}_a\phi$, plus the knowledge that the action is indeed a way to produce ϕ, that is $\mathbb{K}_a[\alpha]_a\phi$. Then, an attempt can be successful or unsuccessful depending on the fact that the intended action is going to be completed by the agent, $[a]done_a(\alpha)$, or not, $\neg[a]done_a(\alpha)$.

A final asset of the present framework is that, using the previous concepts, it is possible to provided a detailed interpretation of the basic levels of responsibility corresponding to the levels of culpability acknowledged in [1]. The Model Penal Code proposes four levels of culpability: purpose, knowledge, recklessness, and negligence, which are referred to as states of mind ascribable to the agent. The definitions of these states of mind, relative to the agent's conduct or a result thereof, can be summed up as follows:[3]

- A person acts *purposely* with respect to her conduct or a result thereof when it is her conscious object to engage in such conduct or to cause such a result;

- A person acts *knowingly* with respect to her conduct or a result thereof when she is aware of being engaged in such conduct or that her conduct will cause such a result;

- A person acts *recklessly* with respect to her conduct or a result thereof when she consciously disregards a substantial and unjustifiable risk that the material element exists or will result from his conduct;

- A person acts *negligently* with respect to her conduct or a result thereof when she should be aware of a substantial and unjustifiable risk that the material element exists or will result from his conduct.

Definition 16. *Basic distinctions concerning the levels of responsibility.*

1. knowingly producing ϕ:
$\mathbb{I}_a done_a(\alpha) \wedge [a]done_a(\alpha) \wedge \mathbb{K}_a[\alpha]_a\phi$

2. intentionally producing ϕ:
$\mathbb{I}_a done_a(\alpha) \wedge [a]done_a(\alpha) \wedge \mathbb{I}_a\phi \wedge \mathbb{K}_a[\alpha]_a\phi$

3. non-intentionally producing ϕ:
$\mathbb{I}_a done_a(\alpha) \wedge [a]done_a(\alpha) \wedge \neg(\mathbb{I}_a\phi \wedge \mathbb{K}_a[\alpha]_a\phi)$

4. precondition of negligently producing ϕ:
$\mathbb{I}_a done_a(\alpha) \wedge [a]done_a(\alpha) \wedge [\alpha]_a\phi \wedge \neg\mathbb{I}_a\phi \wedge \neg\mathbb{K}_a[\alpha]_a\phi$

A fully intentional production of a state of affairs by means of an action is a process where all the objective and subjective elements of responsibility occur. Thus,

---

[3]Note that *purposely* doing is assumed to coincide with *intentionally* doing. See [1], §2.02 for a full description of these states taking into account all the material elements involved in the offense resulting from the agent's conduct.

a fully intentional production of ϕ via α is a process defined by the conjunction of an intention to act, $\mathbb{I}_a done_a(\alpha)$, a successful action, $[a]done_a(\alpha)$, an intention to obtain the consequence by acting, $\mathbb{I}_a\phi$, a causal-like connection, $[\alpha]_a\phi$, and a causal knowledge, $\mathbb{K}_a[\alpha]_a\phi$. If there is no intention to obtain the consequence, or no knowledge, we have a non-intentional production of a state of affairs. Still, in both cases, the production is an instance of knowingly doing, which is the generic concept relative to intentional and non-intentional production. As highlighted in [8]:

> The distinction between purpose and knowledge, then, is that between conscious objective and awareness. It is important to get this distinction straight.

In our terms, the requirement of conscious objective is expressed as $\mathbb{I}_a[a]\phi$, which is the intention to obtain ϕ as a consequence of an action, while the requirement of awareness is expressed as $\mathbb{J}_a[\alpha]_a\phi$, which is implicit is $\mathbb{K}_a[\alpha]_a\phi$. Hence, the distinction between intentionally doing and knowingly doing is made precisely in terms of the fact that the consequence is a conscious objective for the agent. Furthermore, recklessly doing is intended as a kind of non-intentionally doing, that is as the non-intentional production of a practically certain effect, which implies disregarding a substantial and unjustifiable risk that the material element will result from the current conduct. Finally, it is then evident why distinguishing actions and consequences is necessary for introducing the concept of negligent production. In fact, in the absence of such distinction it is not possible to express that the agent is unaware of the fact that her action is causing a certain consequence.

3.3 The Notion of Causality

In characterizing the basic concepts involved in the notion of responsibility a very elementary definition of the concept of causal connection was provided. To be sure, the fact that ϕ is being caused by doing α was identified with the fact that ϕ is a consequence of α, i.e. that ϕ is something that necessarily follows from the fact that α has been performed at a certain stage: $[\alpha]_a\phi$. This definition is admittedly too rough to be assumed as an appropriate characterization of a causal connection, since it allows for the fact that any necessary state of affairs is produced by any action. As it is well-known, a suitable strategy to improve this account consists in adding to the *positive causal condition*, stating that the performance of an action is sufficient for ϕ to be the case, a *negative causal condition*, to the effect that the omission of that action would be sufficient for ϕ not to be the case (see [3, 12, 19]). In the present framework this strategy can be pursued in at least two different ways.

1. Exploiting the possibility of avoiding the specific action On the one hand, we can take as negative condition the avoidability of the performed action, when the performance of the action is both sufficient and necessary to obtain the consequence. Accordingly, an agent is said to be causing ϕ precisely when she is performing an action having ϕ as a consequence in a context where she could have done otherwise and, in doing so, obtain the negation of ϕ.

1. positive condition: $[a]done_a(\alpha) \wedge [\alpha]_a\phi$
2. negative condition: $\Diamond_a[a]done_a(\overline{\alpha}) \wedge [\overline{\alpha}]_a\neg\phi$

In this case, the negative condition is a way to state the *sine qua non* condition adopted as a criterion of ascription of crimes: ϕ is caused by an agent's action α only if ϕ would not have been the case if the agent had not performed α.[4] Still, this account seems to be problematic in so far as an action causing ϕ is to be both sufficient and necessary for ϕ to be the case, while in many cases an agent can produce ϕ by means of many different actions.

2. Exploiting the possibility of acting in a different way On the other hand, we can take as negative condition the possibility of adopting a line of conduct ensuring the negation of the consequence. Accordingly, an agent is said to be causing ϕ precisely when she is acting in such a way that ϕ is a consequence of her action in a context where she could have acted so as to prevent ϕ. Hence

1. positive condition: $[a]done_a(\alpha) \wedge [\alpha]_a\phi$
2. negative condition: $\Diamond_a[a]\neg\phi$

In this case, the negative condition is slightly more general and it is consistent with the intuition that it is not necessary for an action to be both a sufficient and a necessary condition for a consequence to be the case. In fact, what is crucial is not that the agent's *action* is necessary for ϕ to obtain, but that the agent's *activity* constitutes a necessary condition. In the light of this, we introduce the following definition of causing ϕ by doing α:

$C_a(\phi, \alpha) := [a]done_a(\alpha) \wedge [\alpha]_a\phi \wedge \Diamond_a[a]\neg\phi.$

Thus, an agent is causing ϕ via α precisely when she is doing an action, α, which is sufficient for ϕ in a context where she could have acted so as to prevent ϕ.

4 Conclusion

The main contribution of this paper consists in providing a definition of the notion of responsibility and a precise distinction of different levels of responsibility in a system that subsumes both standard action logic and *stit* logic. The system here proposed seems to allow us to model all the elements that are involved in the conduct of an agent in an intuitive and straightforward way. In systems of action logic developed

---

[4]See [1], §2.03: (1) Conduct is the cause of a result when: (a) it is an antecedent but for which the result in question would not have occurred.

in the *stit* tradition, we are typically able to express the fact that an agent can produce a certain effect by making a certain choice, but we are unable to express what kind of action causes a certain effect. The present system overcomes this problem by introducing formulas describing the fact that a state of affairs obtains as a consequence of the instantiation of a certain action type. In systems of action logic developed in the dynamic logic tradition, we are typically able to express what kind of action causes a certain effect, but we are not able to express the fact that an agent has indeed performed a certain action, or made a certain choice. The present system overcomes this problem by introducing formulas describing the fact that a state of affairs obtains as a consequence of the actual conduct of the agent. Furthermore, formulas describing the premeditated effects of the intentional conduct of the agent are introduced in order to distinguishing different ways in which an action can be performed and formulas describing what the agent is justified to believe are introduced in order to identifying what the agent is aware of. Finally, the intentional operator is hyperintensional, thus avoiding standard problems deriving from forms of logical omniscience. The resulting system has proved to be sufficiently strong to define different notions of ability, to distinguish between intentional and non-intentional actions, to distinguish between negligent and non-negligent actions, and to provide an interesting definition of causality, while being based on a semantics that allows for a simple axiomatization. In view of this high expressive power, a promising line of research consists in studying the possibility of interpreting and integrating the most significant systems of ability and action logic (see [10, 12]) in the present one. In a different direction, the possibility to extend its semantics so as to introduce dynamic procedures of model updates is also enticing.

References

[1] ALI. *Model Penal Code*. The American Law Institute, Philadelphia, 1984.

[2] Sergei Artemov. The logic of Justification. *The Review of Symbolic Logic*, 1:477–513, 2008.

[3] Nuel Belnap. Backwards and Forwards in the Modal Logic of Agency. *Philosophy and Phenomenological Research*, 51:777–807, 1991.

[4] Jan Broersen. Deontic Epistemic Stit Logic Distinguishing Modes of Mens Rea. *Journal of Applied Logic*, 9:127–152, 2011.

[5] Jan Broersen. Making a Start with the stit Logic Analysis of Intentional Action. *Journal of Philosophical Logic*, 40:499–530, 2011.

[6] Ilaria Canavotto and Alessandro Giordani. Enriching Deontic Logic. *Journal of Logic and Computation*, first online:1–23, 2018.

[7] Tiago De Lima, Lamber Royakkers, and Frank Dignum. A Logic for Reasoning about Responsibility. *Logic Journal of IGPL*, 18:99–117, 2010.

[8] Markus D. Dubber. *An Introduction to the Model Penal Code*. Oxford University Press, Oxford, 2015.

[9] George P. Fletcher. *Basic Concepts of Criminal Law*. Oxford University Press, Oxford, 1998.

[10] Andreas Herzig. Logics of knowledge and action: critical analysis and challenges. *Autonomous Agents and Multi-Agent Systems*, 29:719–753, 2015.

[11] Andreas Herzig and Emiliano Lorini. A Dynamic Logic of Agency I: STIT, Capabilities and Powers. *Journal of Logic, Language and Information*, 19:89–122, 2010.

[12] Risto Hilpinen. On Action and Agency. In E. Ejerhed and S. Lindström, editors, *Logic, Action and Cognition: Essays in Philosophical Logic*, pages 3–27. Kluwer, Dordrecht, 1997.

[13] John Horty. *Agency and Deontic Logic*. Oxford University Press, Oxford, 2001.

[14] John Horty and Nuel Belnap. The Deliberative Stit: A Study of Action, Omission, Ability, and Obligation. *Journal of Philosophical Logic*, 24:583–644, 1995.

[15] John Horty and Eric Pacuit. Action Types in Stit Semantics. *The Review of Symbolic Logic*, first online:1–21, 2017.

[16] Emiliano Lorini and Andreas Herzig. A logic of Intention and Attempt. *Synthese*, 163:45–77, 2008.

[17] John-Jules C. Meyer. A Different Approach to Deontic Logic: Deontic Logic Viewed as a Variant of Dynamic Logic. *Notre Dame Journal of Formal Logic*, 29:109–136, 1998.

[18] Georg H. von Wright. *Norm and Action: A Logical Inquiry*. Routledge and Kegan Paul, London, 1963.

[19] Georg H. von Wright. *Practical Reason: Philosophical Papers*, volume 1. Blackwell, Oxford, 1983.

[20] Ming Xu. Combinations of Stit with Ought and Know. *Journal of Philosophical Logic*, 44:851–877, 2015.

DIALOGUES ON MORAL THEORIES

GUIDO GOVERNATORI, FRANCESCO OLIVIERI, RÉGIS RIVERET
Data61, CSIRO, Australia

ANTONINO ROTOLO
University of Bologna, Italy

SERENA VILLATA
Université Côte d'Azur, I3S, Inria, CNRS, France

Most ethical systems define how the individuals ought morally act, being part of a society. The process of elicitation of a moral theory governing the agents in a society requires them to express their own norms with the aim to find a moral theory on which all may agree upon. We address this issue by proposing a formal framework that can instantiate in agents' dialogues moral/rational criteria, such as the *maximin principle*, *Pareto efficiency*, and *impartiality*, which were used, e.g., by John Rawls' theory or rule utilitarianism.

1 Introduction

Many conceptions of autonomy have been developed in social science and philosophy [21, 36]. One successful approach in Artificial Intelligence (AI) and Multi-Agent Systems (MAS) sees an autonomous agent as "self-contained, reactive, proactive [...], typically with a central focus of control, that is able to communicate with other agents [...]. A more specific usage is to mean a computer system that is either conceptualised or implemented in term of concepts more usually applied to humans (such as beliefs, desires and intentions)". Autonomy, in particular, is "the assumption that, although we generally intend agents to act on our behalf, they nevertheless act without direct human or other intervention, and have some kind of control over their internal state" [41].

A theoretical contribution to the idea of autonomy comes of course from moral philosophy, which takes this idea as crucial: the moral life of agents, their values, norms, and the related concept of moral responsibility are all meaningless without assuming that agents can deliberate and are decision-makers. Specific moral traditions, such as the Kantian one, claim in addition that precisely the idea of autonomy—conceived of as agent's capability of adopting right and universal norms of behaviour—pertains to the domain of *morality*, in

---

The authors are supported by the EU H2020 research and innovation programme under the Marie Sklodowska-Curie grant agreement No. 690974 for the project MIREL: *MIning and REasoning with Legal texts*.

contrast with other domains of practical reasoning, such as the *law*, where the grounds of agency is *heteronomy* and the binding force of norms is contextual and depends on external factors—typically, coercion [19].

The role of norms to characterise moral autonomy, and autonomous actions, has been differently, but successfully proposed by (competing) moral views like Kantianism and rule utilitarianism. Drawing from these traditions, we formally explore the following intuition:

Intuition. *Autonomous agents take decisions about the moral theory governing their society, and elicit new theories that would improve the welfare. Decision-making is performed through a collective deliberative procedure, called* moral dialogue.

On the basis of this intuition, a question arises: *how to define a generic formal framework to mimic the determination of moral theories in a society of autonomous individuals?* The research question can break down into subquestions, for example: *(i)* how to represent rational criteria which are the basis of several moral theories? *(ii)* how to represent building blocks for grasping at least basic aspects of both Rawls' approach to morality and rule utilitarianism?

To address these open challenges, we propose a generic framework capturing the determination of moral theories as a dialogical process. In our framework, moral views are represented through the form of rule-based theories expressing agents' norms, which are associated with an utility function. In other words, theories are nothing but normative systems on which agents argue in regard to their moral justification: agents have thus to deliberate about which theories should regulate the society through a dialogue. By putting on the table their arguments in the dialogue, agents determine what are the (possibly new) rules that should govern the society and its welfare, depending on a specific moral theory. We considered building blocks and rational criteria for characterising influential moral theories such as

- Rawls' contractualist model of deontological morality [32]; this approach sees moral principles for a society as resulting from a collective deliberative process among the agents belonging to such a society, a process which, if it runs under impartiality, supports the maximin principle; and

- rule utilitarianism [6, 16, 17]; this approach requires agents to follow the rules[1] that maximise utility, and which often connects morality to the theory of rational action, where expected-utility maximisation and Pareto efficiency are fundamental concepts [16].

Such moral theories express opposite views [16], and our approach aims at accommodating these different views (possibly amongst others). To sum up, our contribution is as follows:

- Agents propose in a dialogue the normative (moral) theory that they would prefer for their society;

---

[1]Hereafter, the terms rule and norm will be interchangeably used.

- Each theory is associated with an utility that measures the impact of the proposed norms; the intended reading could be, for example, in terms of the consequence for the society if all agents would conform to such norms (as suggested by rule utilitarianism);

- Agents deliberate in a different way depending on which of the above theories are employed to compute the utility, leading then to the emergence of a society regulated by the selected moral theory.

Drawing ideas from Rawls' theory and rule utilitarianism, we provide a formal framework for moral dialogues which can accommodate the following rational/moral criteria:

- Welfare maximisation [16] and Pareto efficiency [13];

- Maximin principle [38, 32];

- Impartiality [18, 6].

Ethics and moral theories are becoming more and more relevant for AI in general [39, 26, 37, 28, 10, 12, 8], and for autonomous systems in particular [3], as evidenced by initiatives like The Moral Machine by the MIT[2] and the REINS Project [7]. Enhancements in robotics, knowledge representation and reasoning, and cognitive modelling cast a new light on these challenges making their discussion more urgent than in the past. The generic formal approach we propose is, up to our knowledge, the only formal framework devoted to represent and understand the determination of (well known) moral theories in a society of autonomous individuals. Our framework is in line with Artificial Moral Agents (AMA) proposed by Dignum [12]: our intelligent systems incorporate moral reasoning in their deliberation process, and they rely on argumentation theory to explain their behaviour in terms of moral concepts. From a broader perspective, our approach provides a first step towards the so called "beneficial AI" [34], so that agents are designed to stick to a specific moral theory, e.g., Rawls' maximin principle, emerging from a dialogue-based deliberation process and having as a goal the societal welfare.

The reader may argue that often people speak of *discovering* moral rules, rather than deliberating on moral rules. In this case, the appropriate dialogue would be like a mathematical dialogue, where people are engaged in the process of discovering independent truths, beyond agreeing on the rules to live by. The issue of discovering moral rules is out of the scope of this paper. We choose a more contractualist approach to deliberate on the *best* set of moral rules depending on the utility of the agents involved in the deliberation process.

The paper is organised as follows: first, we provide the definition of moral theory in our formal setting, and then we sketch how the utility of theories could be determined. After some problem definitions, we propose moral dialogues to address such problems, and we study some basic properties of these dialogues. We conclude with a comparison with the literature, and some future perspectives.

---

[2] http://moralmachine.mit.edu/

2 Moral Theory Setting

A moral theory defines norms stating what to do; it deals also with forming judgements about what one ought (e.g., morally) to do. We assume a logic language from which it is possible to build moral theories. A moral theory is made of a set rules and a superiority relation over the rules.

Definition 1 (Moral theory). *A **moral theory** is a tuple $\mathcal{T} = \langle \mathcal{R}, \succ \rangle$ where $\mathcal{R}$ is a set of rules, and $\succ \subseteq \mathcal{R} \times \mathcal{R}$ is a superiority relation over the rules.*

In the remainder of the paper, a set of moral theories is denoted $\mathfrak{T}$, and we may just say theory instead of moral theory.

If a moral theory is meant to be applied in some context (represented by for example some facts), then such a moral theory can be embedded within a larger 'contextual' theory. In this paper, possible contexts are left implicit, assuming that everything can be parametrised with respect to them.

When agents argue about theories to govern their own society, they consider the utility springing from these theories.

Definition 2 (Agent theory utility distribution). *Let $\mathfrak{T}$ be a set of theories, $\mathbb{V}$ an ordered set of values (on which the moral utility functions are computed), and Ag a set of agents. An **agent moral theory utility distribution** is a function*

$$U: \mathfrak{T} \to \prod_0^{|Ag|} \mathbb{V}.$$

Given a theory and n agents, the function returns a vector of $n+1$ values, where the first value, conventionally, indicates the total welfare for the set of agents, and the remaining values define the value of the theory for each agent. The value of the theory for agent i corresponds to the projection on the i-th element of the vector, thus $U_i(\mathcal{T}) = \pi_i(U(\mathcal{T}))$. In the remainder, $U_{Ag}(\mathcal{T})$ denotes agents' utility $\pi_0(U(\mathcal{T}))$, and $U_i(\mathcal{T})$ the utility $\pi_i(U(\mathcal{T}))$ of agent i.

3 Theory Utility

In this section, we briefly illustrate two possible approaches for computing theory utility, i.e., from rule utility and from literal utility.

3.1 Theory Utility from Rule Utility

As argued in the context of rule utilitarianism, we can determine what is the value of each rule for each agent (based on the context in which the theory is used) by introducing the following function [16].

Definition 3 (Agent rule valuation). *Let Ag and $\mathbb{V}$ be, respectively, a set of agents and an ordered set of values (on which a moral utility function is computed). An **agent rule valuation** is a function*
$$V\colon Ag \times \mathcal{R} \times \mathfrak{T} \to \mathbb{V}.$$

This function assigns to every rule in every theory a value to be used in a moral utility function. Based on this definition, we establish that the elements of the agent utility distribution are computed based on the utilities of the rules (and the context in which they appear), using the following equation

$$U_i(\mathcal{T}) = F^i_{r \in \mathcal{R}}(V(Ag_i, r, \mathcal{T})) \tag{1}$$

where F^i is a function/operator that agglomerates the individual values for a set of rules into a single value, and Ag_i denotes agent i.

If we move from rule utility, as discussed above, two options are possible for computing *theory utility*: in the first case, this is computed by other elements of the vector based on a predefined function (for instance, the total welfare is the sum of the welfare values for the individual agents); in the second case, there is an individual rule depending function designed for it working on values attributed to rules according to the context in which they appear, namely, we have a function

$$V_{Ag}\colon \mathcal{R} \times \mathfrak{T} \to \mathbb{V} \tag{2}$$

from which the total welfare can be computed as

$$U_{Ag}(\mathcal{T}) = F^0_{r \in \mathcal{R}}(V_{Ag}(r, \mathcal{T})). \tag{3}$$

Functions for rule valuation can be further specified, in particular with respect to some inference mechanisms as proposed later.

3.2 Theory Utility from Literal Utility

A more fine-grained approach to articulate the way in which utility springs from any theory $\mathcal{T}$ is based on the utility of conclusions that follow from arguing on $\mathcal{T}$.

Let us first give a basic language setting. A literal is a propositional atom or the negation of a propositional atom. Given a literal ϕ, its complementary literal is a literal, denoted as $\sim \phi$, such that if ϕ is an atom p then $\sim \phi$ is its negation $\neg p$, and if ϕ is $\neg q$ then $\sim \phi$ is q. If $Prop$ is a set of propositional atoms then $Lit = Prop \cup \{\neg p \mid p \in Prop\}$ is a set of literals.

For each literal l in a set Lit of literals and given a (possibly different) set of literals $\{l_1, \ldots, l_n\}$, we can define a function λ that assigns for each agent i in Ag an utility value, i.e., the utility that the state of affairs denoted by l brings to i in a context described by $l_1, \ldots, l_n$.

Definition 4 (Agent literal valuation). *Let Ag and $\mathbb{V}$ be, respectively, a set of agents and an ordered set of values. An **agent literal valuation** is a function*

$$\lambda : Ag \times Lit \times \text{pow}(Lit) \to \mathbb{V}.$$

If $E(\mathcal{T}) = \{c_1, \ldots, c_m\}$ is the set of conclusions of a theory $\mathcal{T}$, then an individual agent utility can be given by agglomerating the values of all conclusions.

$$U_i(\mathcal{T}) = \underset{\forall l \in E(\mathcal{T})}{F^i} \lambda(Ag_i, l, E(\mathcal{T})). \tag{4}$$

where F^i is a function/operator that agglomerates the individual values into a single value.

As mentioned earlier, the utility of theories can be further specified by considering inference mechanisms to reason on the theories. Different mechanisms are possible, we propose next to use an argument-based reasoning setting.

4 Theory Utility in an Argumentation Setting

In this section, we sketch how the utility of a theory could be determined in an argumentation setting.

4.1 Argumentation Setting

To reason on theories, and in particular to determine justified and rejected conclusions of any theory, we adopt an argumentation setting. Different argumentation settings exist in the literature, see e.g. [14, 31, 20] amongst many others. We adopt an ASPIC$^+$-like setting, and the (internal) logical structure of arguments are specified in such a way that arguments are logical inference trees built out from rules (we adjust the definition in [31] to meet our definition of theory). In the following definition, for a given argument A, Conc returns its conclusion, Sub returns all its sub-arguments, Rules returns the set of rules in the argument and, finally, TopRule returns the last inference rule in the argument.

Definition 5 (Argument). *Let $\mathcal{T} = (\mathcal{R}, \succ)$ be a theory where rules have the form $\psi_1, \ldots, \psi_n \Rightarrow \phi$ ($0 \leq n$), $\psi_1, \ldots, \psi_n, \phi \in Lit$. An **argument** A constructed from $\mathcal{T}$ has the form $A_1, \ldots, A_n \Rightarrow_r \phi$, where*

- A_k *is an argument constructed from $\mathcal{T}$, and*
- $r : \text{Conc}(A_1), \ldots, \text{Conc}(A_n) \Rightarrow \phi$ *is a rule in $\mathcal{R}$.*

With regard to argument A, the following holds:

$$\begin{aligned}
&\text{Conc}(A) = \phi \\
&\text{Sub}(A) = \text{Sub}(A_1) \cup \ldots \cup \text{Sub}(A_n) \cup \{A\} \\
&\text{TopRule}(A) = r : \text{Conc}(A_1), \ldots, \text{Conc}(A_n) \Rightarrow \phi \\
&\text{Rules}(A) = \text{Rules}(A_1) \cup \ldots \cup \text{Rules}(A_n) \cup \{\text{TopRule}(A)\}.
\end{aligned}$$

Arguments may support conflicting conclusions and thus attacks may appear between arguments. Effective attacks between arguments are defined here with respect to the superiority relation over rules which are used to build arguments.

Definition 6 (Attacks). *An argument B attacks an argument A iff $\exists A' \in \mathrm{Sub}(A)$ such that $\mathrm{Conc}(B) = \sim\mathrm{Conc}(A')$, and $\mathrm{TopRule}(A') \not\succ \mathrm{TopRule}(B)$.*

Given a theory from which arguments are built and attacks determined, we can define an argumentation framework, along with standard argumentation semantics.

Definition 7 (Argumentation framework). *Let $\mathcal{T} = (\mathcal{R}, \succ)$ be a theory. The **argumentation framework** $AF_\mathcal{T}$ determined by $\mathcal{T}$ is a tuple $\langle \mathcal{A}, \rightsquigarrow \rangle$ where $\mathcal{A}$ is the set of all arguments constructed from $\mathcal{T}$, and $\rightsquigarrow \subseteq \mathcal{A} \times \mathcal{A}$ is an attack relation.*

Definition 8 (Argumentation semantics).

Conflict-free set: *A set $\mathcal{S}$ of arguments is conflict-free iff there exist no arguments A and B in $\mathcal{S}$ such that B attacks A.*

Argument defence: *Let $\mathcal{S} \subseteq \mathcal{A}$ be a set of arguments. The set $\mathcal{S}$ defends an argument $A \in \mathcal{A}$ iff for each argument B attacking A there is an argument C in $\mathcal{S}$ that attacks B.*

Complete extension: *Let $AF = (\mathcal{A}, \rightsquigarrow)$ and $\mathcal{S} \subseteq \mathcal{A}$. The set $\mathcal{S}$ is a complete extension of AF iff $\mathcal{S}$ is conflict-free and $\mathcal{S} = \{A \in \mathcal{A} \mid \mathcal{S} \text{ defends } A\}$.*

Grounded extension: *A grounded extension $\mathrm{GE}(AF)$ of an argumentation framework AF is the minimal complete extension of AF.*

Justified argument and conclusion: *An argument A and its conclusion are justified w.r.t. an argumentation framework AF iff $A \in \mathrm{GE}(AF)$.*

Rejected argument and conclusion: *An argument A is rejected w.r.t. an argument framework AF iff $A \notin \mathrm{GE}(AF)$; its conclusion is rejected iff it is not justified.*

Other semantics could be employed, see e.g. [1, 2]. For our purposes and the sake of simplicity, we consider the grounded semantics only, and we leave how to deal with other semantics to future investigations.

4.2 Theory Utility from Rule Utility (cont'd)

On the basis of the argumentation setting to reason upon any theory, we can now provide a simple way for determining the utility of any theory from its rules. We recall that the assumption is that any rule r in any theory $\mathcal{T}$ can be associated with an utility value v (see Definition 3), which means that complying with this rule r produces utility v.

Suppose that the only rules that contribute to the utility function for an agent are the rules contributing to justified arguments/conclusions. So, assuming $\mathbb{V} = \mathbb{Z}$ one can create an agent rule valuation as follows:

$$V(Ag_i, r, \mathcal{T}) = \begin{cases} n \neq 0 & \text{if } r \in \text{Rules}(A),\ A \in \text{GE}(AF_{\mathcal{T}}) \\ 0 & \text{otherwise.} \end{cases} \quad (5)$$

The individual utility U_i of agent i is then possibly the sum of the agent rule valuations:

$$U_i(\mathcal{T}) = \sum_{r \in \mathcal{R}} V(Ag_i, r, \mathcal{T}). \quad (6)$$

Hence, following an intuition from rule utilitarianism, this sketches how the utility of theories can be determined as the sum of the single agent rule valuations [17][3].

4.3 Theory Utility from Literal Utility (cont'd)

An alternative to determine theory utility consists in computing the utility from justified conclusions. For any theory $\mathcal{T}$ let us specify the set $E(\mathcal{T})$ of conclusions of $\mathcal{T}$ (Section 3.2) as follows:

$$E(\mathcal{T}) = \{\psi \mid \forall A \in \text{GE}(AF_{\mathcal{T}}),\ \psi = \text{Conc}(A)\}. \quad (7)$$

Hence, the function based on Equation (4) can be easily applied here.

We leave to future investigations the complete development of the two approaches sketched in this section to determine the utility of theories with respect to various contexts in an argumentation setting. Whatever the way a theory is associated with a utility value, we have then the problem of finding the 'right' theory.

5 Problem Definitions

As any theory can be associated with a utility, we may identify particular theories. For example, one may consider agents' utility optimal theories, i.e., theories maximising the agents' utility, or (strong) 'Pareto optimal theories', i.e., theories for which no agent can be made better off by making some agents worse off, or 'maximin optimal theories', i.e., theories maximising the utility of the worst off agents.

Definition 9 (Agents' utility optimal theory)**.** *Let Ag be a set of agents. A theory $\mathcal{T}^*$ is an* **agents' utility optimal theory** *amongst a set of theories $\mathfrak{T}$ iff there is no theory $\mathcal{T} \in \mathfrak{T}$ such that $U_{Ag}(\mathcal{T}) > U_{Ag}(\mathcal{T}^*)$.*[4]

Definition 10 (Pareto optimal theory)**.** *Let Ag be a set of agents. A theory $\mathcal{T}^*$ is a* **Pareto optimal theory** *amongst a set of theories $\mathfrak{T}$ iff there is no theory $\mathcal{T} \in \mathfrak{T}$ such that $U_i(\mathcal{T}^*) \leq U_i(\mathcal{T})$ for all $i \in Ag$ and $U_i(\mathcal{T}^*) < U_i(\mathcal{T})$ for some $i \in Ag$.*

---
[3]We do not commit to any specific utility theory, which would require a more detailed machinery: see [17].

[4]Equivalently, we can say that a theory $\mathcal{T}^*$ is agents' utility optimal amongst a set of theories $\mathfrak{T}$ iff for all $\mathcal{T} \in \mathfrak{T}$ it holds that $U_{Ag}(\mathcal{T}) \leq U_{Ag}(\mathcal{T}^*)$.

Definition 11 (Maximin optimal theory). *Let Ag be a set of agents. A theory $\mathcal{T}^*$ is a **maximin optimal theory** amongst a set of theories $\mathfrak{T}$ iff there is no theory $\mathcal{T} \in \mathfrak{T}$ such that $\min_{i \in Ag} U_i(\mathcal{T}) > \min_{i \in Ag} U_i(\mathcal{T}^*)$.*

We can now formulate the general problem of a moral theory elicitation.

> **Given:** a set of agents Ag and a set of theories $\mathfrak{T}$;
> **Find:** a specific moral theory $\mathcal{T}$ in $\mathfrak{T}$.

The problem can be specified. For instance, one may seek an agents' utility optimal theory, or a Pareto optimal theory.

Definition 12 (Agents' utility theory problem).

> ***Given:*** *a set of agents Ag and a set of theories $\mathfrak{T}$;*
> ***Find:*** *a theory $\mathcal{T}$ in $\mathfrak{T}$ which is agents' utility optimal amongst $\mathfrak{T}$.*

Definition 13 (Pareto theory problem).

> ***Given:*** *a set of agents Ag and a set of theories $\mathfrak{T}$;*
> ***Find:*** *a theory $\mathcal{T}$ in $\mathfrak{T}$ which is Pareto optimal amongst $\mathfrak{T}$.*

Definition 14 (Maximin theory problem).

> ***Given:*** *a set of agents Ag and a set of theories $\mathfrak{T}$;*
> ***Find:*** *a theory $\mathcal{T}$ in $\mathfrak{T}$ which is maximin optimal amongst $\mathfrak{T}$.*

Given a theory, a brute force solution is to compute every sub-theory along with a utility distribution and then retain the specific sub-theory we are after. However, this solution is of course not efficient. In the next section, we initiate possible alternative solutions by means of dialogues.

6 Moral Dialogues

A moral dialogue is the process through which agents propose their normative theories with the aim to improve on the current state-of-the-art theory. The normative system resulting from the dialogue is taken to be morally justified. Moral dialogues are based on dialogues.

Definition 15 (Dialogue). *A **dialogue** is a sequence of theories $(\mathcal{T}_k)_{k=1,\ldots K}$ such that*

- *theory $\mathcal{T}_1$ is an (arbitrary) initial theory;*
- *for every $\mathcal{T}_k$, there is a set of theories $\mathfrak{T}^k = \{\mathcal{T}_1^k, \ldots, \mathcal{T}_n^k\}$ (proposed by some agents);*

- theory $\mathcal{T}_{k+1} = Choice(\mathfrak{T}^k)$, where $Choice$ is a function that selects theory $\mathcal{T}_{k+1}$ out of a non-empty set $\mathfrak{T}^k$;
- theory $\mathcal{T}_K$ is terminal iff $\mathfrak{T}^K = \emptyset$.

Definition 16 (Theories proposed in a dialogue). *The set of theories $\mathfrak{T}^d$ proposed in a dialogue $d = (\mathcal{T}_k)_{k=1,...K}$ is $\bigcup_{k \in \{1,...K\}} \mathfrak{T}^k$.*

We can note that theory $\mathcal{T}_k$ may be included in $\mathfrak{T}^k$, possibly leading to some sort of equilibrium. However, in this paper, we are not interested in computing *equilibria* as, being interested in moral reasoning, we deal with principles and not with *moves* as in standard game theoretic approaches. For this reason, we rely on dialogues and not on games, though our dialogues may be seen as *mirroring* such games.

A dialogue is moral if, and only if, the choice function is moral. We concentrate on a few moral *Choice* functions, each corresponding to a well established ethical/rational criterion: in an impartial choice, theories are drawn at random; following rule utilitarianism, other choices are maximising agents' utility choice, or a Pareto choice.

Definition 17 (Impartial choice). *The choice function of a dialogue $(\mathcal{T}_k)_{k=1,...K}$ is an **impartial choice function** iff any theory $\mathcal{T}_k$ ($2 \leq k$) is drawn at random from $\mathfrak{T} \subseteq \mathfrak{T}^{k-1}$ with a uniform probability.*

Definition 18 (Agents' utility maximising choice). *The choice function of a dialogue $(\mathcal{T}_k)_{k=1,...K}$ is an **agents' utility maximising choice function** iff any theory $\mathcal{T}_k$ ($2 \leq k$) is an agents' utility optimal theory amongst the set of theories $\mathfrak{T}^{k-1}$.*

Definition 19 (Pareto choice). *The choice function of a dialogue $(\mathcal{T}_k)_{k=1,...K}$ is a **Pareto choice function** iff any theory $\mathcal{T}_k$ ($2 \leq k$) is a Pareto optimal theory amongst the set of theories $\mathfrak{T}^{k-1}$.*

Randomising theories is also a basis to mimic an intuition developed by moral philosophers such as John Rawls [32], who argued the importance of the so-called second-order impartiality, according to which norms and principles are impartially evaluated and selected by agents by demonstrating that they would be selected by a group of impartial persons who were choosing the moral rules for their society. In particular, Rawls said that a normative theory of a just society should be chosen by self-interested rational agents in the original position, i.e., a position in which agents are rational and possess broad knowledge about the world, but are denied specific information regarding their own particular identities and personal convenience. On this basis, Rawls argued that rational agents with risk aversion should yield the 'Difference Principle', according to which inequalities are to the greatest benefit of the least advantaged members of society, i.e., a maximin choice [15].

Definition 20 (Maximin choice). *The choice function of a dialogue $(\mathcal{T}_k)_{k=1,...K}$ is a **maximin choice function** iff any theory $\mathcal{T}_k$ ($2 \leq k$) is a maximin optimal theory amongst the set of theories $\mathfrak{T}^{k-1}$.*

Example. *Let us consider a group of three autonomous individuals: agent 1 has high incomes because of its high salary, agent 2 has high incomes because of tax evasion, and agent 3 has low incomes. Suppose we have an initial theory $\mathcal{T}_0$ with utility distribution $[5, 3, 1, 1]$ (where 5 is the global utility), and suppose theories $\mathcal{T}_1$, $\mathcal{T}_2$ and $\mathcal{T}_3$ are proposed such that $U(\mathcal{T}_1) = [8, 2, 0, 6]$ (i.e., taxes slightly raised for upper classes, tax evasion is severely punished, and public subsidies are introduced for lower classes), $U(\mathcal{T}_2) = [6, 2, 2, 2]$ (i.e., taxes slightly raised for upper classes together with public subsidies for lower classes and imprisonment for tax fraud is lowered from 5 years to 3 years), and $U(\mathcal{T}_3) = [7, 3, 3, 1]$ (i.e., a tax evasion amnesty is proposed). If the agents' utility maximising choice is adopted then $\mathcal{T}_1$ is elicited, while the maximin choice yields $\mathcal{T}_2$, and the Pareto choice results into $\mathcal{T}_3$.*

One may complement agents' utility maximising choices, or Pareto choices, or maximin choices with an impartial choice when there exist multiple agents' utility, or Pareto, or maximin theories, respectively, in a set of theories $\mathfrak{T}^k$. For example, a moral dialogue $(\mathcal{T}_k)_{k=1,...,K}$ can have an impartial and agents' utility maximising choice function if any theory $\mathcal{T}_k$ is drawn at random from the set theories maximising agents' utility.

Whatever the moral choice function, a moral theory may not emerge from the theories proposed in a dialogue, especially if theory $\mathcal{T}_{k-1}$ is not included in $\mathfrak{T}^{k-1}$. We further investigate moral dialogues in that regard in the next section.

7 Moral Optimising Dialogues

The use of dialogues and their iterative nature suggests a few different (search) strategies to find an optimal theory in a set of theories. We do not propose any particular strategies in this paper, leaving them to future work. In this section, we simply give some basic properties of moral dialogues that may characterise such strategies.

7.1 Agents' Utility Optimising Dialogue

For the terminal theory of a dialogue to be agents' utility optimal amongst the theories proposed in the dialogue, it is sufficient that the dialogue has an agents' utility maximising choice function whose output theory $\mathcal{T}_k$ is always included in the proposed theories $\mathfrak{T}^k$.

Proposition 1. *The terminal theory of a dialogue $d = (\mathcal{T}_k)_{k=1,...K}$ with an agents' utility maximising choice function is agents' utility optimal amongst the set of theories $\mathfrak{T}^d$ proposed in the dialogue if for any $\mathcal{T}_k$, it holds that $\mathcal{T}_k \in \mathfrak{T}^k$.*

Therefore, such an agents' utility maximising dialogue solves the problem given in Definition 12, where $\mathfrak{T} = \mathfrak{T}^d$.

However, the terminal theory may not be a strict 'improvement' of the initial theory. For this reason, one may consider dialogues to elicit agents' utility optimal theories based on the idea of improving theories.

Definition 21 (Agents' utility improving theory). *Let Ag a set of agents. A theory $\mathcal{T}^*$ is an **agents' utility improvement** of a theory $\mathcal{T}$ iff $U_{Ag}(\mathcal{T}^*) > U_{Ag}(\mathcal{T})$.*

Proposition 2. *A theory is an agents' utility optimal theory amongst a set of theories $\mathfrak{T}$ iff there exist no agents' utility improvements in $\mathfrak{T}$ of the theory.*

Proof. By Definition 9, a theory $\mathcal{T}^*$ is agents' utility optimal amongst a set of theories $\mathfrak{T}$ iff there exists no agents' utility improving theories in $\mathfrak{T}$ of $\mathcal{T}^*$. □

Consequently, the initial theory is not optimal if there exists an improvement.

Proposition 3. *The terminal theory of a dialogue $d = (\mathcal{T}_k)_{k=1,...K}$ with a agents' utility maximising choice function is agents' utility optimal amongst the set of theories $\mathfrak{T}^d$ proposed in the dialogue and it is an agents' utility improvement of the initial theory, if for any $\mathcal{T}_k$, it holds that $\mathcal{T}_k \in \mathfrak{T}^k$, and there exists a theory $\mathcal{T}_k$ which is an agents' utility improvement of $\mathcal{T}_{k-1}$.*

In other words, if there exists no improvement in a dialogue then the initial theory remains the optimal theory, and a moral dialogue is not necessary to find the optimal theories.

7.2 Pareto Optimising Dialogue

Moral dialogues can be similarly tuned to elicit Pareto optimal theories.

Proposition 4. *The terminal theory of a dialogue $d = (\mathcal{T}_k)_{k=1,...K}$ with a Pareto choice function is Pareto optimal amongst the set of theories $\mathfrak{T}^d$ proposed in the dialogue if for any $\mathcal{T}_k$, it holds that $\mathcal{T}_k \in \mathfrak{T}^k$.*

Therefore, a Pareto improving dialogue solves the problem given in Definition 13, where $\mathfrak{T} = \mathfrak{T}^d$.

As the terminal theory may not be an improvement of the initial theory, we can consider Pareto improving theories, i.e., theories leading to a utility gain, without any agents being made worse off.

Definition 22 (Pareto improving theory). *Let Ag a set of agents. A theory $\mathcal{T}^*$ is a **Pareto improvement** of a theory $\mathcal{T}$ iff $U_i(\mathcal{T}^*) \leq U_i(\mathcal{T})$ for all $i \in Ag$ and $U_i(\mathcal{T}^*) < U_i(\mathcal{T})$ for some $i \in Ag$.*

Proposition 5. *A theory is a Pareto optimal theory amongst a set of theories $\mathfrak{T}$ iff there exist no Pareto improvements in $\mathfrak{T}$ of the theory.*

Proof. By Definition 10, a theory $\mathcal{T}^*$ is Pareto optimal amongst a set of theories $\mathfrak{T}$ iff there is no Pareto improving theory of $\mathcal{T}^*$. □

Proposition 6. *The terminal theory of a dialogue $d = (\mathcal{T}_k)_{k=1,...K}$ with a Pareto choice function is Pareto optimal amongst the set of theories $\mathfrak{T}^d$ proposed in the dialogue and it is an agents' utility improvement of the initial theory, if for any $\mathcal{T}_k$, it holds that $\mathcal{T}_k \in \mathfrak{T}^k$, and there exists a theory $\mathcal{T}_k$ which is a Pareto improvement of $\mathcal{T}_{k-1}$.*

7.3 Maxmin Optimising Dialogue

Similarly to agents' utility and Pareto improving choice functions, maximin can be accommodated in dialogues.

Proposition 7. *The terminal theory of a dialogue $d = (\mathcal{T}_k)_{k=1,\ldots K}$ with a maximin choice function is maximin optimal amongst the set of theories $\mathfrak{T}^d$ proposed in the dialogue if for any $\mathcal{T}_k$, it holds that $\mathcal{T}_k \in \mathfrak{T}^k$.*

Therefore, a maximin improving dialogue solves the problem given in Definition 14, where $\mathfrak{T} = \mathfrak{T}^d$.

Definition 23 (Maximin improving theory). *Let Ag a set of agents. A theory $\mathcal{T}^*$ is a **maximin improvement** of a theory $\mathcal{T}$ iff $\min_{i \in Ag} U_i(\mathcal{T}) > \min_{i \in Ag} U_i(\mathcal{T}^*)$.*

Proposition 8. *A theory is a maximin optimal theory amongst a set of theories $\mathfrak{T}$ iff there exist no maximin improvements in $\mathfrak{T}$ of the theory.*

Proof. By Definition 10, a theory $\mathcal{T}^*$ is maximin optimal amongst a set of theories $\mathfrak{T}$ iff there is no maximin improving theory of $\mathcal{T}^*$. □

Proposition 9. *The terminal theory of a dialogue $d = (\mathcal{T}_k)_{k=1,\ldots K}$ with a maximin choice function is maximin optimal amongst the set of theories $\mathfrak{T}^d$ proposed in the dialogue and it is an agents' utility improvement of the initial theory, if for any $\mathcal{T}_k$, it holds that $\mathcal{T}_k \in \mathfrak{T}^k$, and there exists a theory $\mathcal{T}_k$ which is a maximin improvement of $\mathcal{T}_{k-1}$.*

Notice that, while Rawls justifies the maximin choice on the assumptions of agents with risk aversion and an impartial choice (from the 'original position'), we do not aim here at directly covering this justification.

8 Related Work

Autonomy and agency are central properties in robotic systems, assisted living applications, responsible vehicles and many other application domains. As the complexity of the situations faced by such autonomous agents is increasing, agents' decision-making has to take into account new elements like ethical and moral considerations. For these reasons, the last years have seen a raising number of projects tackling this issue, e.g., the REINS project about responsibility [7], the ETHICAA project about defining regulation modes to manage ethical conflicts within socio-technical systems [3], and the MIT Moral Machine about autonomous self-driving cars.

New approaches and position papers are also appearing in the literature: Charisi *et al.* [8] report and discuss the immediate challenges faced by the problem of the engineering of machine ethics, and Dignum [12] describes the leading ethics theories, and proposes alternative ways to ensure ethical behaviour by autonomous agents. Our approach is closer to the goals of the ETHICAA project regarding the *reasoning perspective* where a representation of ethical principles is provided together with decision-making models [3]. Following [25], our agents converge on a moral theory by assessing the arguments and values at stake under specific ethical principles.

Our approach can also be related to the work on computational justice in self-organising electronic institutions proposed by Pitt et al. [30], where agents can agree on a set of rules to self-organise and self-regulate a distribution of resources [29]. We take a more abstract stance, and provide a dialogical framework for agents' deliberation about moral theories.

An argumentation-based perspective to ethical systems design is proposed by Verheij [37]. Based on the assumptions that ethical system's decisions depend on the *values* and the rules embedded in the system, Verheij [37] studies the issue of the comparison of values in value-guided argumentation, through techniques connecting qualitative and quantitative primitives from evidential argumentation applied to value-guided argumentation. The problem tackled in this paper is different from our goal, as well as the methodology proposed to address it. The only shared point is the fact of relying on rules to represent ethical decisions and theories, and on (two different kinds of) formal argumentation frameworks.

Another approach to ethics in a multi-agent scenario has been proposed by Cointe et al. [9], where the authors studied the problem of ethical judgement, i.e., the assessment of the appropriateness of agents' behaviours with respect to moral convictions and ethical principles. No specific ethical theories are considered. Again, the goal and methodology to address it differ from our paper, even if the two approaches may be seen as complementary, as after the elicitation of a new moral theory to ensure the welfare the society then the agents need to judge whether the behaviour of the other agents complies or not with it.

Dennis et al. [11] propose a theoretical framework for ethical plan selection that can be formally verified. The authors formally verify that the agent chooses to execute the most ethical available plan, given its belief set. This approach focuses on the formal verification of ethical decision-making within autonomous agents. This goal differs from ours, as we are interested in a dialogical assessment of moral theories. Moreover, further differences arise, i.e., ethical principles are considered as abstract while we take into account three well known moral theories. The common point is the representation of moral theories using (ethical) rules.

Other contributions have been presented about formal models of ethics but with different goals than the one we addressed in this paper. A logic-based approach to model moral reasoning with deontic constraints is presented by Wiegel [40]. It is an approach to represent the theory of good, and ethical reasoning is addressed in the meta-level with the aim to support the adoption of a less restrictive model of behaviour. Other approaches like [4, 35] provide a direct translation of some well-known ethical principles as Kant's Categorical Imperative or Thomas Aquinas' Doctrine of Double Effect into logic programming.

Other approaches aim at formalising, with the help of logic, classical game theory and evolutionary game theory, the influence of social preferences and moral values on the decision-making process of autonomous agents, with special emphasis on fairness values based on Rawls' maxmin criterion. More precisely, Lorini [22] proposes a logical formalisation of social preferences based on Rawls' fairness principle. Lorini [23] again proposes a logical formalisation of the relationship between moral values and preferences. Lorini and Muhlenbernd [24] provide a game-theoretic and evolutionary analysis of Rawls' fairness principle and its connection with the concepts of responsibility and guilt. The latter work focuses on the integration of moral and ethical aspects into the decision-making processes of rational autonomous agents and it is related with some existing economic theories of morality [5]. Another recent game-theoretic account of morality and its connection with guilt has been proposed in the area of multi-agent systems by Pereira *et al.* [27].

9 Conclusion

Autonomous agents raise the problem of eliciting moral norms governing the behaviour of the agents in a society. We addressed the elicitation problem of moral norms as normative theories by proposing a generic dialogical framework. We showed how the framework can accommodate different and well-established moral choices (such as utility maximising, Pareto and maximin choices), possibly leading to different moral theories. By doing so, different ethical criteria can be compared. To the best of our knowledge, there exists no other formal comparative framework designed to deal with the elicitation of moral theories in MAS showing the above mentioned features.

Future work directions are multiple. They include a deeper comparative investigation of moral principles. From the philosophical point of view, we also aim to extend this framework to cope with the theory of the *ideal observer* (as a condition for ensuring impartiality in ethics) [18, 6]. The idea is that the moral judgement is well founded if it is accepted by the set of fully-informed agents in the society based on well founded rational arguments. A goal is to prove that these conditions are equivalent to a situation where moral rules are selected by an infinite set of agents who know everything. Eventually, given the attention paid to computational self-organising institutions (see e.g. [30, 33]), it would be interesting to investigate how dialogues on moral theories can fit with these implementations.

References

[1] Pietro Baroni, Martin Caminada, and Massimiliano Giacomin. An introduction to argumentation semantics. *Knowledge Eng. Review*, 26(4):365–410, 2011.

[2] Pietro Baroni, Guido Governatori, and Régis Riveret. On labelling statements in multi-labelling argumentation. In *Proc. of the 22nd Euro. Conf. on Artificial Intelligence*, volume 285 of *Frontiers in Artificial Intelligence and Applications*, pages 489–497. IOS Press, 2016.

[3] Aline Belloni, Alain Berger, Olivier Boissier, Grégory Bonnet, Gauvain Bourgne, Pierre-Antoine Chardel, Jean-Pierre Cotton, Nicolas Evreux, Jean-Gabriel Ganascia, Philippe Jaillon, Bruno Mermet, Gauthier Picard, Bernard Rever, Gaële Simon, Thibault de Swarte, Catherine Tessier, François Vexler, Robert Voyer, and Antoine Zimmermann. Dealing with ethical conflicts in autonomous agents and multi-agent systems. In *Papers from the 2015 AAAI Workshop on Artificial Intelligence and Ethics*. AAAI Press, 2015.

[4] Fiona Berreby, Gauvain Bourgne, and Jean-Gabriel Ganascia. Modelling moral reasoning and ethical responsibility with logic programming. In *Proc. of 20th Inter. Conf. on Logic for Programming, Artificial Intelligence, and Reasoning*, pages 532–548. Springer, 2015.
[5] K. Binmore. *Natural Justice*. Oxford University Press, 2005.
[6] Richard Brandt. *Ethical Theory*. Prentice Hall, 1959.
[7] Jan M. Broersen. Responsible intelligent systems - the REINS project. *KI*, 28(3):209–214, 2014.
[8] Vicky Charisi, Louise A. Dennis, Michael Fisher, Robert Lieck, Andreas Matthias, Marija Slavkovik, Janina Sombetzki, Alan F. T. Winfield, and Roman Yampolskiy. Towards moral autonomous systems. *CoRR*, abs/1703.04741, 2017.
[9] Nicolas Cointe, Grégory Bonnet, and Olivier Boissier. Ethical judgment of agents' behaviors in multi-agent systems. In *Proc. of the 15th Inter. Conf. on Autonomous Agents & Multiagent Systems*, pages 1106–1114. ACM, 2016.
[10] Boudewijn de Bruin and Luciano Floridi. The ethics of cloud computing. *Science and Engineering Ethics*, 23(1):21–39, 2017.
[11] Louise A. Dennis, Michael Fisher, Marija Slavkovik, and Matt Webster. Formal verification of ethical choices in autonomous systems. *Robotics and Autonomous Systems*, 77:1–14, 2016.
[12] Virginia Dignum. Responsible autonomy. In *Proc. of the 26th Inter. Joint Conf. on Artificial Intelligence*, pages 4698–4704. ijcai.org, 2017.
[13] Allan M. Feldman. welfare economics. In Steven N. Durlauf and Lawrence E. Blume, editors, *The New Palgrave Dictionary of Economics*. Palgrave Macmillan, Basingstoke, 2008.
[14] Guido Governatori, Michael J. Maher, Grigoris Antoniou, and David Billington. Argumentation semantics for defeasible logic. *J. Log. Comput.*, 14(5):675–702, 2004.
[15] John C. Harsanyi. Can the maximin principle serve as a basis for morality? a critique of john rawls's theory. *American Political Science Review*, 69(2):594–606, 1975.
[16] John C. Harsanyi. Morality and the theory of rational behavior. *Social Research*, 44:623–656, 1977.
[17] John C. Harsanyi. Rule utilitarianism and decision theory. *Erkenntnis*, 11(1):25–53, 1977.
[18] Troy Jollimore. Impartiality. In Edward N. Zalta, editor, *The Stanford Encyclopedia of Philosophy*. Metaphysics Research Lab, Stanford University, spring 2017 edition, 2017.
[19] I. Kant and M.J. Gregor. *Practical Philosophy*. Kant, Immanuel, 1724-1804. Works. Engl. 1992. Cambridge University Press, 1999.
[20] Ho-Pun Lam, Guido Governatori, and Régis Riveret. On ASPIC[+] and Defeasible Logic. In *Proc. of 6th Conf. on Computational Models of Argument*, volume 287 of *Frontiers in Artificial Intelligence and Applications*, pages 359–370. IOS Press, 2016.
[21] Richard Lindley. *Autonomy*. Atlantic Highlands, NJ: Humanities Press International, 1986.
[22] Emiliano Lorini. From self-regarding to other-regarding agents in strategic games: a logical analysis. *Journal of Applied Non-Classical Logics*, 21(3-4):443–475, 2011.
[23] Emiliano Lorini. A logic for reasoning about moral agents. *Logique & Analyse*, 58:177–218, 2016.
[24] Emiliano Lorini and Roland Mühlenbernd. The long-term benefits of following fairness norms under dynamics of learning and evolution. *Fundam. Inform.*, 158(1-3):121–148, 2018.
[25] Michel Meyer. *Principia Moralia*. Fayard, 2013.

[26] Brent Daniel Mittelstadt and Luciano Floridi. The ethics of big data: Current and foreseeable issues in biomedical contexts. *Science and Engineering Ethics*, 22(2):303–341, 2016.

[27] Luís Moniz Pereira, Tom Lenaerts, Luis A. Martinez-Vaquero, and The Anh Han. Social manifestation of guilt leads to stable cooperation in multi-agent systems. In Kate Larson, Michael Winikoff, Sanmay Das, and Edmund H. Durfee, editors, *Proceedings of the 16th Conference on Autonomous Agents and MultiAgent Systems, AAMAS 2017, São Paulo, Brazil, May 8-12, 2017*, pages 1422–1430. ACM, 2017.

[28] Luís Moniz Pereira and Ari Saptawijaya. *Programming Machine Ethics*, volume 26 of *Studies in Applied Philosophy, Epistemology and Rational Ethics*. Springer, 2016.

[29] Jeremy Pitt, Dídac Busquets, and Régis Riveret. Procedural justice and fitness for purpose of self-organising electronic institutions. In *Proc. of the 16th Inter. Conf. on Principles and Practice of Multi-Agent Systems*, pages 260–275. Springer, 2013.

[30] Jeremy Pitt, Dídac Busquets, and Régis Riveret. The pursuit of computational justice in open systems. *AI Soc.*, 30(3):359–378, 2015.

[31] Henry Prakken. An abstract framework for argumentation with structured arguments. *Argument & Computation*, 1(2):93–124, 2010.

[32] John Rawls. *A Theory of Justice*. London: Oxford University Press, 1971.

[33] Régis Riveret, Alexander Artikis, Jeremy V. Pitt, and Erivelton G. Nepomuceno. Self-governance by transfiguration: From learning to prescription changes. In *Proc. of the 8th IEEE Inter. Conf. on Self-Adaptive and Self-Organizing Systems*, pages 70–79. IEEE Computer Society, 2014.

[34] Stuart J. Russell. Provably beneficial artificial intelligence. *Exponential Life, The Next Step*, 2017.

[35] Ari Saptawijaya and Luís Moniz Pereira. Logic programming for modeling morality. *Logic Journal of the IGPL*, 24(4):510–525, 2016.

[36] J. B. Schneewind. *The Invention of Autonomy*. Cambridge University Press, 1998.

[37] Bart Verheij. Formalizing value-guided argumentation for ethical systems design. *Artif. Intell. Law*, 24(4):387–407, 2016.

[38] Abraham Wald. *Statistical Decision Functions*. Wiley, 1950.

[39] Toby Walsh, editor. *Artificial Intelligence and Ethics, Papers from the 2015 AAAI Workshop*, volume WS-15-02 of *AAAI Workshops*, 2015.

[40] Vincent Wiegel and Jan van den Berg. Combining moral theory, modal logic and mas to create well-behaving artificial agents. *I. J. Social Robotics*, 1(3):233–242, 2009.

[41] Michael Wooldridge and Nicholas R. Jennings. Agent theories, architectures, and languages: A survey. In *Proc. of the Workshop on Agent Theories, Architectures, and Languages on Intelligent Agents*, ECAI-94, pages 1–39. Springer, 1995.

Epistemic Oughts in STIT Semantics (Abbreviated Version)

John Horty
University of Maryland, USA

1 Introduction

This paper explores some of the ways in which agentive, deontic, and epistemic concepts intertwine to yield ought statements—or simply, *oughts*—of different characters. Consider an example. Suppose I place a coin on the table, either heads up or tails up though the coin is covered and you do not know which. And suppose you are then asked to bet whether the coin is heads up or tails up, with $10 to win if you bet correctly. If the coin is heads up but you bet tails, there is a sense in which we could naturally say that you ought to have made the other choice—at least, things would have turned out better for you if you had. But an ought statement like this does not involve any suggestion that you should be criticized for your actual choice. Nobody could blame you, in this situation, for betting incorrectly. By contrast, imagine that the coin is placed in such a way that you can see that it is heads up, and so know that it is heads up, but you bet tails anyway. Again we would say that you ought to have made the other choice, but this time it seems that you could legitimately be criticized for your choice.

These two scenarios have much in common: the coin is placed heads up but you bet tails, so that, in both cases, we could naturally say that you ought to have done otherwise. All that differs between the two scenarios is your knowledge—whether or not you know that the coin is heads up. Yet this difference is enough to influence the character of the resulting oughts, inviting criticism in one case but not the other.

The primary goal of the paper is to investigate agentive ought statements of the sort found in the second of these scenarios, where violation of the ought seems to invite criticism of the agent. Since an appeal to knowledge seems to play such an important role in the characterization of these statements, I refer to them as *epistemic oughts*. This investigation is not carried out in general, but in the particular setting of *stit semantics*, a logical framework for the analysis of agentive statements originating with a series of papers by Nuel Belnap, Michael Perloff, and Ming Xu, culminating in their [1].

Although the standard framework of stit semantics contains entities that can be regarded as action tokens—particular, concrete actions—it makes no appeal to general, repeatable kinds of actions, or action types. There is, for example, no action type of "betting tails." There are only particular instances of betting tails—by particular individuals at particular moments in particular games—with nothing to group them together as actions of the same kind. In recent work, Eric Pacuit and I [6] have argued that, in order to represent an epistemic sense of ability, it is helpful to enrich the standard framework of stit semantics with an explicit set of action types, in addition to the action tokens already present; the result is the new framework that we refer to as *labeled stit semantics*, where each action token is assigned a label, indicating the type of which it is a token. What the current paper shows is that an appeal to action types is likewise helpful in the analysis of epistemic oughts.

The paper is organized as follows: The next section summarizes the standard framework of stit semantics and then reviews the approach to agentive oughts in stit semantics that was set out earlier in [4], which relies on a preference ordering among the action tokens available to an individual. Section 3 explores the idea that epistemic oughts might by analyzed by combining this earlier approach with epistemic information in a particularly straightforward way, and points out problems with this initial proposal. This discussion motivates the introduction of action types in Section 4, which reviews the new framework of labeled stit semantics. Within this new framework, Section 5 suggests a new account of epistemic oughts that avoids the problems with the initial proposal; the account is similar in spirit to that of [4], but is based on an ordering of action types, rather than action tokens.

2 Background

2.1 Branching time

Stit semantics is cast against the background of a theory of indeterministic time, first set out by A. N. Prior [7] and developed in more detail by Richmond Thomason [10], according to which moments are ordered into a treelike structure, with forward branching representing the indeterminacy of the future and the absence of backward branching representing the determinacy of the past.

This picture leads to a notion of *branching time frames* as structures of the form $\langle Tree, < \rangle$, in which $Tree$ is a nonempty set of moments and $<$ is a strict partial ordering of these moments without backward branching: for any m, m', and m'' from $Tree$, if $m' < m$ and $m'' < m$, then either $m' = m''$ or $m'' < m'$ or $m' < m''$. A maximal set of linearly ordered moments from $Tree$ is a *history*, representing some complete temporal evolution of the world. If m is a moment and h is a history, then the statement that $m \in h$ can be taken to mean that m occurs at some point in the course of the history h, or that h passes through m. Because of indeterminism, a number of different histories might pass through a single moment. We let $H^m = \{h : m \in h\}$ represent the set of histories passing through m; and when h belongs to H^m, we speak of a moment/history pair of the form m/h as an *index*.

A *branching time model* is a structure that supplements a branching time frame with a *valuation function* v mapping each propositional constant from some background language into the set of m/h indices at which, intuitively, it is thought of as true. If we suppose that formulas are formed from truth functional connectives as well as the usual temporal operators P and F, representing past and future, the satisfaction relation $\models$ between indices and formulas true at those indices is defined as follows.

Definition 1 (Evaluation rules: basic operators). Where m/h is an index and v is the evaluation function from a branching time model $\mathcal{M}$,

- $\mathcal{M}, m/h \models A$ if and only if $m/h \in v(A)$, for A a propositional constant,
- $\mathcal{M}, m/h \models A \wedge B$ if and only if $\mathcal{M}, m/h \models A$ and $\mathcal{M}, m/h \models B$,
- $\mathcal{M}, m/h \models \neg A$ if and only if $\mathcal{M}, m/h \not\models A$,
- $\mathcal{M}, m/h \models \mathsf{P}A$ if and only if there is an $m' \in h$ such that $m' < m$ and $\mathcal{M}, m'/h \models A$,
- $\mathcal{M}, m/h \models \mathsf{F}A$ if and only if there is an $m' \in h$ such that $m < m'$ and $\mathcal{M}, m'/h \models A$.

In addition to the usual temporal operators, the framework of branching time allows us to define the concept of historical necessity, along with its dual concept of historical possibility: the formula $\Box A$ is taken to mean that A is historically necessary, while $\Diamond A$, defined as $\neg \Box \neg A$, means that A is still open as a possibility.

Definition 2 (Evaluation rule: $\Box A$). Where m/h is an index from a branching time model $\mathcal{M}$,

- $\mathcal{M}, m/h \models \Box A$ if and only if $\mathcal{M}, m/h' \models A$ for each history $h' \in H^m$.

The notion of historical necessity can be registered in the metalanguage by defining a formula A as *settled true* at a moment m from a model $\mathcal{M}$ just in case $\mathcal{M}, m/h \models A$ for each h from H^m; likewise A can be defined as *settled false* just in case $\mathcal{M}, m/h \models \neg A$ for each h from H^m. A formula is *moment determinant* just in case it is either settled true or settled false.

The set of possible worlds accessible at a moment m can be identified with the set H^m of histories passing through that moment. The propositions at m can thus be identified with sets of accessible histories, subsets of H^m. And the particular proposition expressed by a sentence A at a moment m in a model $\mathcal{M}$ can be identified with the set $|A|_{\mathcal{M}}^m = \{h \in H^m : \mathcal{M}, m/h \models A\}$ of histories from H^m in which that sentence is true. Here and elsewhere, we will omit reference to the background model when context allows, writing $|A|^m$, for example, to refer to the proposition expressed by A in some model that can be identified by the context, or in an arbitrary model.

2.2 Stit semantics

Within stit semantics, the idea that an agent α sees to it that A is taken to mean that the truth of A is guaranteed by an action performed by α. In order to capture this idea, we must be able to speak of individual agents and of their actions; and so the basic framework of branching time is supplemented with two additional primitives.

The first is simply a set *Agent* of agents, individuals thought of as acting in time. Now, what is it for one of these agents to act? Setting aside vagueness, probability, and many of the richer components of human action, stit semantics is based on the idea that acting, at a moment, is nothing more than constraining the course of future events to lie within some definite subset of the histories still available at that moment. These constraints are encoded through our second primitive: a function *Choice*, mapping each agent α and moment m to a partition $Choice_\alpha^m$ of the set H^m of histories through m.[1] The idea is that, by acting at m, the agent α selects a particular one of the equivalence classes from $Choice_\alpha^m$ within which the history to be realized must then lie, but that this is the extent of the agent's influence.

If K is a choice cell from $Choice_\alpha^m$, we speak of K as an action—or more precisely, an *action token*—available to the agent α at the moment m, and we say that α *performs* the action token K at the index m/h just in case h is a history belonging to K. We let $Choice_\alpha^m(h)$ (defined only when $h \in H^m$) stand for the particular equivalence class from $Choice_\alpha^m$ that contains the history h; $Choice_\alpha^m(h)$ thus represents the particular action token performed by the agent α at the index m/h.

With these new primitives, a *stit frame* can be defined as a structure of the form $\langle Tree, <, Agent, Choice \rangle$, supplementing a branching time frame with the additional components *Agent* and *Choice*, as specified above, and a *stit model* as a model based on a stit frame. Although stit frames and models can be very general, we simplify here in two ways. First, we suppose that, at any given moment, at most one agent faces a nontrivial choice. Second, we suppose that any choice facing an agent involves only finitely many options.

We can now introduce a standard stit operator—written [... *stit*: ...]—and allowing for statements of the form [α *stit*: A], with the intuitive meaning that the agent α sees to it that A. A statement of this form is defined as true at an index m/h just in case the action token performed by α at that index guarantees the truth of A. Formally, we can say that some action token K available to an agent at the moment m guarantees the truth of A just in case A holds at m/h for each history h from K—just in case, that is, $K \subseteq |A|^m$. Since the action token performed by α at the index m/h is $Choice_\alpha^m(h)$, our semantic analysis can be captured through the following evaluation rule.[2]

Definition 3 (Evaluation rule: [α *stit*: A]). Where α is an agent and m/h is an index from a stit model $\mathcal{M}$,

- $\mathcal{M}, m/h \models [\alpha\ stit\!:\ A]$ if and only if $Choice_\alpha^m(h) \subseteq |A|_\mathcal{M}^m$.

2.3 Agentive oughts

We begin by supplementing stit frames with a function *Value* mapping each history h into a numerical value $Value(h)$, representing its overall worth, or desirability. The resulting *deontic stit frames* can be defined as structures of the form $\langle Tree, <, Agent, Choice, Value \rangle$, and *deontic stit models* as models based on a deontic stit frame.

---

[1] The *Choice* function is subject to two technical constraints that are not discussed here.

[2] Those familiar with stit logics will recognize this particular operator as the "Chellas stit," first introduced by Horty and Belnap [5], but drawing on ideas from Chellas [2].

The idea explored in [4] was that an agentive deontic operator could be defined, not directly in terms of the ordering on histories provided by the *Value* function, but in terms of a dominance ordering on the action tokens available to an agent—where this later ordering is itself defined in terms of the value ordering on histories.

Definition 4 (Dominance orderings on action tokens; $\leq, <$). Let α be an agent and m a moment from a deontic stit frame, and let K and K' belong to $Choice_\alpha^m$. Then $K \leq K'$ (K' weakly dominates K) if and only if $Value(h) \leq Value(h')$ for each $h \in K$ and $h' \in K'$; and $K < K'$ (K' strongly dominates K) if and only if $K \leq K'$ and it is not the case that $K' \leq K$.

It is easy to see that both the weak and strong dominance orderings on action tokens are transitive, and that the strong ordering is, in addition, irreflexive.

With this dominance ordering in place, we can now represent what an agent ought to do. Syntactically, the idea is carried by an *agentive ought* operator $\odot[\ldots \mathit{stit} \colon \ldots]$, allowing for construction of statements of the form $\odot[\alpha \; \mathit{stit} \colon A]$, with the intuitive meaning that α ought to see to it that A.

In order to describe the semantics of statements like these, we first define the optimal action tokens available to an agent at a moment as those that are not strongly dominated by any others.

Definition 5 (Optimal action tokens; K-$Optimal_\alpha^m$). Where α is an agent and m a moment from a deontic stit frame, the optimal action tokens available to α at m are those belonging to the set

$$K\text{-}Optimal_\alpha^m = \{K \in Choice_\alpha^m : \text{there is no } K' \in Choice_\alpha^m \text{ such that } K < K'\}.$$

Because of our second simplifying assumption on stit models, that any choice involves only finitely many options—and because the strong dominance relation is transitive and irreflexive—the set of optimal action tokens available to an agent at a moment is guaranteed to be nonempty. The meaning of our agentive ought operator can therefore be defined very simply, through the stipulation that an agent ought to see to it that A just in case the truth of A is guaranteed by each optimal action tokens available to that agent.

Definition 6 (Evaluation rule: $\odot[\alpha \; \mathit{stit} \colon A]$). Where α is an agent and m/h is an index from a deontic stit model $\mathcal{M}$,

- $\mathcal{M}, m/h \models \odot[\alpha \; \mathit{stit} \colon A]$ if and only if $K \subseteq |A|_\mathcal{M}^m$ for each $K \in K\text{-}Optimal_\alpha^m$.

It is easy to see that this definition leads to a normal modal logic, with statements of the form $\odot[\alpha \; \mathit{stit} \colon A]$ moment determinant, and with the characteristic deontic principle that ought implies can holding in the form of the validity of $\odot[\alpha \; \mathit{stit} \colon A] \supset \Diamond[\alpha \; \mathit{stit} \colon A]$.

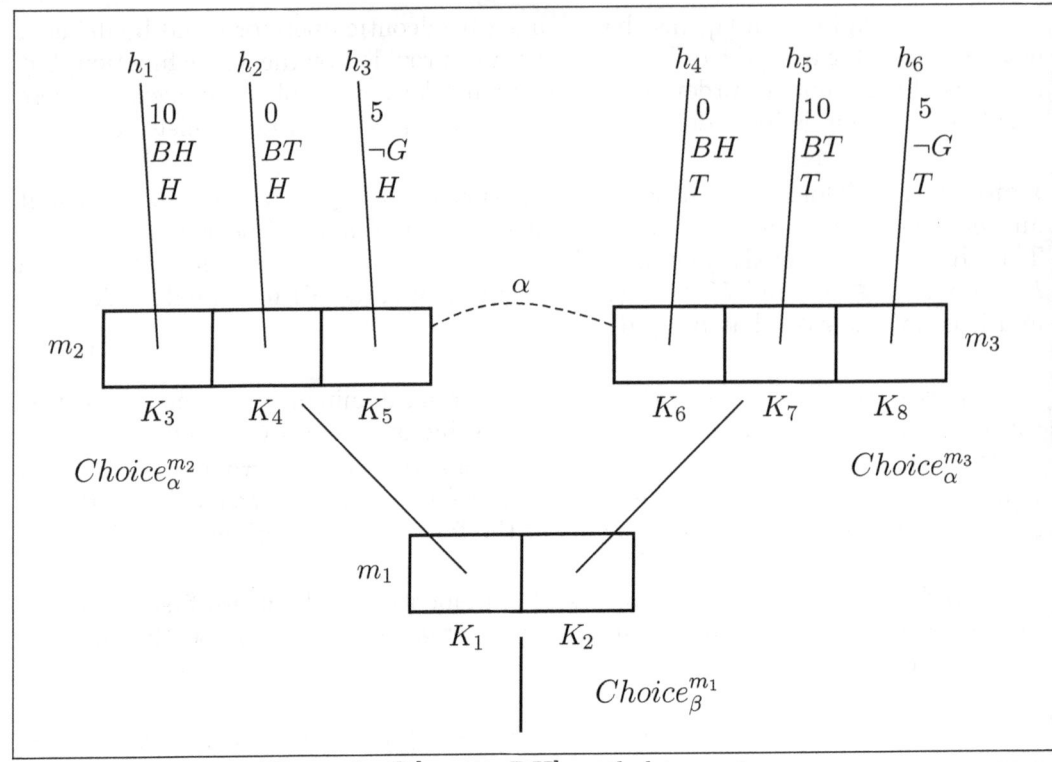

Figure 1: $\odot[\alpha \; stit: BH]$ settled true at m_2

3 Knowledge and oughts

3.1 An initial proposal

The logic just sketched, built around an ordering on action tokens, is useful in many ways. But it is less helpful in situations in which our evaluation of oughts is influenced by epistemic considerations.

To see this, we first incorporate epistemic information into the framework of stit semantics by adapting techniques that are, by now, standard in logic and game theory: we posit, for each agent α, an equivalence relation $\sim_\alpha$ among the moments from a stit frame, where $m \sim_\alpha m'$ is taken to mean that the moments m and m' are epistemically indistinguishable for α, or that nothing α knows allows her to distinguish m from m'.[3] Our previous deontic stit frames can now be supplemented with the additional component $\{\sim_\alpha\}_{\alpha \in Agent}$, a set containing indistinguishability relations for the various agents from $Agent$, leading to frames of the form $\langle Tree, <, Agent, Choice, Value, \{\sim_\alpha\}_{\alpha \in Agent}\rangle$, which are both deontic and *epistemic*; models can be defined, as usual, through the addition of a valuation function.

In this epistemic setting, let us now consider a situation very similar to the initial

---

[3] Although this idea of analyzing epistemic ideas by positing an indistinguishability relation is indeed standard in logic and game theory, it was not until Herzig and Troquard [3] that anyone even thought to explore the idea in the context of stit semantics.

example from this paper, where I first place a coin on the table in such a way that you cannot see whether it is heads up or tails up, and then you bet heads or tails. In the new situation, however, you are faced with a true gamble, not just an innocent bet: you must risk five dollars for the opportunity to bet whether the coin is heads up or tails up, with ten dollars to win if you bet correctly and your original five dollars to lose if you bet incorrectly; or you can choose not to gamble, preserving your original sum of five dollars.

This new situation is depicted in Figure 1. Here, α represents you, β represents me, and m_1 is the moment at which I place the coin on the table, either heads up by performing the action token K_1, or tails up by performing K_2. Next, you choose whether to guess heads or tails. This action occurs at one of the later moments m_2 or m_3 in the branching time structure, depending on my initial choice at m_1. If I have placed the coin heads up, your choice occurs at the moment m_2, where you can bet heads by performing K_3, tails by performing K_4, or refrain from gambling by performing K_5. If I have placed the coin tails up, then your choice occurs at m_3, where you can bet heads by performing K_6, tails by performing K_7, or refrain from gambling by performing K_8. Of course, since you do not know, at the time of your choice, whether I have placed the coin heads up or tails up, the moments m_2 and m_3 are indistinguishable for you. We thus have $m_2 \sim_\alpha m_3$, indicated by an α-arc these two moments in the diagram.[4] The histories h_1 and h_5, in which you bet correctly, have a value of 10, while h_2 and h_4, in which you bet incorrectly, have a value of 0; the histories h_3 and h_6, in which you refrain from gambling, have a value of 5.

Finally, the statements H and T stand for the respective propositions that I placed the coin heads up, or tails up; the first holds at the indices m_2/h_1, m_2/h_2, and m_2/h_3, while the second holds at m_3/h_4, m_3/h_5, and m_3/h_6. The statements BH and BT stand for the respective propositions that you bet heads or tails; the first holds at m_2/h_1 and m_3/h_4, while the second holds at m_2/h_2 and m_3/h_5. The statement G, equivalent to the disjunction $BH \vee BT$, stands for the proposition that you gamble; this statement is true at any index where either BH or BT is true, and false at the indices m_2/h_3 and m_3/h_6, where you refrain from gambling.

Now suppose what actually happens at m_1 is that I place the coin heads up, so that, at the time of your choice, you occupy the moment m_2. What ought you to do? According to the theory summarized in the previous section, the answer is unequivocal. Since K_3, the unique optimal action available to you at m_2, guarantees that you bet heads, that is what you ought to do: $K\text{-}Optimal_\alpha^{m_2} = \{K_3\}$ and $K_3 \subseteq |BH|^{m_2}$, so that $\odot[\alpha \; stit: BH]$ is settled true at m_2. And indeed, as we noted earlier, there does seem to be a sense in which it is right to say, in this situation, that you ought to bet heads, since betting heads will result in a value of 10, the greatest value available. But again, an ought statement like this is not an epistemic ought—there is no suggestion that you should be criticized if you violate the ought.

How, then, can we represent epistemic oughts, which seem to be sensitive to an agent's knowledge, and invite criticism when violated? One very natural reaction to this situation is that, although it may in fact be the case that you ought to

---

[4]Since indistinguishability is an equivalence relation, the actual indistinguishability relation at work in any particular stit frame is the closure of the relation explicitly depicted in the diagram of that frame under reflexivity, transitivity, and symmetry.

bet heads, the reason you would not be criticized for failing to bet heads is simply that you did not know this—you did not know that you ought to bet heads. This reaction suggests an initial proposal about the way in which epistemic and deontic concepts might interact to yield epistemic ought statements, whose violation invites criticism of the agent of the ought. According to this proposal, criticism is tied, not to violations of what an agent in fact ought to do, but only to violations of what an agent knows she ought to do.

In order to capture this proposal formally, we must be able to speak explicitly of what an agent knows, or does not know. We therefore introduce, for each agent α, an operator K_α representing that agent's knowledge, so that a statement of the form $\mathsf{K}_\alpha A$ means that α knows that A. This knowledge operator is defined here in a standard fashion, adapted only slightly to fit the framework of branching time, through the stipulation that an agent knows that A at an index m/h whenever A holds at every index m'/h' based on a moment m' that the agent cannot distinguish from m.

Definition 7 (Evaluation rule: $\mathsf{K}_\alpha A$). Where α is an agent and m/h an index from an epistemic stit model $\mathcal{M}$,

- $\mathcal{M}, m/h \models \mathsf{K}_\alpha A$ if and only if $\mathcal{M}, m'/h' \models A$ for all m'/h' such that $m' \sim_\alpha m$ and $h' \in H^{m'}$.

Once this knowledge operator has been introduced into the language, our initial proposal can be set out as follows: the ought statements that matter in terms of criticism are not statements of the form $\odot[\alpha\ \mathit{stit}\colon A]$, describing what the agent in fact ought to do, whether she knows it or not, but statements of the form $\mathsf{K}_\alpha \odot[\alpha\ \mathit{stit}\colon A]$, describing what the agent knows she ought to do.

This initial proposal provides us with a formal solution to the difficulty raised by our current example, from Figure 1. What we need to understand is why, when situated at m_2, you would not be criticized for failing to bet heads, even though you ought to bet heads—even though, that is, the statement $\odot[\alpha\ \mathit{stit}\colon BH]$ holds at m_2. And the answer provided by the initial proposal is that criticism is not warranted because, even though you ought to bet heads, you do not know that you ought to bet heads—the statement $\mathsf{K}_\alpha \odot[\alpha\ \mathit{stit}\colon BH]$ fails. We can see this by verifying that not every optimal action token available to you at every moment indistinguishable from m_2 guarantees that you bet heads. Since you do not know whether I have placed the coin heads up or tails up, the moments you cannot distinguish from m_2 are m_2 itself and m_3, with the optimal action tokens available to you at these moments calculated as: $K\text{-}Optimal_\alpha^{m_2} = \{K_3\}$ and $K\text{-}Optimal_\alpha^{m_3} = \{K_7\}$. And while your unique optimal action at m_2 guarantees that you bet heads, your unique optimal action at m_3 does not: while $K_3 \subseteq |BH|^{m_2}$, we do not have $K_7 \subseteq |BH|^{m_3}$.

3.2 Problems with the initial proposal

Our initial proposal—that criticism is tied to knowledge of oughts, rather than oughts themselves—has a good deal of intuitive appeal, and seems to offer a satisfying solution to the difficulty raised by our example. But the proposal fails, as we can see by considering three further problems.

To understand the first of these, we need only look a bit more closely at the situation depicted in Figure 1, supposing again that I have placed the coin heads up, so that you occupy m_2. As we have just seen, the optimal action tokens available to you at the moments indistinguishable from the moment you occupy are $K\text{-}Optimal_\alpha^{m_2} = \{K_3\}$ and $K\text{-}Optimal_\alpha^{m_3} = \{K_7\}$, with the result that the statement $\mathsf{K}_\alpha\odot[\alpha\ stit\colon BH]$ is settled false at m_2. You do not know that you ought to bet heads, since not all of these optimal action tokens guarantee that you bet heads. In the same way, we can see that the statement $\mathsf{K}_\alpha\odot[\alpha\ stit\colon BT]$ is settled false. You do not know that you ought to bet tails, since not all of these optimal action tokens guarantee that you bet tails: while $K_7 \subseteq |BT|^{m_3}$, we do not have $K_3 \subseteq |BT|^{m_2}$. But now, recall the statement letter G, equivalent to $BH \lor BT$, representing the proposition that you gamble. It turns out that the statement $\mathsf{K}_\alpha\odot[\alpha\ stit\colon G]$ is settled true at m_2. You do know that you ought to gamble, on the current analysis, since each of the optimal action tokens available to you at any moment indistinguishable from m_2 guarantees that you either bet heads or bet tails, and both betting heads and betting tails are ways of gambling: $K_3 \subseteq |G|^{m_2}$ and $K_7 \subseteq |G|^{m_3}$.

This is already bad enough, since it does not seem right to say, in this situation, that you know you ought to gamble—the fact that the statement $\mathsf{K}_\alpha\odot[\alpha\ stit\colon G]$ is true suggests that it does not even properly capture the intuitive idea that you know you ought to gamble. And things get even worse when we recall that, according to our initial proposal, criticism is tied to violation of the statement $\mathsf{K}_\alpha\odot[\alpha\ stit\colon G]$. Even though this statement is true in the current situation, it does not seem that you could be criticized if you choose not to gamble.

This second problem is illustrated in Figure 2, which depicts a situation nearly identical to that from Figure 1, with agents, action tokens, and statement letters interpreted in the same way, differing only in the value assigned to histories. In this case, however, the gamble you face is peculiar: although you must risk five dollars to gamble, all you stand to gain if you gamble and win is five dollars, your original stake. As a result, the histories h_2 and h_4, in which you bet incorrectly, have a value of 0, and the histories h_3 and h_6, in which you refrain from gambling, have a value of 5. But in this case, the histories h_1 and h_5, in which you bet correctly, also have a value of 5.

As before, suppose that I have placed the coin heads up, so that you occupy the moment m_2, though you do not know this, since you cannot distinguish m_2 from m_3. The optimal action tokens available to you at the moments you might occupy, as far as you know, can be calculated as: $K\text{-}Optimal_\alpha^{m_2} = \{K_3, K_5\}$ and $K\text{-}Optimal_\alpha^{m_3} = \{K_7, K_8\}$. And even though some of these optimal action tokens guarantee that you refrain from gambling, others do not: while $K_5 \subseteq |\neg G|^{m_2}$ and $K_8 \subseteq |\neg G|^{m_3}$, we have neither $K_3 \subseteq |\neg G|^{m_2}$ nor $K_7 \subseteq |\neg G|^{m_3}$. It therefore follows that the statement $\mathsf{K}_\alpha\odot[\alpha\ stit\colon \neg G]$ is settled false at m_2—you do not know that you ought not to gamble.

As before, this result is objectionable from the start, since it seems natural to conclude that you do know that you ought not to gamble, or at least that an ideal reasoner would know that. And once again, it becomes even more problematic when we consider that, according to the initial proposal, criticism is tied to a violation of the statement $\mathsf{K}_\alpha\odot[\alpha\ stit\colon \neg G]$. Since this statement is false, you would not

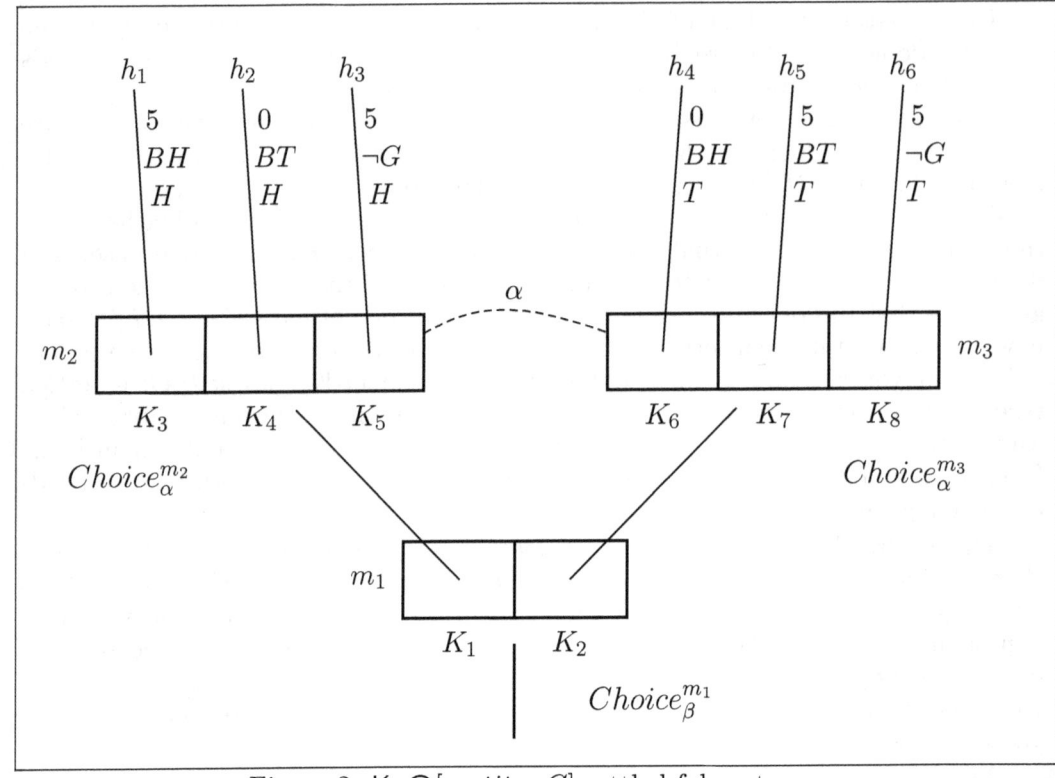

Figure 2: $K_\alpha \odot [\alpha \text{ stit}: \neg G]$ settled false at m_2

run afoul of what it requires by gambling, so that, according to the initial proposal, criticism would not be appropriate. But it does seem, in this situation, that criticism of gambling is appropriate.

To understand the third problem with the initial proposal, we return to the situation depicted in Figure 1, once more supposing that I have placed the coin heads up, so that you occupy m_2. Let us now take the new statement letter W, equivalent to the formula $(BH \wedge H) \vee (BT \wedge T)$, to represent the proposition that you win, or bet correctly; this statement is true at the indices m_2/h_1 and m_3/h_5 and nowhere else. As we have seen, the optimal action tokens available to you at the moments you cannot distinguish from m_2 are the members of $K\text{-}Optimal_\alpha^{m_2} = \{K_3\}$ and $K\text{-}Optimal_\alpha^{m_3} = \{K_7\}$, and both of these optimal actions guarantee that you win: $K_3 \subseteq |W|^{m_2}$ and $K_7 \subseteq |W|^{m_3}$. As a result, the statement $K_\alpha \odot [\alpha \text{ stit}: W]$ is settled true at m_2—you know that you ought to win.

The truth of this statement points to a different kind of problem for the initial proposal, according to which criticism is tied to violation of statements like this. Here, it is not so much that the results of the proposal are evidently incorrect. Perhaps you do know that you ought to win—perhaps this statement reflects some sort of conceptual truth about gambling. The problem in this case is that, all the same, it does not seem that you could legitimately be criticized for failing to win. Why not? Well, you can legitimately be criticized for failing to do something only if it is something you are able to do. This idea is often captured with the slogan that ought implies can, or in the presence of agency, that ought implies ability. But in the current example, winning simply does not seem to be something that lies within your abilities—it does not seem to be an outcome that you are able to guarantee.

4 Labeled stit semantics

4.1 Action types and ability

The argument just offered—that you cannot be criticized for failing to win because you are not able to win—may seem to be too quick. One might object that, at the moment m_2 in the situation from Figure 1, you are, in fact, able to win. You could perform the action token K_3, in which case you would win—indeed, the statement $\Diamond[\alpha$ stit: $W]$, which is taken in [4] to represent the proposition that you have the ability to win, is settled true at m_2. But this objection turns on an ambiguity. There is a sense, captured by the truth of $\Diamond[\alpha$ stit: $W]$, in which you do have the ability to win—Pacuit and I refer to this in [6] as the *causal* sense of ability. But since you do not know whether the coin is heads up or tails up, there is another sense of ability—the *epistemic* sense—in which you do not have this ability, and it is this latter sense that seems to be crucial for assessing the legitimacy of criticism.

In trying to understand this epistemic sense of ability, we run up against a limitation of the standard stit framework: its restriction to action tokens. From an intuitive standpoint, what you face in the situation from Figure 1 are three options: betting heads, betting tails, or refraining from gambling. These three options must be thought of as action types, rather than tokens, since their execution at different moments results in the performance of different action tokens.

The remainder of this section summarizes the approach developed in [6], extending the framework of stit semantics to include types as well as tokens, defining a new epistemic stit operator that draws on these action types, and also a new formula to capture the epistemic sense of ability.

4.2 The *kstit* operator

We begin by explicitly postulating a set $Type = \{\tau_1, \tau_2, \ldots, \tau_n\}$ of action types—general kinds of action, as opposed to the concrete action tokens already present in stit logics. We assume here, for simplicity only, that there are a finite number of action types and that all action types are primitive. In contrast to action tokens, action types are repeatable. A robot might execute the action type of raising its left arm four inches twice during the day, once at the lab in the morning and once at home in the evening, resulting in two concrete action tokens of the same type; a gambler might execute the action type of betting heads in two different games, or at two different points in the same game

Once action types have been introduced into stit semantics, it is most natural to assume that it is the execution of these action types, rather than the performance of concrete action tokens, that falls most directly within the agent's control. This point can be illustrated by returning to Figure 1. It is hard to see, in this situation, how you could actually choose to perform the action token K_3, for example, since that action token is not available to you unless you occupy the moment m_2, and you do not know whether you occupy m_2 or m_3. All you can do is execute the action type of betting heads, which will then result in the performance of the token K_3 if you are at m_2 and K_6 if you are at m_3.

Formally, the new action types introduced here are related to the action tokens already present in stit semantics through two functions. The first is a partial *execution function*—written, [ ]—mapping each action type τ into the particular action token $[\tau]_\alpha^m$ that results when τ is executed by the agent α at the moment m. Of course, the action token $[\tau]_\alpha^m$ must be one of those available to α at m—that is, we must have $[\tau]_\alpha^m \in Choice_\alpha^m$. The execution function is partial because it seems best to assume that not every action type is available for execution by every agent at every moment.

Just as the execution function maps the action type τ executed by an agent α at a moment m into a particular action token $[\tau]_\alpha^m$ from $Choice_\alpha^m$, we postulate, in addition, a one-one *label function*—written, $Label$—mapping each action token K from $Choice_\alpha^m$ into a particular action type $Label(K)$ from $Type$, where the label assigned to the action token K is the type of action under which this particular token falls. The interaction between the execution and label functions is governed by two *execution/label constraints*:

If $K \in Choice_\alpha^m$, then $[Label(K)]_\alpha^m = K$,

If $\tau \in Type$ and $[\tau]_\alpha^m$ is defined, then $Label([\tau]_\alpha^m) = \tau$.

The first of these requires that, if K is an action token available to α at m whose type is $Label(K)$, then the execution of that action type by α at m results in the performance of K itself; the second requires that, if τ is an action type whose execution by α at m results in the performance of the action token $[\tau]_\alpha^m$, then the type of that action token is τ itself.

Our previous definition of the action tokens available to an agent at a moment, as well as our definition of the particular action token performed by an agent at an index, can now be lifted from tokens to types in the natural way. Since $Choice_\alpha^m$ is the set of action tokens available to the agent α at the moment m, we can take $Type_\alpha^m = \{Label(K) : K \in Choice_\alpha^m\}$ as the set of action types available to α at m; and since $Choice_\alpha^m(h)$ is the particular action token performed by α at the index m/h, we can take $Type_\alpha^m(h) = Label(Choice_\alpha^m(h))$ as the action type executed by α at that index.

Putting these various ideas together, we can define a *labeled deontic stit frame* as a structure of the form $\langle Tree, <, Agent, Choice, Value, \{\sim_\alpha\}_{\alpha \in Agent}, Type, [\ ], Label\rangle$, containing all the components introduced earlier, as well as $Type$, [ ], and $Label$ as specified above; a *labeled deontic stit model* results when such a frame is supplemented with a valuation.

And we can then introduce a new epistemic stit operator—written, [... *kstit*: ...]—allowing for statements such as [α *kstit*: A]. As with our earlier stit statements, a statement of this new form can likewise be interpreted to mean that α sees to it that A, but in a different, epistemic sense. While the earlier [α *stit*: A] was

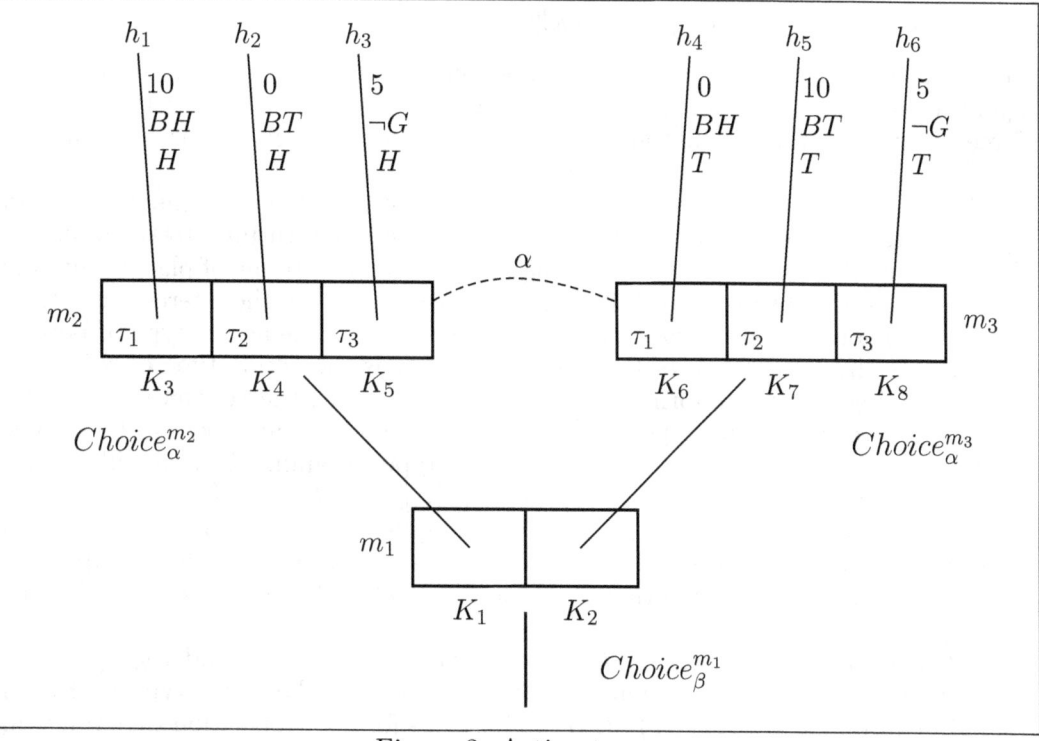

Figure 3: Action types

taken to mean that α performs an action token guaranteeing the truth of A, what $[\alpha \text{ kstit}: A]$ means, somewhat roughly, is that α executes an action type that she knows to guarantee the truth of A. More precisely, this statement will be defined as true at an index m/h just in case the action type executed by α at that index guarantees the truth of A at every moment m' that is indistinguishable for α from m. The action type executed by α at the index m/h is $Type_\alpha^m(h)$, as we have seen, and so the execution of this action type by α at another moment m' is $[Type_\alpha^m(h)]_\alpha^{m'}$. The evaluation rule for our new operator, therefore, is as follows.

Definition 8 (Evaluation rule: $[\alpha \text{ kstit}: A]$). Where α is an agent and m/h an index from a labeled deontic stit model $\mathcal{M}$,

- $\mathcal{M}, m/h \models [\alpha \text{ kstit}: A]$ if and only if $[Type_\alpha^m(h)]_\alpha^{m'} \subseteq |A|_\mathcal{M}^{m'}$ for all m' such that $m' \sim_\alpha m$.

This rule introduces a complication, which we can see by noting that it begins with an action type $Type_\alpha^m(h)$ executed by the agent α at the index m/h, and then considers the effects arising from an execution of that same action type by the same agent at a different moment m', where m and m' are linked only by being indistinguishable for the agent. In order for this procedure to make sense, and so for the evaluation rule to be well-defined, we need to ensure that the action type executed by the agent at m/h is available for execution also at m'. We therefore stipulate that labeled stit frames must satisfy the *type/indistinguishability* constraint

If $m' \sim_\alpha m$, then $Type_\alpha^{m'} = Type_\alpha^m$,

according to which the same action types must be available for execution by an agent at any two moments that are indistinguishable for that agent; the intuitive force of this constraint is that an agent must know which action types are available for execution.

The new *kstit* operator can be illustrated, and contrasted with the original *stit* operator, representing agency only in a causal sense, by returning to the situation from Figure 1. Ignoring the uninteresting actions available to me of placing the coin on the table either heads up or tails up, and considering only the interesting actions available to you, we take $Type = \{\tau_1, \tau_2, \tau_3\}$, where τ_1 is the action type of betting heads, τ_2 is the action type of betting tails, and τ_3 is the action type of refraining from gambling. As our informal description makes clear, the concrete actions K_3 and K_6 are tokens of the type betting heads, K_4 and K_7 are tokens of the type betting tails, and K_5 and K_8 are tokens of the type refraining from gambling. We therefore have $[\tau_1]_\alpha^{m_2} = K_3$ and $[\tau_1]_\alpha^{m_3} = K_6$, $[\tau_2]_\alpha^{m_2} = K_4$ and $[\tau_2]_\alpha^{m_3} = K_7$, and $[\tau_3]_\alpha^{m_2} = K_5$ and $[\tau_2]_\alpha^{m_3} = K_8$. This information appears in Figure 3, with the action types implicit in our informal description of the situation now displayed explicitly, in accord with the convention that the type of an action token is written inside the rectangle indicating that token.

Let us focus on the index m_2/h_1, where $Choice_\alpha^{m_2}(h_1)$ is K_3 and $Type_\alpha^{m_2}(h_1)$ is τ_1—you are performing the action token K_3 by executing the action type τ_1. Recall that W, equivalent to $(BH \wedge H) \vee (BT \wedge T)$, stands for the proposition that you win, and that G, equivalent to $BH \vee BT$, stands for the proposition that you gamble. Because $K_3 \subseteq |W|^{m_2}$, the statement $[\alpha \; stit: W]$ is true at this index—you see to it that you win in the causal sense captured by the ordinary *stit* operator. But the statement $[\alpha \; kstit: W]$ is false—you do not see to it that you win in the epistemic sense captured by the *kstit* operator, since there are moments indistinguishable for you from the one you occupy at which τ_1, the action type you execute, results in the performance of an action token that does guarantees the truth of W. In particular, you cannot distinguish m_3 from m_2, and $[\tau_1]_\alpha^{m_3} = K_6$, as we have seen, but we do not have $K_6 \subseteq |W|^{m_3}$. On the other hand, the statement $[\alpha \; kstit: G]$ is true at m_2/h_1—you do see to it that you gamble even in the epistemic sense, since the action type you execute at this index results, at each moment indistinguishable for you from m_2, in the performance of an action token that guarantees the truth of G.

5 Epistemic oughts

5.1 Ordering the action types

We now turn to our central topic: the definition of an epistemic ought operator based on a preference ordering of action types, rather than action tokens. The definition proceeds relative to the notion of an *information set bearing on an agent* α—or where clarity allows, simply an *information set*—defined as a nonempty set I of moments subject to the *type/information* constraint

If $m, m' \in I$, then $Type_\alpha^m = Type_\alpha^{m'}$.

A set of this kind can be thought of as representing the information that the agent α occupies some moment belonging to the set; the constraint tells us that the same action types are available for execution by the agent at each moment from the set. Of course, if an information set represents a body of information, the question naturally arises exactly whose information this is supposed to be. In evaluating oughts pertaining to an agent α at a moment m, we will concentrate in this section on information sets of the form $I_\alpha^m = \{m' : m \sim_\alpha m'\}$, representing the information available to the agent herself, at that moment. Our initial definitions, however, will be developed in terms of an arbitrary information set.

How, then, should we define a preference ordering on the action types available to an agent at the moments from an information set, on the basis of the information only that the agent occupies one of those moments? The definition proposed here lifts our previous ordering on action tokens to an ordering on action types, relative to an information set, by quantifying over the moments from that information set.

Definition 9 (Dominance orderings on action types; $\preceq_\alpha^I, \prec_\alpha^I$). Let α be an agent from a labeled deontic stit frame, I an information set bearing on α, and τ and τ' action types belonging to $Type_\alpha^m$ for each moment m from I. Then $\tau \preceq_\alpha^I \tau'$ (τ' weakly dominates τ, on the basis of I) if and only if $[\tau]_\alpha^m \leq [\tau']_\alpha^m$ for each moment m from I; and $\tau \prec_\alpha^I \tau'$ (τ' strongly dominates τ, on the basis of I) if and only if $\tau \preceq_\alpha^I \tau'$ and it is not the case that $\tau' \preceq_\alpha^I \tau$.

And in the same way, the current orderings on types inherit the properties of the earlier orderings on tokens: both the strong and weak orderings on types are transitive, and the strong ordering is irreflexive.

We can now, at last, introduce a new *epistemic ought* operator $\odot[\ldots \; kstit: \ldots]$, allowing for statements of the form $\odot[\alpha \; kstit: A]$. Just like our earlier agentive oughts, an epistemic ought statement of this form can also be taken to mean that the agent α ought to see to it that A, but now in an epistemic sense.

Following the route mapped out in our earlier treatment of agentive oughts, the semantics of epistemic oughts relies on the idea of optimality—but now of optimal action types, rather than tokens, where the action types that are optimal on the basis of an information set are those that are not strongly dominated, on the basis of that information set.

Definition 10 (Optimal action types; $T\text{-}Optimal_\alpha^I$). Where α is an agent from a labeled deontic stit frame, I is an information set bearing on α, and m is a moment from I,

$$T\text{-}Optimal_\alpha^I = \{\tau \in Type_\alpha^m : \text{there is no } \tau' \in Type_\alpha^m \text{ such that } \tau \prec_\alpha^I \tau'\}.$$

As with tokens, since the set of types is finite, and because the strong dominance relation is transitive and irreflexive, the set of optimal action types available to an agent must be nonempty.

An epistemic ought statement $\odot[\alpha \; kstit: A]$ can now be defined as holding at a moment from an information set whenever A is guaranteed, at each moment from that information set, by the execution of each action type that is optimal on the basis of that information set. But which information set? In evaluating an epistemic ought of this form, at a moment m, we focus on the information set I_α^m, representing the information available to the agent α herself at the very moment of evaluation.

Definition 11 (Evaluation rule: $\odot[\alpha\ kstit:\ A]$). Where α is an agent and m/h an index from a labeled deontic stit model $\mathcal{M}$,

- $\mathcal{M}, m/h \models \odot[\alpha\ kstit:\ A]$ if and only if $[\tau]_\alpha^{m'} \subseteq |A|_\mathcal{M}^{m'}$ for each $\tau \in T\text{-}Optimal_\alpha^{I_\alpha^m}$ and for each $m' \in I_\alpha^m$.

It is easy to verify that this epistemic ought is a normal modal operator, that epistemic ought statements are moment determinant, and that it satisfies the very strong deontic principle that, if an agent ought to see to it that A, then that agent has the ability to see to it that A even in the epistemic sense: the formula $\odot[\alpha\ kstit:\ A] \supset \Diamond[\alpha\ kstit:\ A]$ is valid.

5.2 Exploring the epistemic ought

The epistemic ought just introduced is proposed as an operator that combines agentive, deontic, and epistemic ideas in the right way, through a formula of the form $\odot[\alpha\ kstit:\ A]$, to yield an agentive ought statement whose violation invites criticism of the agent. Earlier, we considered the proposal that this idea could be captured simply by combining the agent's knowledge with an ordinary agentive ought, through a formula of the form $\mathsf{K}_\alpha\odot[\alpha\ stit:\ A]$. This initial proposal was rejected on the grounds that it offered problematic predictions in three representative cases. We now return to these three cases to confirm that the current suggestion yields better results.

The first problem was based on the example from Figure 1, later reproduced with action types rendered explicit in Figure 3. As we saw, the problem presented by this example for the initial proposal was that, supposing I place the coin heads up, so that you occupy m_2, the proposal predicts that you know you ought to gamble: the statement $\mathsf{K}_\alpha\odot[\alpha\ stit:\ G]$ is settled true at m_2. Yet this does not seem like the right result—it does not seem like you know you ought to gamble, or that you could reasonably be criticized for failing to gamble.

The current suggestion avoids this first problem, since it does not predict that you ought to gamble, at least in the epistemic sense: the statement $\odot[\alpha\ kstit:\ G]$ is settled false at m_2. This statement holds just in case, at each member of your information set $I_\alpha^{m_2} = \{m_2, m_3\}$, the execution of each action type that is optimal on the basis of this information guarantees that you gamble. The set of action types that are optimal on the basis of this information is the entire set $T\text{-}Optimal_\alpha^{I_\alpha^{m_2}} = \{\tau_1, \tau_2, \tau_3\}$; each of these action types is optimal because none is even weakly dominated by another. But it is not the case that the execution of each of these optimal action types guarantees that you gamble at each moment from your information set. In particular, the execution of τ_3, the action type of refraining from gambling, at either m_2 or m_3 does not guarantee that you gamble: we have neither $[\tau_3]_\alpha^{m_2} \subseteq |G|^{m_2}$ nor $[\tau_3]_\alpha^{m_3} \subseteq |G|^{m_3}$.

The second problem for the initial proposal was based on the example from Figure 2, exactly like that depicted in Figures 1 and 3 except that the histories carry different values: this time, since the gamble is peculiar, the histories that result from betting correctly carry no more value than the histories that result from refraining from the gamble. Although action types are not represented explicitly in Figure 2, we can assume that τ_1, τ_2, and τ_3 again represent the action types of betting heads, betting tails, and refraining, and also that, as in Figure 3, the execution of these types result in the respective action tokens K_3, K_4, or K_5 at m_2, and K_6, K_7, or K_8 at m_3.

The problem presented by this example for the initial proposal was that, supposing again that you occupy m_2, the proposal fails to predict that you know you ought not to gamble: the statement $\mathsf{K}_\alpha \odot [\alpha \text{ stit}: \neg G]$ is settled false at m_2. But contrary to this prediction, it does seem that, in light of your information, you ought not to gamble, that you know this or at least that an ideal reasoner would know it, and that you could reasonably be criticized for gambling.

The current suggestion avoids this second problem by correctly predicting that you ought not to gamble: the statement $\odot[\alpha \text{ kstit}: \neg G]$ is settled true at m_2. On the basis of your information $I_\alpha^{m_2} = \{m_2, m_3\}$, the unique member of the set $T\text{-}Optimal_\alpha^{I_\alpha^{m_2}} = \{\tau_3\}$ is your only available optimal action type, since it strongly dominates each of the others—the execution of τ_3 always yields an action token that weakly dominates the execution of τ_1 or τ_2, and for each, there is some moment in your information set at which the execution of τ_3 strongly dominates. And the execution of this optimal action type guarantees that you refrain from gambling at each moment from your information set: we have both $[\tau_3]_\alpha^{m_2} \subseteq |\neg G|^{m_2}$ and $[\tau_3]_\alpha^{m_3} \subseteq |\neg G|^{m_3}$.

The third problem for the initial proposal was based, once again, on the example from Figures 1 and 3, with the statement letter W, equivalent to the formula $(BH \wedge H) \vee (BT \wedge T)$, now representing the proposition that you win, or bet correctly. In this case, the problem posed for the initial proposal is that, supposing that you occupy m_2, the proposal predicts that you know you ought to win: $\mathsf{K}_\alpha \odot [\alpha \text{ stit}: W]$ is settled true at m_2. As we noted, you may indeed know that you ought to win, but not in a sense in which failure to win would invite criticism. You could not legitimately be criticized for failing to win, because winning is not something you are, in the epistemic sense, able to do.

The current suggestion avoids this third problem as well, since it does not predict that you ought to win, at least in the epistemic sense in which failure to win would invite criticism: the statement $\odot [\alpha \text{ kstit}: W]$ is settled false at m_2. This statement does not hold since the set of optimal action types available to you on the basis of your information set $I_\alpha^{m_2} = \{m_2, m_3\}$ is again the entire set $T\text{-}Optimal_\alpha^{I_\alpha^{m_2}} = \{\tau_1, \tau_2, \tau_3\}$, and it is not the case that the execution of each of these optimal action types guarantees winning at each moment from your information set.

Having verified that the epistemic ought operator resolves the problems posed for the initial proposal, it is worth exploring the relations between this new operator and the agentive ought defined earlier. Even though the epistemic *kstit* operator is strictly stronger than the familiar causal *stit*, it turns out that the epistemic ought operator is neither stronger nor weaker than the familiar agentive ought: neither $\odot[\alpha \text{ kstit}: A] \supset \odot[\alpha \text{ stit}: A]$ nor $\odot[\alpha \text{ stit}: A] \supset \odot[\alpha \text{ kstit}: A]$ is valid. A countermodel to the first formula is provided by the example from Figure 2, where, as we have seen, $T\text{-}Optimal_\alpha^{I_\alpha^{m_2}} = \{\tau_3\}$, so that $\odot[\alpha\ kstit\colon \neg G]$ is settled true at m_2, since $[\tau_3]_\alpha^{m_2} \subseteq |\neg G|^{m_2}$, but $K\text{-}Optimal_\alpha^{m_2} = \{K_3, K_5\}$, so that $\odot[\alpha\ stit\colon \neg G]$ is settled false, since it is not the case that $K_3 \subseteq |\neg G|^{m_2}$. What this example shows is that action tokens can be optimal even if they do not result from the execution of optimal action types: here, K_3 is an optimal action token even though it results from the execution of the non-optimal action type τ_1. A countermodel to the second formula is provided by the example from Figures 1 and 3, where $K\text{-}Optimal_\alpha^{m_2} = \{K_3\}$, so that $\odot[\alpha\ stit\colon BH]$ is settled true at m_2, since $K_3 \subseteq |BH|^{m_2}$, but $T\text{-}Optimal_\alpha^{I_\alpha^{m_2}} = \{\tau_1, \tau_2, \tau_3\}$, so that $\odot[\alpha\ kstit\colon BH]$ is settled false, since it is not the case that $[\tau_2]_\alpha^{m_2} \subseteq |BH|^{m_2}$, for example. What this example shows is that an action type can be optimal on the basis of an agent's information even if its execution at some moment consistent with that information does not result in an optimal action token: here, τ_2 is an optimal action type even though the action token K_2 resulting from its execution at m_1 is not optimal.

Although there are not, then, any general connections between the epistemic ought operator defined here and the ordinary agentive ought, the two operators are equivalent in models satisfying the *perfect information* constraint, mentioned earlier, which tells us that an agent always knows which moment she occupies. In this case, the information set for an agent α occupying the moment m is simply $I_\alpha^m = \{m\}$, from which we can conclude that $K\text{-}Optimal_\alpha^m = \{[\tau]_\alpha^m : \tau \in T\text{-}Optimal_\alpha^{I_\alpha^m}\}$, or that the optimal action tokens are exactly those resulting from the execution of optimal action types.[5] Given this identity, it follows at once that the two oughts coincide: the formula $\odot[\alpha\ kstit\colon A] \equiv \odot[\alpha\ stit\colon A]$ is valid.

There is one further logical point worth noting: the formula $\odot[\alpha\ kstit\colon A] \supset K_\alpha \odot[\alpha\ kstit\colon A]$ is valid, so that it follows, from the fact that an agent ought to do something in the epistemic sense, that she knows she ought to do it. The current suggestion can therefore be seen as respecting the intuition underlying our initial proposal—that we can be criticized for failing to do what we ought to do only if we know we ought to do it.

6 Conclusion

This paper proposes one way in which agentive, deontic, and epistemic concepts might interact to yield ought statements whose violations seem to invite criticism. The account is similar to that presented in earlier work, but based on an ordering of action types rather than tokens.

A longer version of the paper considers generalizations both to assessment sensitive, or relativistic, oughts, and to conditional oughts. And much future work remains. The most pressing matter, to my mind, involves relaxing the current simplifying assumption that, at any moment, at most one agent faces a non-trivial choice. Not only would this allow us to analyze individual epistemic oughts in a richer setting, it would also allow us to epistemic oughts involving groups of agents

---

[5]Or equivalently, looked at from the other side, we have $T\text{-}Optimal_\alpha^{I_\alpha^m} = \{Label(K) : K \in K\text{-}Optimal_\alpha^m\}$, so that the optimal action types are exactly those that are labels of optimal action tokens.

and the relation between epistemic oughts bearing on groups and those bearing on the individual agents belonging to those groups. In a standard stit framework, without epistemic information, the relation between group and individual oughts was addressed in a preliminary way in [4], and has recently received renewed attention.[6] Moving the issue to an epistemic setting could be very rewarding.

References

[1] Nuel Belnap, Michael Perloff, and Ming Xu. *Facing the Future: Agents and Choices in Our Indeterministic World*. Oxford University Press, 2001.

[2] Brian Chellas. *The Logical Form of Imperatives*. PhD thesis, Philosophy Department, Stanford University, 1969.

[3] Andreas Herzig and Nicolas Troquard. Knowing how to play: uniform choices in logics of agency. In *Proceedings of the Fifth International Joint Conference on Autonomous Agents and Multi-agent Systems (AAMAS-06)*, pages 209–216. The Association for Computing Machinery Press, 2006.

[4] John Horty. *Agency and Deontic Logic*. Oxford University Press, 2001.

[5] John Horty and Nuel Belnap. The deliberative stit: a study of action, omission, ability, and obligation. *Journal of Philosophical Logic*, 24:583–644, 1995.

[6] John Horty and Eric Pacuit. Action types in stit semantics. *Review of Symbolic Logic*, 10:17–37, 2017.

[7] Arthur Prior. *Past, Present, and Future*. Oxford University Press, 1967.

[8] Frederik Van De Putte. Choosing the right concept of right choice. Unpublished manuscript, 2018.

[9] Allard Tamminga and Hein Duijf. Collective obligations, group plans, and individual actions. *Economics and Philosophy*, 33:187–214, 2017.

[10] Richmond Thomason. Indeterminist time and truth-value gaps. *Theoria*, 36:264–281, 1970.

---

[6]See, for example, Tamminga and Duijf [9] and Van De Putte [8].

S5 as a Deontic Logic

Fengkui Ju
School of Philosophy, Beijing Normal University, Beijing
fengkui.ju@bnu.edu.cn

This paper presents a simple dynamic deontic logic, SimDDL, to formalize conditional obligations and the way new prohibitions and permissions change our obligations. SimDDL presupposes finitely many good, bad and neutral colors and a betterness relation among them. It has two parts. The static part of it is a variant of the modal logic S5, called $S5_c$. $S5_c$ has a universal modality and a number of propositional constants denoting colors. Two types of obligations can be defined in $S5_c$: ideal and practical ones. The dynamic part of SimDDL consists of three dynamic operators: the first one represents "*if* ϕ" as a restrictor; the second indicates the action of issuing new prohibitions; the third denotes the action of offering new permissions. A couple of deontic puzzles can be solved in SimDDL. It is shown that the three dynamic operators do not contribute any expressive power to SimDDL. So the completeness of SimDDL can be reduced to that of $S5_c$.

1 Introduction

Problems Smith lives in a community of a town. The town have a rule: *all the fences must be white*. The community also have a rule: *there must be no fence*. In this situation, *Smith ought to have no fence*, but *if he is going to have a fence, it ought to be white*. A vicious dog often comes to Smith's yard and scares his daughter. To avoid this happening again, Smith decides to have a fence. In this case, *Smith ought to have a white fence*. As white fences are fences, *Smith ought to have a fence*.

This is strange, as the prohibition to have a fence and the obligation to have a fence do not seem to coexist. This puzzle is from [17] and is an ought-to-be version of the Paradox of Gentle Murder presented in [5]. We call it *the problem of white fence* in this paper.

Due to much complaint, the town issue a new regulation: *there may be black or white fences*. Now *Smith may have a fence*. How do new permissions change our

---

Thanks go to Rou Gao, Fenrong Liu, Yanjing Wang, the audience of the 5th PIOTR project meeting in The John Paul II Catholic University of Lublin, and the anonymous referees, for their useful comments and suggestions. The research has been supported by the National Social Science Foundation of China (No. 12CZX053) and the National Science Centre, Poland (Project: Permissions, Information and Institutional Dynamics, Obligations, and Rights, UMO-2014/15/G/HS1/04514).

obligations? Here is a specific problem. It seems that the previous regulation implies *there may be black fences*. By classical logics, this inference is not valid. This is the so-called Free Choice Permission Paradox identified by [19].

To avoid dogs scare people again, the community introduce a new rule: *there must be no dog*. Jones, Smith's neighbor, has a dog. Now *he may not have a dog any more*. How do new prohibitions change our obligations?

To avoid being lonely, Jones decides to keep his dog. It is a social norm that *if there is a dog, there ought to be a warning sign*, and *if there is no dog, there ought to be no sign*. It is not a norm that *if there is no dog, there ought to be a sign*. Then *Jones ought to put a sign near his house*. How to express Jones's present obligations is a problem.

Let d denote "*there is a dog*" and s "*there is a sign*". In classical logics, there are two natural and uniform ways to express Jones's duties: (a) $O\neg d$, $O(d \to s)$, $O(\neg d \to \neg s)$ and $\neg O(\neg d \to s)$; (b) $O\neg d$, $d \to Os$, $\neg d \to O\neg s$ and $\neg(\neg d \to Os)$.

Assume the former. From d, that is, *Jones is going to keep his dog*, Od does not follow. Assume the latter. Then a few problems arise. Firstly, $\neg(\neg d \to Os)$ implies $\neg d$; this is strange. Secondly, $\neg(\neg d \to Os)$ implies $\neg Os$; given d, Os follows; then we have a contradiction. Thirdly, given d, $\neg d \to Os$ follows; then we have another contradiction.

This example is also from [17] and is a variant of Chisholm's Paradox presented in [4]. We call it *the problem of warning sign* in this paper.

Views The problem of white fence and the problem of warning sign are essentially the same problem, that is, how to correctly express conditional obligations "*if ϕ then it is obligatory that ψ*".

There have been a lot of formalizations of conditional obligations in the literature. See [3] and [15] for brief surveys. Generally speaking, these formalizations respectively adopt two approaches. Works following the first approach view conditional obligations as a special type of obligations, contrasting with regular ones; they introduce dyadic deontic operators $O(\psi|\phi)$ to express conditional obligations. Works following the second approach think that a proper theory of conditional obligations should be the product of a theory of obligations and a theory of conditionals; they use non-material conditionals and unary deontic operators to express conditional obligations.

We prefer the second approach. The reason is as follows. Besides simple conditional obligations "*if ϕ, then it is obligatory that ψ*", there are complex ones such as "*if ϕ, then it is obligatory that ψ or it is obligatory that χ*". In order to handle complex obligations, we have to introduce more types of obligations if we follow the first approach. Then we would have the problem of type explosion. The second approach does not have this problem.

Here are our views on obligations. Our world have different possible futures.

Some of them are good and others are bad. Among good futures, some are relatively better than others. This is also the case for bad futures. We *should* realize good futures but not bad ones. Of course, we *can* freely choose which good future to realize. In some situations, there is no good future. Then we *should* realize an optimal one. The idea of introducing priorities to normative contexts can be found in various works such as [7] and [21].

In the scenario of white fence, the futures where Smith has no fence are good and others are bad. Making decisions excludes possible futures. When Smith decides to have a fence, those futures where he has no fence disappear. Then there is no good future any more and the optimal ones are those where Smith has a white fence. In the example of warning sign, the futures where there is no dog and sign are good and others are bad. After Jones decides to keep his dog, only bad futures are left and the optimal ones are those where there is a sign.

There are two types of obligations: *ideal* and *practical* ones[1]. The former concerns good futures and the latter concerns optimal ones. In the scenario of white fence, before Smith decides to have a fence, having no fence is both ideally and practically obligatory for him. After he makes the decision, having no fence is still ideally obligatory, in a trivial way, but not practically obligatory any more. Now it is practically obligatory to have a white fence. So conflicting ideal and practical obligations can coexist.

In the example of warning sign, before Jones decides to keep his dog, having no dog and putting no sign are both ideal and practical obligations. When Jones makes the decision, neither of them is still practically obligatory. Now putting a sign is practically obligatory.

Ideal and practical obligations are identical in normal situations where some future is good, as good and optimal futures are identical in normal situations. They differ in abnormal situations where no future is good. Ideal obligations do not really work in abnormal situations in the following sense: everything is ideally obligatory in abnormal situations. Practical obligations always work.

Concerning conditionals, we adopt the so-called *restrictor view*, which is proposed in many works such as [14] and [13] and can be traced back to Ramsey Test [18]. By this view, the conditional "*if ϕ then ψ*" is not a connective relating two sentences. Instead, the if-clause "*if ϕ*" is a device for restricting discourse domains, which are usually classes of possibilities. "*if ϕ then ψ*" is true with respect to a domain iff ψ is true with respect to the restricted domain by "*if ϕ*".

The discourse domain of the scenario of white fence consists of possible futures. The conditional "*if Smith is going to have a fence, it ought to be white*" means that Smith ought to have a white fence with respect to the class of possible futures where he has a fence.

The discourse domain of the example of warning sign is also a class of possible

---

[1]The two terms are borrowed from [8].

futures. The conditionals *"if there is a/no dog, there ought to be a/no warning sign"* mean that there ought to be a/no sign with respect to the class of possible futures where there is a/no dog.

We consider issuing permissions/prohibitions as actions which tend to make possible futures good/bad. Every issued permission/prohibition has a *weight*, which determines how good/bad it would make possible futures.

The Free Choice Permission Paradox concerns the implication relation between issuing the permission to have black or white fences and issuing the permission to have black fences. This relation can be defined as follows: an action implies another if the effects of performing the former subsume the effects of performing the latter. Under this definition, offering the permission to have black or white fences implies offering the permission to have black fences. The idea of defining the implication relation between actions in terms of the effects of performing them can be found in [8] and [23].

Contribution of the paper In this paper, we present a logic SimDDL to make the preceding ideas precise. SimDDL presupposes that there are finitely many good colors, the same number of bad colors and a neutral color. There is a preference order over good colors and also a preference order over bad colors.

The language of SimDDL has a number of propositional constants, indicating colors, a universal modality, and three dynamic operators, respectively denoting if-clauses, issuing permissions and issuing prohibitions. Each model of SimDDL contains a class of states, representing possible futures, and a coloring function.

By goodness and badness of colors and coloring functions, we can identify good and bad futures. By the preference orders over good and bad colors, we can prioritize good and bad futures. Then with the universal modality, we can define ideal and practical obligations.

The dynamic operator denoting if-clauses exclude futures. The one indicating giving permissions lets futures have good colors. The one indicating giving prohibitions make futures have bad colors. As a result, the problems mentioned above get solved.

We show that the three dynamic operators do not contribute any expressive power to SimDDL. So the completeness of SimDDL can be reduced to the fragment of it containing no dynamic operator, which is just the modal logic S5 plus a simple axiom governing those propositional constants indicating colors.

Structure of the paper In Section 2 we define the language and semantics of SimDDL. In Section 3 we define ideal and practical obligations in a derivative way and show that the problems stated previously are solved. The completeness of SimDDL is proved in Section 4. In Section 5 we point out some connections with other works. The paper is concluded by Section 6.

2 The logic SimDDL

Colors Let k be a positive integer and $C = \{c_{-k}, \ldots, c_0, \ldots, c_k\}$ be a set of colors. Let $\leq$ be an order on C such that $c_i \leq c_j$ iff $i \leq j$. $c_i \leq c_j$ means that c_j is at least as good as c_i. c_i is a *bad* color if $i < 0$, a *neutral* one if $i = 0$, and a *good* one if $i > 0$. c_i is a *fine* color if it is good or neutral. $|i|$ is called the *weight* of c_i. $\leq$ is called a preference order and $(C, \leq)$ a preference structure.

Language Let Φ_0 be a countable set of atomic propositions and p range over it. Let i range over $\{-k, \ldots, 0, \ldots, k\}$ and n over $\{0, \ldots, k\}$. Define a language Φ_{SimDDL} as follows:

$$\phi ::= p \mid \top \mid \mathfrak{c}_i \mid \neg\phi \mid (\phi \wedge \phi) \mid \mathbf{A}\phi \mid [!\psi]\phi \mid [\mathcal{I}_n\psi]\phi \mid [\heartsuit_n\psi]\phi$$

where ψ is a formula in Φ_{PC}, the language of propositional calculus, which is defined in the following way:

$$\psi ::= p \mid \top \mid \neg\psi \mid (\psi \wedge \psi)$$

Due to technical reason, we define a sub-language Φ_{S5_c} of Φ_{SimDDL}:

$$\phi ::= p \mid \top \mid \mathfrak{c}_i \mid \neg\phi \mid (\phi \wedge \phi) \mid \mathbf{A}\phi$$

The featured formulas of Φ_{SimDDL} are read as follows:

1. $\mathfrak{c}_i$: this state has the color c_i.
2. $\mathbf{A}\phi$: ϕ is the case everywhere.
3. $[!\psi]\phi$: given ψ, ϕ is the case.
4. $[\mathcal{I}_n\psi]\phi$: after issuing the prohibition to make ψ true with the weight n, ϕ is the case.
5. $[\heartsuit_n\psi]\phi$: after issuing the permission to make ψ true with the weight n, ϕ is the case.

$\mathbf{A}$ is the universal modality. $\mathbf{E}\phi$, the dual of $\mathbf{A}\phi$, indicates that ϕ is the case somewhere. The other usual propositional connectives and the falsum $\bot$ are defined in the usual way.

Assume that the agent's decision of making ψ true will always make ψ true[2]. Then $!\psi$ can be understood as the action of deciding to make ψ true. So $[!\psi]\phi$ can also be read as that ϕ is the case after the agent decides to make ψ true.

In an intuitive sense, weight of prohibitions/permissions is determined by many factors, for instance, the violation/interference consequence, the authority of the givers, and so on.

---

[2] In this paper, we always maintain this assumption.

Models Note that C is a set of colors. A tuple $\mathfrak{M} = (W, \sigma, V)$ is a model of Φ_{SimDDL} if W is a nonempty set of states, σ is a function from W to C, and V is a function from Φ_0 to 2^W. σ is called the coloring function. For any state w, $\sigma(w)$ is the color of w. w is a good/neutral/bad/fine state if it has a good/neutral/bad/fine color. w is *optimal* if it has the best color among the members of W. w is *acceptable* if it is fine or optimal. Note that optimal states might not be fine, as there might be no fine state. A model is normal if it has a fine state, otherwise abnormal.

Every state of a model represents a possible future of the present world, which is implicit. The agent can realize different futures by performing different actions. He should realize fine futures if there is any, otherwise he should realize optimal futures. Figure 1 indicates a model.

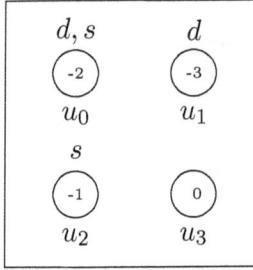

Figure 1: This figure depicts the model for the scenario of warning sign mentioned in Section 1. Here d denotes "*there is a dog*" and s "*there is a warning sign*". u_0 has the color c_{-2}, and so on. Jones can control the truth value of d and s and realize any of the four futures. However, he is only allowed to realize u_3.

Semantics Let $\mathfrak{M} = (W, \sigma, V)$ be a model and ψ a formula in Φ_{PC}. Define (i) $\mathfrak{M}, w \Vdash \phi$, ϕ *being true at w in $\mathfrak{M}$*, (ii) $\mathfrak{M}^{!\psi}$, *the update of $\mathfrak{M}$ with the decision to make ψ true*, (iii) $\mathfrak{M}^{\sharp n \psi}$, *the update of $\mathfrak{M}$ with the prohibition to make ψ true with the weight n*, and (iv) $\mathfrak{M}^{\heartsuit n \psi}$, *the update of $\mathfrak{M}$ with the permission to make ψ true with the weight n*, as follows by mutual induction:

Definition 1.

$$\mathfrak{M}, w \Vdash p \Leftrightarrow w \in V(p)$$
$$\mathfrak{M}, w \Vdash \top$$
$$\mathfrak{M}, w \Vdash \mathbf{c}_i \Leftrightarrow \sigma(w) = c_i$$
$$\mathfrak{M}, w \Vdash \neg \phi \Leftrightarrow \text{not } \mathfrak{M}, w \Vdash \phi$$
$$\mathfrak{M}, w \Vdash \phi \wedge \psi \Leftrightarrow \mathfrak{M}, w \Vdash \phi \text{ and } \mathfrak{M}, w \Vdash \psi$$
$$\mathfrak{M}, w \Vdash \mathbf{A}\phi \Leftrightarrow \text{for any } u \text{ in } W, \mathfrak{M}, u \Vdash \phi$$

$\mathfrak{M}, w \Vdash [!\psi]\phi \Leftrightarrow \mathfrak{M}^{!\psi}, w \Vdash \phi$ if $\mathfrak{M}^{!\psi}$ is defined and w is in it

$\mathfrak{M}, w \Vdash [\maltese_n \psi]\psi \Leftrightarrow \mathfrak{M}^{\maltese_n \psi}, w \Vdash \phi$

$\mathfrak{M}, w \Vdash [\heartsuit_n \psi]\psi \Leftrightarrow \mathfrak{M}^{\heartsuit_n \psi}, w \Vdash \phi$

Definition 2. *Suppose $\mathfrak{M}, x \Vdash \psi$ for some x in W. Let $W^{!\psi} = \{x \in W \mid \mathfrak{M}, x \Vdash \psi\}$. Define $\mathfrak{M}^{!\psi}$ as $(W^{!\psi}, \sigma', V')$, where $\sigma' = \sigma \upharpoonright W^{!\psi}$ and $V'(p) = V(p) \cap W^{!\psi}$ for any $p \in \Phi_0$. $\mathfrak{M}^{!\psi}$ is undefined if there is no x in W such that $\mathfrak{M}, x \Vdash \psi$.*

Definition 3. *Define $\sigma^{\maltese_n \psi} : W \to C$ as the following coloring function: $\sigma^{\maltese_n \psi}(w) = c_{-n}$ if $\mathfrak{M}, w \Vdash \psi$ and n is not less than the weight of $\sigma(w)$, otherwise $\sigma^{\maltese_n \psi}(w) = \sigma(w)$. Define $\mathfrak{M}^{\maltese_n \psi}$ as $(W, \sigma^{\maltese_n \psi}, V)$.*

Definition 4. *Define $\sigma^{\heartsuit_n \psi} : W \to C$ as the following coloring function: $\sigma^{\heartsuit_n \psi}(w) = c_n$ if $\mathfrak{M}, w \Vdash \psi$ and n is not less than the weight of $\sigma(w)$, otherwise $\sigma^{\heartsuit_n \psi}(w) = \sigma(w)$. Define $\mathfrak{M}^{\heartsuit_n \psi}$ as $(W, \sigma^{\heartsuit_n \psi}, V)$.*

We say that $(\mathfrak{M}^{!\psi}, w)$ is well-defined if $\mathfrak{M}^{!\psi}$ is defined and w is a state of it. Note that $\mathfrak{M}, w \Vdash [!\psi]\phi$ holds trivially if $(\mathfrak{M}^{!\psi}, w)$ is not well-defined. It can be verified that $(\mathfrak{M}^{!\psi}, w)$ is well-defined iff $\mathfrak{M}, w \Vdash \psi$.

We say that w is a χ-state if $\mathfrak{M}, w \Vdash \chi$. Deciding to make ψ true removes the $\neg\psi$-states. Technically the operator $[!\psi]$ works in the same way as the so-called public announcement operator in Public Announcement Logic [22].

Prohibiting to make ψ true with the weight n changes the color of the ψ-states w to c_{-n} unless n is less than the weight of the color of w. In that case, the color of w does not change.

Permitting to make ψ true with the weight n changes the color of the ψ-states w to c_n if n is not less than the weight of the color of w, otherwise leaves the color of w unchanged.

A formula ϕ is valid if for any $\mathfrak{M}$ and w, $\mathfrak{M}, w \Vdash \phi$. We in the sequel use SimDDL to denote the set of valid formulas of Φ_{SimDDL}. A set of formulas Γ entails ϕ, $\Gamma \vDash \phi$, if for any $\mathfrak{M}$ and w, if $\mathfrak{M}, w \Vdash \Gamma$, then $\mathfrak{M}, w \Vdash \phi$.

Truth at models A formula ϕ is true at a model $\mathfrak{M}$, $\mathfrak{M} \Vdash \phi$, if $\mathfrak{M}, w \Vdash \phi$ for any w of $\mathfrak{M}$. Models can be viewed as discourse domains consisting of colorful possible futures. So being true at a model means being true with respect to the discourse domain represented by the model. Let d denote "*there is a dog*". We look at the model depicted in Figure 1. Clearly $d \vee \neg d$ is true at this model. This means that there will be a dog or not in the future. But neither d nor $\neg d$ is true at the model. So it it not the case that there will be a dog in the future and it is also not the case that there will be no dog in the future.

$[!\phi]\psi$ is the formalization of the conditional "*if ϕ then ψ*". Here are some features of it concerned with truth at models. Firstly, $\mathfrak{M} \Vdash \phi$ and $\mathfrak{M} \Vdash [!\phi]\psi$ trivially imply

$\mathfrak{M} \Vdash \psi$, as $\mathfrak{M}^{!\phi} = \mathfrak{M}$ if $\mathfrak{M} \Vdash \phi$. Secondly, $\mathfrak{M} \nVdash \phi$ might not imply $\mathfrak{M} \Vdash [!\phi]\psi$. For a counter-example, let $\phi = p$ and $\psi = \mathbf{A}\neg p$. Assume $\mathfrak{M}$ is a model containing both p-states and $\neg p$-states. Then $\mathfrak{M} \nVdash p$ and $\mathfrak{M} \nVdash [!p]\mathbf{A}\neg p$. Thirdly, $\mathfrak{M} \Vdash \neg\phi$ trivially implies $\mathfrak{M} \Vdash [!\phi]\psi$, as $\mathfrak{M}^{!\phi}$ is not defined if $\mathfrak{M} \Vdash \neg\phi$.

3 Derived normativity

Let $\mathfrak{f}$ denote $\mathfrak{c}_0 \vee \cdots \vee \mathfrak{c}_k$, meaning that this is a fine state. Define $O^+\phi$ as $\mathbf{A}(\mathfrak{f} \to \phi)$, indicating that the agent ideally should make ϕ true. Define ideal permissions $P^+\phi$ as $\mathbf{E}(\mathfrak{f} \wedge \phi)$ and ideal prohibitions $F^+\phi$ as $\mathbf{A}(\mathfrak{f} \to \neg\phi)$. $O^+\phi$ is true at a state in a model iff ϕ is true at all fine states of this model. Truth of $O^+\phi$ at a state has nothing to do with this state and is global. This is also the case for $P^+\phi$ and $F^+\phi$.

For any propositional constant $\mathfrak{c}_i$, let $\Delta^>_{\mathfrak{c}_i} = \{\mathfrak{c}_j \mid j > i\}$. $\bigvee \Delta^>_{\mathfrak{c}_i}$, the disjunction of the elements of $\Delta^>_{\mathfrak{c}_i}$, says that this state has a better color than $\mathfrak{c}_i$. Note that $\bigvee \Delta^>_{\mathfrak{c}_i}$ is equivalent to $\bot$ if $\Delta^>_{\mathfrak{c}_i}$ is empty. The formula $\mathbf{E} \bigvee \Delta^>_{\mathfrak{c}_i}$ indicates that some state has a better color than $\mathfrak{c}_i$. Define $\mathfrak{o}$ as $\mathfrak{c}_k \vee (\mathfrak{c}_{k-1} \wedge \neg \mathbf{E} \bigvee \Delta^>_{\mathfrak{c}_{k-1}}) \vee \cdots \vee (\mathfrak{c}_{-k} \wedge \neg \mathbf{E} \bigvee \Delta^>_{\mathfrak{c}_{-k}})$, which says that this is an optimal state. $\mathfrak{a}$, defined as $\mathfrak{f} \vee \mathfrak{o}$, means that this is an acceptable state.

Define $O\phi$ as $\mathbf{A}(\mathfrak{a} \to \phi)$, meaning that the agent practically should make ϕ true. Define practical permissions $P\phi$ as $\mathbf{E}(\mathfrak{a} \wedge \phi)$ and practical prohibitions $F\phi$ as $\mathbf{A}(\mathfrak{a} \to \neg\phi)$. $O\phi$ is true at a state in a model iff ϕ is true at all acceptable states. Truth of $O\phi$ at states is also global.

Note that ideal and practical notions are identical in normal models. In abnormal models, nothing is ideally permitted, which means that ideal notions are not really working.

Define the implication relation between actions of offering permissions as follows: $\heartsuit_n\phi$ implies $\heartsuit_n\psi$ iff for any model $\mathfrak{M}$, $(\mathfrak{M}^{\heartsuit_n\phi})^{\heartsuit_n\psi} = \mathfrak{M}^{\heartsuit_n\phi}$. The implication from $\heartsuit_n\phi$ to $\heartsuit_n\psi$ means that it is useless for you to offer the permission to make ψ true if you already offered the permission to make ϕ true. It can be verified that $\heartsuit_n(\phi \vee \psi)$ implies $\heartsuit_n\phi$. So offering the permission to have black or white fences implies offering the permission to have black fences. The Free Choice Permission Paradox is solved in a way.

Conditional ideal/practical obligations, "*if ϕ then it is ideally/practically obligatory that ψ*" are expressed as $[!\phi]O^+\psi/[!\phi]O\psi$. Figure 2 shows how the problem of white fence is solved and illustrates how new permissions change the agent's obligations. Figure 3 illustrates how new prohibitions change the agent's obligations and show how the problem of warning sign is solved.

Remarks The idea of defining obligations by finitely many prioritized colors can be found in many works. [12] and [1] introduce two colors, that is, a bad one and a fine one, and define obligations in the following way: a proposition is obligatory if

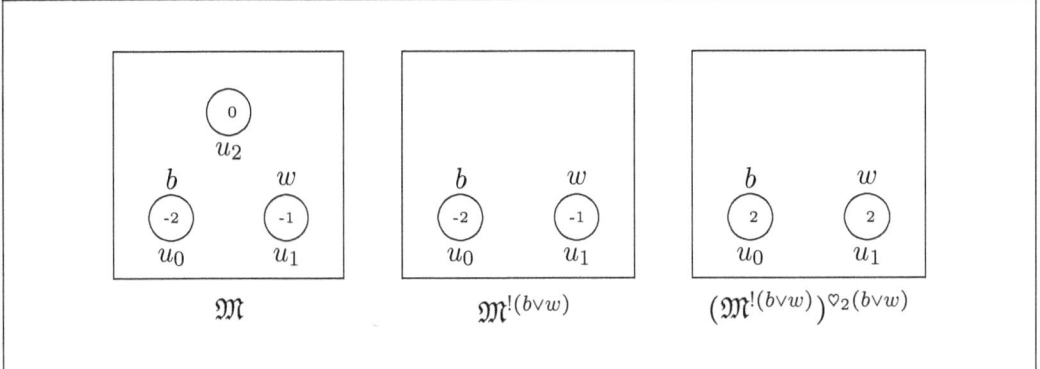

Figure 2: This figure explains what happens in the scenario of white fence. Let b indicate "*there is a black fence*" and w "*there is a white fence*". The model $\mathfrak{M}$ is the starting situation of this scenario. It is a normal situation containing a fine future, that is, u_2. At this model, $O^+\neg(b \vee w)$, "*Smith ideally ought to have no fence*", and $[!(b \vee w)]Ow$, "*if he is going to have a fence, he practically ought to have a white one*", are true. Then Smith decides to have a fence. Now the possible future u_2 disappears and $\mathfrak{M}$ evolves to $\mathfrak{M}^{!(b \vee w)}$ which is an abnormal model. At $\mathfrak{M}^{!(b \vee w)}$, both $O^+\neg(b \vee w)$ and Ow are true. Then the town issue a new regulation, *there may be black or white fences*, and $\mathfrak{M}^{!(b \vee w)}$ becomes to $(\mathfrak{M}^{!(b \vee w)})^{\diamond_2(b \vee w)}$ that is a normal model. Now *Smith ideally can have a black fence* and *ideally can have a white fence*.

it is true at all fine states. [2] uses finitely many linearly ordered colors and defines obligations as follows: a proposition is obligatory if it is true at all optimal states. [21] also contains a similar definition.

In this work, we use three kinds of colors: good, bad and neutral. The reason that we make a distinction between good and bad colors is two-fold: firstly, we want to define two types of obligations: ideal and practical ones; secondly, we want to handle issuing permissions. The introduction of the neutral color is just for conceptual reason: there might be some futures which are neither good nor bad.

As mentioned previously, there are two approaches in the literature on formalizing *simple* conditional obligations "*if ϕ, then it is obligatory that ψ*": one introduces dyadic deontic operators $O(\psi|\phi)$ and one uses non-material implications and unary deontic operators. This work follows the second approach and expresses conditional obligations as $[!\phi]O\psi$. Actually, if we do not consider the distinction between good and bad states, then $[!\phi]O\psi$ shares the same semantics with $O(\psi|\phi)$ in many works including [7], [2] and [21]: ψ is true at all optimal states of the class of ϕ-states.

There have been some works such as [16] in the literature which is in the same research line with this work. Our work adopts the restrictor view on conditionals. To the best of our knowledge, this makes a difference from other works.

The dynamic operator $[!\phi]$ updates models by removing the $\neg\phi$-states. As men-

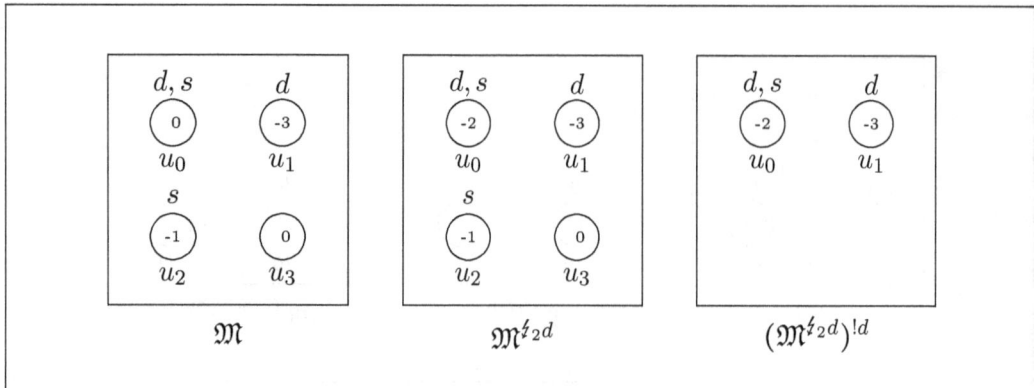

Figure 3: This figure reflects the scenario of warning sign. Let d denote "*there is a dog*" and s "*there is a warning sign*". $\mathfrak{M}$ is the starting point of the scenario, where P^+d is true. So *Jones ideally can have a dog*. Then the community issue a new rule, *there ought to be no dog*, and $\mathfrak{M}$ becomes to $\mathfrak{M}^{\sharp 2d}$. $[!d]Os$ and $[!\neg d]O^+\neg s$ are true at $\mathfrak{M}^{\sharp 2d}$, but $[!\neg d]O^+s$ is false there. So *if there is a dog, there ought to be a sign practically* and *if there is no dog, there ought to be no sign ideally*, but it is not that *if there is no dog, there ought to be a sign ideally*. Then Jones decides to keep his dog and $\mathfrak{M}^{\sharp 2d}$ evolves to $(\mathfrak{M}^{\sharp 2d})^{!d}$, which is an abnormal model. Now *Jones practically ought to put a sign near his house.*

tioned above, technically this is not something new: it already occurs in dynamic epistemic logic. But we think that our understanding of it from the perspective of restricting possible futures by actions such as making decisions is novel. Following this understanding and working in a richer language, we can develop logics beyond dynamic epistemic logic [11].

4 Completeness by reduction

In what follows, we prove that the three dynamic operators do not contribute any expressive power to Φ_{SimDDL}. So the completeness of SimDDL can be reduced to that of $\mathsf{S5_c}$, the logic of the language $\Phi_{\mathsf{S5_c}}$. Let $\veebar$ be the exclusive disjunction. The axiomatization of $\mathsf{S5_c}$ is the axiomatization of S5 plus the axiom $c_{-k} \veebar \cdots \veebar c_k$, saying that every state has a unique color.

What follow are the reduction axioms for $[!\phi]$. They are the same as the reduction axioms for the public announcement operator in Public Announcement Logic.

Lemma 1. *The following formulas are valid:*

1. $[!\phi]p \leftrightarrow (\phi \to p)$
2. $[!\phi]\top \leftrightarrow \top$

3. $[!\phi]\mathfrak{c}_i \leftrightarrow (\phi \to \mathfrak{c}_i)$
4. $[!\phi]\neg\psi \leftrightarrow (\phi \to \neg[!\phi]\psi)$
5. $[!\phi](\psi \wedge \chi) \leftrightarrow ([!\phi]\psi \wedge [!\phi]\chi)$
6. $[!\phi]\mathbf{A}\psi \leftrightarrow (\phi \to \mathbf{A}[!\phi]\psi)$

Define $\Delta_n^{|\leq|} = \{\mathfrak{c}_i \mid |i| \leq n\}$, where $|i|$ is the absolute value of i. Given $\mathfrak{M}, w \Vdash \phi$, $\mathfrak{M}, w \Vdash \bigvee \Delta_n^{|\leq|}$ iff the color of w is changed to c_{-n}/c_n by $\mathit{\mbox{\textsterling}}_n\phi/\Diamond_n\phi$.

The reduction axioms for $[\mathit{\mbox{\textsterling}}_n\phi]$ are as follows:

Lemma 2. *The following formulas are valid:*

1. $[\mathit{\mbox{\textsterling}}_n\phi]p \leftrightarrow p$
2. $[\mathit{\mbox{\textsterling}}_n\phi]\top \leftrightarrow \top$
3. (a) $[\mathit{\mbox{\textsterling}}_n\phi]\mathfrak{c}_i \leftrightarrow ((\neg\phi \wedge \mathfrak{c}_i) \vee (\phi \wedge \bigvee \Delta_n^{|\leq|}))$, where $i = -n$
   (b) $[\mathit{\mbox{\textsterling}}_n\phi]\mathfrak{c}_i \leftrightarrow (\mathfrak{c}_i \wedge (\neg\phi \vee \neg \bigvee \Delta_n^{|\leq|}))$, where $i \neq -n$
4. $[\mathit{\mbox{\textsterling}}_n\phi]\neg\psi \leftrightarrow \neg[\mathit{\mbox{\textsterling}}_n\phi]\psi$
5. $[\mathit{\mbox{\textsterling}}_n\phi](\psi \wedge \chi) \leftrightarrow ([\mathit{\mbox{\textsterling}}_n\phi]\psi \wedge [\mathit{\mbox{\textsterling}}_n\phi]\chi)$
6. $[\mathit{\mbox{\textsterling}}_n\phi]\mathbf{A}\psi \leftrightarrow \mathbf{A}[\mathit{\mbox{\textsterling}}_n\phi]\psi$

Proof. We only consider the cases (3a) and (3b) and skip the others.

(3a) Let $i = -n$.

$\Leftarrow$. Assume $\mathfrak{M}, w \Vdash \neg\phi \wedge \mathfrak{c}_i$. Then $\sigma^{\mathit{\mbox{\textsterling}}_n\phi}(w) = \sigma(w) = c_i$. Assume $\mathfrak{M}, w \Vdash \phi \wedge \bigvee \Delta_n^{|\leq|}$. Then $\sigma^{\mathit{\mbox{\textsterling}}_n\phi}(w) = c_{-n} = c_i$. In both cases, we have $\mathfrak{M}, w \Vdash [\mathit{\mbox{\textsterling}}_n\phi]\mathfrak{c}_i$.

$\Rightarrow$. Assume $\mathfrak{M}, w \nVdash (\neg\phi \wedge \mathfrak{c}_i) \vee (\phi \wedge \bigvee \Delta_n^{|\leq|})$, that is, $\mathfrak{M}, w \nVdash (\neg\phi \wedge c_{-n}) \vee (\phi \wedge \bigvee \Delta_n^{|\leq|})$. Then $\mathfrak{M}, w \Vdash (\phi \vee \neg c_{-n}) \wedge (\neg\phi \vee \neg \bigvee \Delta_n^{|\leq|})$. Then $\mathfrak{M}, w \Vdash (\phi \wedge \neg\phi) \vee (\phi \wedge \neg \bigvee \Delta_n^{|\leq|}) \vee (\neg c_{-n} \wedge \neg\phi) \vee (\neg c_{-n} \wedge \neg \bigvee \Delta_n^{|\leq|})$. It is impossible $\mathfrak{M}, w \Vdash \phi \wedge \neg\phi$. Assume $\mathfrak{M}, w \Vdash \phi \wedge \neg \bigvee \Delta_n^{|\leq|}$. Then $\mathfrak{M}, w \Vdash \neg \bigvee \Delta_n^{|\leq|}$ and $\mathfrak{M}, w \nVdash c_{-n}$. Then $\sigma^{\mathit{\mbox{\textsterling}}_n\phi}(w) = \sigma(w) \neq c_{-n} = c_i$. Assume $\mathfrak{M}, w \Vdash \neg c_{-n} \wedge \neg\phi$. Then $\sigma^{\mathit{\mbox{\textsterling}}_n\phi}(w) = \sigma(w) \neq c_n = c_i$. Assume $\mathfrak{M}, w \Vdash \neg c_{-n} \wedge \neg \bigvee \Delta_n^{|\leq|}$. Then $\sigma^{\mathit{\mbox{\textsterling}}_n\phi}(w) = \sigma(w) \neq c_n = c_i$. In all the three cases, we have $\mathfrak{M}, w \nVdash [\mathit{\mbox{\textsterling}}_n\phi]\mathfrak{c}_i$.

(3b) Let $i \neq -n$.

$\Leftarrow$. Assume $\mathfrak{M}, w \Vdash \mathfrak{c}_i \wedge (\neg\phi \vee \neg \bigvee \Delta_n^{|\leq|})$. Then $\sigma(w) = c_i$ and $\mathfrak{M}, w \Vdash \neg\phi$ or $\mathfrak{M}, w \Vdash \neg \bigvee \Delta_n^{|\leq|}$. In both cases of $\mathfrak{M}, w \Vdash \neg\phi$ and $\mathfrak{M}, w \Vdash \neg \bigvee \Delta_n^{|\leq|}$, we have $\sigma^{\mathit{\mbox{\textsterling}}_n\phi}(w) = \sigma(w)$. Then $\mathfrak{M}, w \Vdash [\mathit{\mbox{\textsterling}}_n\phi]\mathfrak{c}_i$.

$\Rightarrow$. Assume $\mathfrak{M}, w \nVdash \mathfrak{c}_i$. Then $\sigma(w) \neq c_i$. Since either $\sigma^{\mathit{\mbox{\textsterling}}_n\phi}(w) = c_{-n}$ or $\sigma^{\mathit{\mbox{\textsterling}}_n\phi}(w) = \sigma(w)$, $\sigma^{\mathit{\mbox{\textsterling}}_n\phi}(w) \neq c_i$. Assume $\mathfrak{M}, w \nVdash \neg\phi \vee \neg \bigvee \Delta_n^{|\leq|}$. Then $\mathfrak{M}, w \Vdash \phi \wedge \bigvee \Delta_n^{|\leq|}$. Then $\sigma^{\mathit{\mbox{\textsterling}}_n\phi}(w) = c_{-n} \neq c_i$. In both cases, we have $\mathfrak{M}, w \nVdash [\mathit{\mbox{\textsterling}}_n\phi]\mathfrak{c}_i$. $\square$

The reduction axioms for $[\Diamond_n\phi]$ are similar with the ones for $[\mathit{\mbox{\textsterling}}_n\phi]$ and the only difference lies in that "$i = -n/i \neq -n$" is replaced by "$i = n/i \neq n$". The proofs of their validity are also similar.

Lemma 3. *The following formulas are valid:*

1. $[\heartsuit_n \phi] p \leftrightarrow p$
2. $[\heartsuit_n \phi] \top \leftrightarrow \top$
3. (a) $[\heartsuit_n \phi] c_i \leftrightarrow ((\neg \phi \wedge c_i) \vee (\phi \wedge \bigvee \Delta_n^{|\leq|}))$, where $i = n$
   (b) $[\heartsuit_n \phi] c_i \leftrightarrow (c_i \wedge (\neg \phi \vee \neg \bigvee \Delta_n^{|\leq|}))$, where $i \neq n$
4. $[\heartsuit_n \phi] \neg \psi \leftrightarrow \neg [\heartsuit_n \phi] \psi$
5. $[\heartsuit_n \phi](\psi \wedge \chi) \leftrightarrow ([\heartsuit_n \phi] \psi \wedge [\heartsuit_n \phi] \chi)$
6. $[\heartsuit_n \phi] \mathbf{A} \psi \leftrightarrow \mathbf{A} [\heartsuit_n \phi] \psi$

By the three lemmas, we can show the following theorem:

Theorem 1. *For any $\phi \in \Phi_{\mathsf{SimDDL}}$, there is a $\phi' \in \Phi_{\mathsf{S5_c}}$ that is equivalent to ϕ.*

5 Connections

We say that a binary relation T on a nonempty set X has *the black hole property* if for any x and y in X, $T(x) = T(y)$, where $T(x) = \{z \in X \mid Txz\}$ and $T(y) = \{z \in X \mid Tyz\}$. The structure (X, T) looks like a black hole if R has that property, as illustrated by Figure 4. A black hole relation T is *serial* if $T(x) \neq \emptyset$ for any x in X.

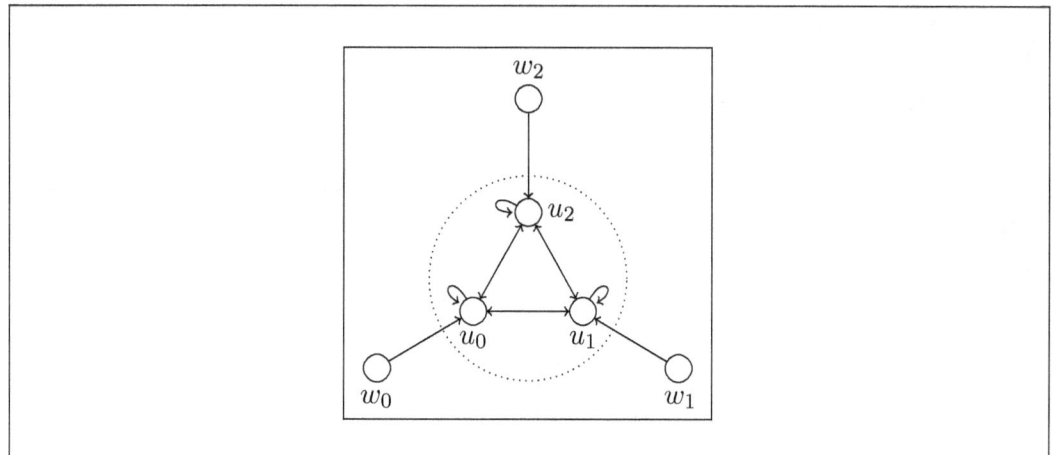

Figure 4: This figure depicts a relational structure (X, T) where T has the black hole property. Note that T is transitive and some arrows are omitted.

Let $\mathfrak{M} = (W, \sigma, V)$ be a model of Φ_{SimDDL}. Note that $\mathrm{O}^+ \phi$ is true at a state w in $\mathfrak{M}$ iff ϕ is true at all fine states, and $\mathrm{O}\phi$ is true at w iff ϕ is true at all acceptable states. Define a relation R on W as follows: Rwu iff u is a fine state. O^+ can be viewed as the box modality related with R: $\mathfrak{M}, w \Vdash \mathrm{O}^+ \phi$ iff $\mathfrak{M}, u \Vdash \phi$ for any u such that Rwu. In a similar way, O is related with the following relation S on W: Swu iff u is an acceptable state.

Each Φ_{SimDDL}-model determines a relation R and a relation S in the previous way. The class of Φ_{SimDDL}-models determines a class of relations R and a class of relations S. The following theorem tells that the two classes of relations are characterized by the black hole property and the serial black hole property respectively.

Theorem 2. *Let W be a nonempty set. A relation R on W has the black hole property iff there is a coloring function σ from W to the set of colors C such that for any x and y in W, Rxy iff y is a fine state. A relation S on W has the serial black hole property iff there is a coloring function σ such that for any x and y in W, Rxy iff y is an acceptable state.*

Proof. We consider only the first result and only the direction from left to right. Let R be a black hole relation on W. Define a coloring function σ as follows: $\sigma(w) = c_k$ if Rxw for some x, otherwise $\sigma(w) = c_{-k}$. As k is a positive integer, c_k is a fine color and c_{-k} is not. We claim that for any x and y, Rxy iff y is a fine state. Suppose Rxy. Then $\sigma(y) = c_k$ and y is a fine state. Suppose not Rxy. Then there is no z such that Rzy, otherwise $R(x) \neq R(z)$ and R is not a black hole relation. Then $\sigma(w) = c_{-k}$ and y is not a fine state. $\square$

It follows that O$^+$ as a modal operator *corresponds to* the black hole property and O to the serial black hole property.

The modal logic K45 is over transitive and euclidean relational structures and KD45 is over serial, transitive and euclidean relational structures. The following result can be easily shown:

Theorem 3. *The class of black hole relational structures and the class of transitive and euclidean relational structures determine the same modal logic. The class of serial black hole relational structures and the class of serial, transitive and euclidean relational structures determine the same modal logic.*

So, O$^+$ is a K45 operator and O a KD45 operator.

The featured theorems of KD45 are $\Box\phi \to \Diamond\phi$, $\Box\phi \to \Box\Box\phi$ and $\Diamond\phi \to \Box\Diamond\phi$, which respectively define seriality, transitivity and the euclidean property. The featured theorems of K45 are $\Box\phi \to \Box\Box\phi$ and $\Diamond\phi \to \Box\Diamond\phi$[3].

As moral principles, $\Box\phi \to \Box\Box\phi$ and $\Diamond\phi \to \Box\Diamond\phi$ are considered too strong by some logicians. However, as pointed out by [9], $\Box\phi \to \Box\Box\phi$ follows from $\Box\phi \to \mathbf{A}\Box\phi$ and $\mathbf{A}\psi \to \Box\psi$, and $\Diamond\phi \to \Box\Diamond\phi$ follows from $\Diamond\phi \to \mathbf{A}\Diamond\phi$ and $\mathbf{A}\psi \to \Box\psi$, where $\mathbf{A}$ is the universal modality. $\Box\phi \to \mathbf{A}\Box\phi$ and $\Diamond\phi \to \mathbf{A}\Diamond\phi$ result from the fact that

---

[3] K45 and KD45 are commonly viewed as doxastic logics: $\Box\phi$ is read as that the agent believes ϕ. The difference between K45 and KD45 is that the former allows inconsistent beliefs while the latter does not. In the literature, KD45 is also viewed as a deontic logic: it is also called the Deontic S5.

fine/acceptable futures are model determinate. This fact is fine, as which futures are fine/acceptable should be judged from the perspective of the present world, not from the perspective of futures. $\mathbf{A}\psi \to \square\psi$ says whatever is inevitable is obligatory, which sounds plausible.

One typical way to handle the deontic operator O in branching-time models is as follows. Among the futures starting at a point, some are acceptable. $O\phi$ is true at a point if ϕ is true relative to this point and all the acceptable futures starting at this point. This approach can be found in many works such as [20], [9], [6] and [10]. The O operator in all of them is a **KD45** operator.

6 Concluding remarks

In this work, the preference order $\leq$ over colors is linear. What if it is just a preorder? In this case, we can put two constraints on $\leq$: for any i, j such that $0 < i, j \leq k$, (i) $c_{-i} < c_0 < c_j$; (ii) $c_{-i} < c_{-j}$ iff $c_i > c_j$. In principle, all the results in this work would go through.

The intuitive background of this work has a temporal dimension: every state in a model represents a possible future of the present world; the agent should realize a fine future if there is any, otherwise should realize an optimal one. However, the temporal dimension is missing in the formal settings: in models, futures do not have internal structure and the present world is implicit; in the language, there is no temporal operator. It would be a natural further step to introduce a temporal dimension to this work based on branching-time models.

References

[1] A. R. Anderson. Some nasty problems in the formal logic of ethics. *Noûs*, 1(4):345–360, 1967.

[2] L. Åqvist. Systematic frame constants in defeasible deontic logic. In D. Nute, editor, *Defeasible Deontic Logic*, pages 59–77. Springer Netherlands, 1997.

[3] J. Carmo and A. Jones. Deontic logic and contrary-to-duties. In D. Gabbay and F. Guenthner, editors, *Handbook of Philosophical Logic: Volume 8*, pages 265–343. Springer Netherlands, 2002.

[4] R. Chisholm. Contrary-to-duty imperatives and deontic logic. *Analysis*, 24(2):33–36, 1963.

[5] J. Forrester. Gentle murder, or the adverbial Samaritan. *The Journal of Philosophy*, 81(4):193–197, 1984.

[6] T. French, J. McCabe-Dansted, and M. Reynolds. Axioms for obligation and robustness with temporal logic. In G. Governatori and G. Sartor, editors, *Deontic Logic in Computer Science*, pages 66–83. Springer Berlin Heidelberg, 2010.

[7] B. Hansson. An analysis of some deontic logics. *Noûs*, 3(4):373–398, 1969.

[8] R. Hilpinen. Deontic logic. In L. Goble, editor, *The Blackwell Guide to Philosophical Logic*, pages 159–182. Blackwell Publishing, 2001.

[9] J. Horty. *Agency and Deontic Logic*. Oxford University Press, 2001.

[10] F. Ju and G. Grilletti. A dynamic approach to temporal normative logic. In A. Baltag, J. Seligman, and T. Yamada, editors, *Proceedings of the 6th International Workshop of Logic, Rationality, and Interaction*, pages 512–525. Springer Berlin Heidelberg, 2017.

[11] F. Ju, G. Grilletti, and V. Goranko. A logic for temporal conditionals and a solution to the Sea Battle Puzzle. Manuscript, 2018.

[12] S. Kanger. New foundations for ethical theory. In R. Hilpinen, editor, *Deontic Logic: Introductory and Systematic Readings*, pages 36–58. Springer Netherlands, 1971.

[13] A. Kratzer. Conditionals. *Chicago Linguistics Society*, 22(2):1–15, 1986.

[14] D. Lewis. Adverbs of quantification. In E. Keenan, editor, *Formal Semantics of Natural Language*, pages 178–188. Cambridge University Press, 1975.

[15] P. McNamara. Deontic logic. In E. N. Zalta, editor, *The Stanford Encyclopedia of Philosophy*. Metaphysics Research Lab, Stanford University, 2014.

[16] P. Mott. On Chisholm's Paradox. *Journal of Philosophical Logic*, 2(2):197–211, 1973.

[17] H. Prakken and M. Sergot. Contrary-to-duty obligations. *Studia Logica*, 57(1):91–115, 1996.

[18] F. Ramsey. General propositions and causality. In D. Mellor, editor, *Philosophical Papers*, pages 145–163. Cambridge University Press, 1990.

[19] A. Ross. Imperatives and logic. *Philosophy of Science*, 11(1):30–46, 1944.

[20] R. Thomason. Deontic logic as founded on tense logic. In R. Hilpinen, editor, *New Studies in Deontic Logic: Norms, Actions, and the Foundations of Ethics*, pages 165–176. Springer Netherlands, 1981.

[21] J. van Benthem, D. Grossi, and F. Liu. Priority structures in deontic logic. *Theoria*, 80(2):116–152, 2014.

[22] H. van Ditmarsch, W. van der Hoek, and B. Kooi. *Dynamic Epistemic Logic*. Springer, 2007.

[23] F. Veltman. Imperatives at the borderline of semantics and pragmatics. Manuscript, 2010.

Towards a Formal Ethics for Autonomous Cars

Piotr Kulicki, Robert Trypuz[*], Michael P. Musielewicz[†]

The John Paul II Catholic University of Lublin,
Al. Racławickie 14, 20-950 Lublin, Poland
{kulicki,trypuz,michael.musielewicz}@kul.pl

Autonomous cars are one of the emerging technologies that will have a significant impact on society in the upcoming years. Although the predictions estimate that the traffic safety will be significantly improved, many people are afraid and prefer a human driver's control over vehicles or at least human driver's possibility to take control over the car. One of the reasons is that people want to be sure that in case of *hazardous situation or accident* a self-driving car will behave in a proper way. What does it mean "proper way"? There are several levels that can be considered, but at the end there is a level of values, especially moral values.

In this paper we move towards a formal ethics for autonomous vehicles, which will allow people to understand the values influencing a self-driving car. To accomplish this, we first address philosophical concerns for the possibility of ethics for driverless cars, by paying particular attention to the issue of their capacity of being a normative agent. We then discuss a formal ontology for these vehicles and the possibility of the use of such an ontology as a basis for a normative system. The lack of expressive power of ontological tools leads to the conclusion that the formal ethics for autonomous cars requires a more powerful logic. The logic should be able to take into account norms on actions and states, and handle normative conflict and preferences on norms.

1 Introduction

Autonomous cars (also called driverless or self-driving) are one of the emerging technologies that is expected to have a significant impact on society in the upcoming years. The number of companies preparing to manufacture fully autonomous cars is growing and includes major car manufacturers like Toyota, IT companies like

---

A special thanks to the anonymous reviewers for their helpful suggestions.

[*]Robert Trypuz and Piotr Kulicki acknowledge that their research has been supported by the National Science Centre, Poland (grant UMO-2015/17/B/HS1/02569).

[†]Michael P. Musielewicz's research has been supported by the National Science Centre, Poland (BEETHOVEN, UMO-2014/15/G/HS1/04514).

Google, Intel or Nvidia, and transport companies like Uber, thereby making the economic and social expectations in that matter considerable.

Designers and producers of self-driving cars use a variety of technologies to develop the software for controlling their vehicles. While there are particular differences, every one of them use some variant of statistical methods (applying tools such as artificial neural networks, deep learning, reinforcement learning, etc.) as their main tool.

These techniques work very well in typical situations, and at times can even outperform humans[1]. This process, however, leads to a black box algorithm of car control. The system acts successfully but we do not really know why and how (in the sense that the car's choices cannot be explained in a way that would be understood by an average person). As a result, the technology becomes suspicious to many parties involved in the use of autonomous cars.

Moreover, documents from regulatory authorities, *e.g.* [22, 4, 2], and researchers working in the field, *viz.* [6, 10, 5], recognize the need for ethical considerations concerning the behaviour of autonomous vehicles. However, they do not provide any complex and well defined theory in that matter. We also believe that an ethics for self-driving cars is indeed necessary. Additionally, we think that the ethics in consideration should be formalized. That allows for a precise expression of ethical intuitions, which is important for the success of a social debate regarding the subject and may also be useful for self-driving control software specification and development.

We begin the remaining part of this paper with a justification for the need of transparency of the decisions made by autonomous cars and a discussion of factors relevant for those decisions. Then, we approach the postulated formal ethics for autonomous cars. We address some philosophical concerns for the possibility of ethics for driverless cars. We pay particular attention to the issue of the capacity of a driverless car, or its controlling software, to be a normative agent. After this section the paper discusses an ontology for these vehicles which concludes with a proposal of a formal normative reasoning that will be useful for building a formal ethics for autonomous cars.

2 Social Acceptance of Driverless Cars

2.1 Benefits and Threats

It is widely agreed that driverless cars, once introduced, are going to have a major impact upon society. Both the United States of America and the European Union have taken it upon themselves to be ready for this change. In the GEAR 2030 report

---

[1]For example see the outcomes of Microsoft's Bejing team in 2015 ImageNet Large Scale Visual Recognition Competition where they were able to have a 3.57% error rate surpassing human average error rate of 5% for the first time. [13, pp.223–5]

[2] for the European Commission, a sketch of the impact that driverless cars are expected to have upon society is provided. Here, the expected impact ranges from a 90% reduction in human error related road accidents to increased social mobility and even to a reduction of pollution in the environment [2, p.40]. Likewise, the US federal government sees safety as the paramount feature of this new technology and hopes to see a reduction of up to 94% of traffic accidents in the US, along with increased mobility for disabled persons [22, p.5].

The European Commission's report mentions that autonomous vehicles pose "new challenges for regulators and policy makers concerning e.g. road safety, environmental, societal and ethical issues, cybersecurity protection of personal data, competitiveness and jobs, etc. which need to be addressed" [2, p.40]. Solving these issues is needed to build up the social acceptance of driverless cars.

A psychological factor also has to be considered. Although the predictions estimate that traffic safety will be significantly improved, many people are afraid and prefer human driver's control over vehicles or at least the possibility of a human driver to take the control over the car. These fears surface in instances where self-driving cars, that are currently being tested, have failed to avoid serious collisions. Tesla's car in 2016 failed to detect a large white 18-wheel truck and trailer crossing the highway. The car drove full speed under the trailer, causing the collision that killed the 40-year-old behind the wheel inside the Tesla. Recently, an autonomous Uber car killed a woman walking across the street in the State of Arizona[2]. From these examples we can see that the use of autonomous cars is not free from serious risks.

Even specialists in the area remain skeptical about the technology they create. Raj Rajkumar, a leading expert on robotics, who cooperates with General Motors in the construction of autonomous cars, describes the current status of the technology in the following way:

> We as humans understand the situation. We are cognitive, sentient beings. We comprehend, we reason, and we take action. When you have automated vehicles, they are just programmed to do certain things for certain scenarios.[3]

So the users of autonomous vehicles want to know and understand (at some level of generality) how the vehicles are programmed to "do certain things for certain scenarios". They want to be sure that in case of a hazardous situation or an accident a self-driving car will behave in a proper way. Yet we must consider, what do we mean when we say "proper way"?

---

[2]See https://www.theguardian.com/technology/2018/mar/19/uber-self-driving-car-kills-woman-arizona-tempe (retrieved March 20, 2018)

[3]See https://www.technologyreview.com/s/602492/what-to-know-before-you-get-in-a-self-driving-car/ (retrieved March 1, 2018).

2.2 The Need for Transparency

In most cases it is possible to avoid damage to property, health, and the life of passengers and other participants of traffic. Moreover, it seems credible that a well trained algorithm will perform far better in driving than the average human driver or even a very good driver, and so it would seem that ethical considerations for driverless cars is relegated to only extreme situations. But this is not necessarily the case. The effects that these devices have upon their users may differ depending upon how its program is made or trained. The US Federal Government's policy for driverless cars indicates that, "even in instances in which no explicit ethical rule or preference is intended, the programming of an HAV [(highly automated vehicles)] may establish an implicit or inherent decision rule with significant ethical consequences"[22, p.26].

However, the very ascription of values to these objects, resting upon implicit ethical values, must be made clear so that all stakeholders can ensure that these "ethical judgments and decisions are made consciously and intentionally" [22, p.26]. This claim for transparency is mirrored in the report made by the ethics commission of the *Bundesministerium für Verkehr und digitale Infrastruktur* (hereinafter BMVI) made June of 2017. Here the BMVI underscores the importance of maintaining the autonomy of people in making ethical decisions and the prospect of some programmer or commission deciding how a driverless car should act on our behalf is, in and of itself, problematic [4, p.16].

Hod Lipson and Melba Kurman write in their book *Driverless* [13] that drivers make countless calculations and risk assessments of their behavior and of the road as it unfolds around them. When drivers are thrown into a situation where life is at risk they must react accordingly. Do they swerve right and hit a wall, or hit some other vehicle? When it is people making these choices there is an air of spontaneity which allows for us to forgive poor decisions, however the same does not apply for autonomous vehicles. As they say, "those of us fortunate enough never to have had a sever traffic accident have not had to perform the uncomfortable task of publicly articulating why we reacted the way we did when faced with an unavoidable traffic accident. Driverless cars stir up consternation since they force us to publicly reveal this calculation. Even more challenging, driverless cars will require that, as a society, we agree on a uniform set of ethical codes that will guide the decision-making process of artificial intelligence software when faced with an emergency" [13, p.252]. But it is precisely this sort of "digging out" of our ethical calculations that will allow for transparency in this public debate.

We concur that it is crucial for autonomous vehicles' designers, and moreover for all stakeholders in these decisions, to make clear what hierarchy of values they embed in their vehicles. This clarification will enable the potential owners and users of self-driving cars, other traffic participants, the public in general, and regulatory authorities to accept or reject the underlying ethics in the vehicle's decision making algorithms before the wide scale usage of such vehicles.

2.3 Possible Factors Influencing Self-Driving Cars' Expected Conduct

What kind of factors should be taken into account when the "ethical" behavior of self-driving cars is considered? Let us refer to some statements that can illustrate the breadth of possibilities.

Patrick Lin argues that the chief safety feature of driverless cars, that is its "crash-optimization", implicitly means targeting which object to hit in order to optimize a crash [12, pp. 72 – 73]. He notes that if we adopt a preference for minimizing harm to our property the car would need to target objects of a lesser weight than the vehicle; yet if we wish to minimize the harm to other people's property we ought to target an object of greater weight than the vehicle.

Michael Taylor from Car and Driver magazine, reported in [19] that according to Christoph von Hugo, Head of Active Safety in Mercedes-Benz Passenger Cars, all of Mercedes-Benz's future self-driving cars will *prioritize saving the people they carry*. Mercedes-Benz, after a public criticism, soon retracted the statement and indicated they would follow whatever the law proscribes [16]. That highlights the difficulties in pinning down the best response.

In general, can or should an autonomous car value one life more than another on the basis of their relation to that car (value the passenger or owner over other persons), age, sex, status or by applying some other criteria? These difficulties in our (in)ability to choose who to save is seen in the often discussed trolley problems.

On this precise point there are many different points of view. Take for example the report made by the BMVI. There they lay forth 20 ethical rules for automated and connected vehicular traffic. In the 9th rule they proscribe

> In the event of unavoidable accident situations, any distinction based on personal features (age, gender, physical or mental constitution) is strictly prohibited. It is also prohibited to offset victims against one another. General programming to reduce the number of personal injuries may be justifiable. Those parties involved in the generation of mobility risks must not sacrifice non-involved parties. [4, p.11]

These are fairly strong claims and are further supported by first three articles of the *Grundgesetz für die Bundesrepublik Deutschland*, and raises questions if such "targeting" of objects that happen to be people could even be constitutionally permitted within Germany. These claims also seen in important associations in civil society. The IEEE (the Institute of Electrical and Electronics Engineers) also commit their members to these very same standards. Therefore, it would seem to answer our questions concerning whether a driverless car can value one life more than another.

Notwithstanding that apparent answer, there is more to the story than that. If we look at MIT's Moral Machine (http://moralmachine.mit.edu), we see that people do in fact have preferences and seem capable of choosing between two bad options; and when they are given a series of choices of how to act in various dilemmas general trends emerge. For an informal example we can see that enforcing the law, preferring women to men, humans to animals, fit people to fat people are some of the preferences that are noticeable. A more formal example of this is also seen the work of Bonnefon et al. [9] where they noticed a strong preference for cars that minimize harm as such (*i.e.* by choosing self-sacrifice or the sacrifice of even loved ones) but it is conjoined with a general reluctance to buy such a car for themselves or even to have that sort of ethics enforced by legal means.

3 Ethics and Self-driving Cars – Foundational Problems

3.1 What is Meant by Ethics?

Various regulatory institutions like the U.S. Federal Government in its policy [22, p.26], the European Union in its GEAR 2030 report [2, p.40] and press releases [3] in and the German ethics commission of the BMVI in their report [4], all emphatically assert the need for ethics for driverless cars. However, *there is no one clear understanding what is meant by ethics*. Rather, there are some common themes presented within these various works.

First and foremost, they note that ethics covers the decision making processes within the vehicle both in terms of legal and moral reasoning (or in some cases a conflation of these two modes of reasoning). Additionally there needs to be a balancing of moral actions, legal actions, and goal oriented actions made in light of moral and legal reasoning. Furthermore, it is clear that while ethics is important in all stages of automation, it is most important for when the vehicles are at higher stages of autonomy. In these stages, the human agent takes a lesser (to even non-existent) role in the operation of the device. The common convention that these various institutions use is the SAE (Society of Automotive Engineers) levels of automation. Notably, the BMVI's document calls into question whether or not the removal of the human factor is a good thing from an ethical perspective [4, p.20].

In addition to these texts various philosophers have also ventured some essential features of an ethics for autonomous vehicles. Neil McBride, for example, offers the *A.C.T.I.V.E.* (Autonomy, Community, Transparency, Identity, Value, and Empathy) formulation of ethics for autonomous cars, within which the rules that govern the both human-human and human-machine relationships need to be addressed [15]. Lin, believes that ethics has a crucial role in establishing what underlying values we ascribe to objects, which has a direct bearing on how the driverless car's crash optimization will function. Additionally, both McBride and Lin underscore the importance of broader ethical considerations for these new devices. One such example "conservative driving" where the autonomous car is overly cautious and other drivers will try to "'game' it, *e.g.*, by cutting it in front of it knowing that the automated car will slow down or swerve to avoid an accident" [12, p.51]. Another example, found in McBride, poses the question of what rules should govern the community – ranging from regulators, mechanics, the supply chain, etc. – that is formed to support these new devices [15, p.182].

In all of these works, certain aspects emerge. We are living in an age where new technologies are introducing new agents and stakeholders, and while there is a host of benefits resulting from these changes, it calls into question how we should act. Ethics, in this context, then seems to be the establishment of norms that govern the actions of and between these various agents, both in terms of the human – human relationship and the human – robot relationship (to borrow from McBride). It is this understanding of ethics that we wish to use for this paper. Nevertheless, there are still some serious philosophical questions that first need to be addressed in order to build a proper ethics for these new devices.

3.2 Cars as Normative Agents

A poignant problem in designing an ethics for driverless cars is the establishment of these devices as normative agents that operate within a given normative system. If we are to do this, there are several factors that need to be considered. First, we need to see that they are agents. Then if they are agents, we must see if they are normative agents. To establish that autonomous vehicles are normative agents requires first that they are agents that are capable of bearing norms as such and second they are placeable inside of a "normative system." This, however, is no small feat and will depend greatly upon one's conception of norms. It is only once we have established this, that the movement towards a formal ethics for driverless cars makes sense. For once we have the driverless car *qua* normative agent, we can flush out a subsequent ontology and normative reasoning kinds to model it. Although attributing normative agency to computer programs seems to be quite natural for computer science oriented logicians, for many legal theorists and philosophers (ethicist) it is still strange, so in this section we will argue for the aforementioned points.

To begin we need to establish that autonomous vehicles are in fact agents. There are various senses of agency that are used in various fields. In a plain sense, being and agent simply means having the capacity to act. There are, however, other more technical uses of the term. The most natural place to start is with a consideration of agency within computer science, where White and Chopra say (citing another author) that in this field an agent is "a piece of software that acts on behalf of its user and tries to meet certain objectives or complete tasks without any direct input or direct supervision from its user" [17, p.6].

Trypuz, in *Formal Ontology of Action*, furthers this definition and provides a good list of features that artificial agents have as found in the literature, having the following attributes: is autonomous, is situated (embodied in or inhabits an environment), is reactive – senses its environment and is responsive to changes in the environment, acts upon is environment, is proactive – has a set of goals or tasks, contains inner representations of itself and its world, is rational –"acts in its own best interest, given the beliefs that it has about the world", has the ability to perform domain-oriented reasoning, is a persistent (software) entity, and has social ability – interacts (negotiates and cooperates) with other agents (and possibly humans) via some king of agent-communication language: it engages in dialogues and negotiates and coordinates transfer of information [20, p.40]. Given these notions driverless cars seem to meet the well established criteria for being agents with the computer science community. Yet to be an agent is one thing to be a normative agent is quite another.

The problem of normative agency comes to the foreground when we reflect upon the nature of norms in themselves on a philosophical level. When we consider norms, as rules that govern the behavior of various agents, we notice that they have two related aspects. First they set the bounds of obligated, forbidden, or permitted actions and second these actions are ascribable to agents within the system, (*i.e.* a normative agent, who is beholden to these rules by virtue of being in the system).

To make these features clear let us consider an example provided by Ota Weinberger in *Law, Institution and Legal Politics: Fundamental Problems of Legal Theory and Social Philosophy*, where he offers an example of a game of chess to describe what he calls the institutional nature of "social normative systems".

> The rules of the game of chess are defined by its basic conditions: chessboard, figures, starting positions, rules of operation etc. We might ask whether these rules should be regarded as normative rules or as definitions. If they were mere definitions the person who does not adhere to the rules would not be seen as infringing the 'duty of the chess-player', but simply as not playing chess.[footnote omitted] It is true that nobody is obliged to play chess; the rules of chess apply to the players not as a system imposed by society but only as a result of a voluntary participation in the game; but they are relevant for the possibility of setting acts since they lay down a behaviour in accordance with a duty and define the class of possible results of the game: the game which is won (or lost) [24, p.193].

If two players sit down to play chess, they voluntarily enter into a sort of norm-governed activity constrained by the normative rules of the game. Their moves are permitted (such as a pawn may move two spaces in its first move), obliged (a pawn may only take other pieces that are in its diagonals and one space away), or forbidden (a pawn may not capture a unit directly in front of it) in respect to the rules of the game being played.

Weinberger expands this conception of normative systems as a game into broader considerations of law and into other norm based systems. For our purposes we can see how traffic fits within this framework. The driver (and drivers in general) are duty-bound to obey traffic norms, that is to say that the drivers – by the very act of driving – become the "players", the traffic norms constitute the "rules of the game" that they are "playing", and the current state of affairs of the road are much like the game board. The key difference consisting in the complexity of the system, the content and number of norms (described in the previous section as combining moral, legal, and social rules) and the price of failures (in terms of tickets or even possible damage to persons and property). While this is quite clear, what remains a question is whether or not driverless cars are in fact agents, in a normative sense, that can fit into this system.

While the above schema is convenient in that it allows people who wish to study norms (or rights broadly construed) by using more analytic tools, the topic of building an ethics for driverless cars poses a unique problem for it. Normally it is rather simple to use this framework when we apply it to various normative systems, whether in a game of chess or driving a car. The agents are well defined, and so are the rules, and problems are typically introduced when there are normative conflicts or moral dilemmas. When applying this theory to driverless cars we first need to confront, the question "are they really normative agents?" How can they be bearers of rights in the broad sense? Or to use Hohfeldian terminology [8, p.30], can they be part of the duty – claim, privilege – non-claim, liability – power, and disability – immunity relationships?

To underscore this issue let us provide two examples. The first example is common place. A person is trying to cross the road at an uncontrolled intersection. The pedestrian has a claim to cross the road unimpeded which places a duty upon a driver to slow down and allow the pedestrian to cross. Is the driverless car beholden to the duty to "let pedestrians cross the street!"?

A more drastic example can also be taken from the well know trolley problem. An unmanned driverless car is going down the street and it is faced with a dilemma. Its breaks have failed, and now its controlling algorithm needs to make a choice of hitting a person in its lane or two people in the lane next to it. Now a dilemma is introduced if the car is beholden to the rule "Thou shalt not kill!" or say "Maximize the good!" or "Don't commit a forbidden act!" But is this the case? Against whom can the people in this perilous situation invoke a duty not to kill them? People in general? Sure, but in this case that's vapid. The programmer? Of course, but s/he programmed it not to hit people already. How about the driverless car itself? That's not clear. If the answer is no, then it would seem that the car does not have a duty to "not kill!" nor a duty to "maximize the good!" nor even a duty to "avoid committing a forbidden act!" that corresponds to any person's right in such a situation, in much the same way as we would not ascribe normative agency to a raging bull or a falling rock. Yet, if it lacks normative agency, then it falls outside of the normative system. Leaving us in a *de facto* situation were nothing is forbidden for it, and therefore everything is implicitly permitted. But surely that cannot be the case, can it?

If we are to avoid this we first need to dig even further into the theory of rights. There are presently two prominent theories of rights "will theory" and "interest theory" see e.g. [14, p.62]. These views of rights differ in what is required of an agent in order to ascribe to that agent rights as such, or in particular claims – duties, non-claims – privileges etc., and make them normative agents within a particular system of rights.

The key difference rests in the importance of the right bearers' interests and wills in the matter. For interest theorists, the bearers of these rights need only to be a beneficiary (or have some interest in the claim – duty etc. relationship), and the will is not needed. Will theorists, however, maintain that the bearer of these rights, need to be able to take an active role in the fulfillment of these rights, or put otherwise, be able to actualize them, to demand or to waive their right, and their interests need not be protected.

Both of these theories capture some of our basic intuitions on what rights are in relation to their bearers. Will theory maintains the idea that we are little sovereigns over our rights and can dispense or invoke them as we please. In interest theory we maintain the notion that rights ought to somehow be to our benefit. There is, however, a problem with will theory, namely the criterion of the necessity of the will excludes certain classes from bearing rights, even among human beings, that intuitively should have rights. These classes of persons would include the unborn, infants, the invalid, and the senile among others, who not having the capacity to use their will to demand or enforce their rights. For example they do not have the capacity to demand or waive claim to not be arbitrarily killed against some other person who is capable of fulfilling the adjoining duty in the Hohfeldian sense. So as they are incapable of having or exercising their wills they would then would have no rights, which is not the case.

This leads us to the consideration of interest theory. When considering driverless cars they certainly have interests when operating within the context of driving on the street. For example, they have an interest in crossing a busy intersection so they may have a claim of "right of way" against some other driver and that other driver has a duty to yield to the driverless car under that rule. Additionally, they may have in interest in being properly maintained and have a claim on their owner to service them. What is perhaps most important, is that given this conception of rights driverless cars may be bearers of rights (broadly conceived) and that right is granted by the normative system within which it is operating, and thereby are normative agents.

4 Towards an Ontology-Based Normative Reasoning for a Self-Driving Car

In [1] we read that a logical-based control implemented in a self-driving car would contribute to its *self-explanatory capacity* by which the author means a justification and explanation, in human understandable way, what the car "has done, is doing, and will be doing, and why". In the same paper it is emphasized that the self-driving car should have implemented an *operating-system-rooted ethical control* by which the author means "logics that are connected to the operating-system level of [...] cars, and that ensure these cars meet all of their moral and legal obligations, never do what is morally or legally forbidden, invariably steer clear of the invidious, and, when appropriate, perform what is supererogatory".

In section 4.1 we present, in our opinion, one of the most promising modeling framework that proposes a logic-based modeling for autonomous vehicles. It possesses the *self-explanatory capacity* but does not provide *operating-system-rooted ethical control*. In section 4.2 we discuss what should be added to the framework to fill that gap.

4.1 Advanced Driving Assistant System Ontologies

In this section we shall present and discuss an example of ontology-based normative reasoning for a self-driving car taken from the most resent research in the field. We shall refer to the works of Lihua Zhao and her colleagues [27, 25, 26] where they propose Advanced Driving Assistant System Ontologies (hereinafter ADAS Ontologies) and some interesting ideas on how to combine ontologies with logical rules of reasoning that is expressed in the Semantic Web Rule Language (henceforward SWRL). The works, however, do not propose insights on how to incorporate ethical rules into that system. But the frameworks they present are flexible enough to be a good starting point for our purpose.

The authors say that a self-driving car should be able to "infer driving behavior by processing the knowledge" [26, p.1427]. The knowledge they have in mind comes from a mapping of the sensor data—collected by the car in real time—onto the categories of a machine-understandable ontology. So the car's "raw" perception data is transformed into ontological facts inside of the car's knowledge base. The knowledge base is a source of information that is used for making driving decisions. The car's knowledge in ADAS ranges from its spatio-temporal knowledge (maps, driving paths at any given time) through knowledge about itself (its type, size, current speed, etc) to its knowledge about rules (traffic and others) it must follow. That knowledge is to be considered when a decision is taken at any given time and position. For instance, the self-driving car monitors its speed in real-time, so it can be said that it is aware of its speed. If it also knows the speed limit on the road that it is currently driving on, then it can make a decision to accommodate its speed up to that limit. It can also take into account the current weather conditions or other conditions connected to traffic, so other rules (like various safety rules *i.e.* to drive at a slower speed when the roads are icy) can have an impact on its behavior.

The SWRL rules they propose are formulated by means of categories taken from the ADAS Ontologies. The rules are conditional and trigger the execution of actions according to the car's current conditions. It is worth noting that the driving decisions are on the level of basic driving actions such as *Stop*, *TurnRight*, or *Give Way*.

In [26], the authors propose 14 SWRL rules to model "Right-of-Way rules at uncontrolled intersections and on narrow two-way roads." They identified the following three situations that may occur: first "Before an intersection: Give way or move forward in comply with Right-of-Way rules", second "At an intersection: Stop and give way to the other cars when upcoming collisions are detected" (see formula (1)) and third "On a two-way lane: Move to the left side and give way to the other cars coming from the opposite side of the two-way lane."

$$MyCar(?car1) \land isRunningOn(?car1, ?int) \land$$
$$Intersection(?int) \land collisionWarningWith(?car1, ?car2) \Rightarrow \qquad (1)$$
$$Stop(?car1) \land giveWay(?car1, ?car2)$$

These rules are triggered only when the self-driving car receives a collision warning signal from a collision detection system. It should be stressed that both detecting and preventing collisions belong to the most important category of the self-driving car's tasks.

The SWRL rule reasoner performs reasoning on a fragment of the whole knowledge base containing the ontological description of the current driving situation.

Rules proposed in [26], like (1), although they do not contain deontic qualifications, they can still express norms. (1) says that if my car is approaching an intersection and receives a collision warning with another car, then my car is obliged to stop and give a way to the other car. One should notice that $Stop(?car1)$ in (1) means in fact that $?car1$ is classified as an object that should stop. The ontological modelling proposed in [26] contains no action tokens. This is because there is no need in that framework for modeling possible actions and describing possible choices in a given situation. It is assumed that after a collision warning there is a unique rule that, by taking into account the car's conditions, will property classify the car as being obliged to carry out an action of certain type.

4.2 Towards Ethical Control

Let us start with the following example and attempt of its modeling using the framework described above. A self-driving car is driving down a two-lane street, where there is an approaching truck in the opposing lane and a pedestrian on the sidewalk to the right. All of a sudden the truck swerves into the self-driving car's lane setting off the car's collision detection system. The driverless car must now prevent the collision by swerving right and out of the way onto the sidewalk, which triggers another collision warning. Now that it has two warnings, the car must decide whether is should collide with the truck and avoid hitting the pedestrian or swerve out of the way and avoid hitting the truck but then hit the pedestrian.

$$MyCar(?car1) \land collisionWarningWith(?car1, ?car2) \Rightarrow \\ preventCollision(?car1, ?car2) \quad (2)$$

$$preventCollision(?car1, ?car2) \Rightarrow TurnRight(?car1) \quad (3)$$

$$TurnRight(?car1) \Rightarrow collisionWarningWith(?car1, ?person1) \quad (4)$$

$$MyCar(?car1) \land collisionWarningWith(?car1, ?person1) \Rightarrow \\ preventCollision(?car1, ?person1) \quad (5)$$

$$preventCollision(?car1, ?person1) \Rightarrow TurnLeft(?car1) \quad (6)$$

After triggering all of the rules one by one, *i.e.* after triggering the rule (6), the car will come back to the initial conditions and will start triggering the rules from the beginning, starting from (2). So in consequence the car will keep turning left and right until it will eventually hit either the car ?*car2* or the person ?*person1*. It is clear that "deliberating by doing" is not always the best option.

In section 2.3 we discussed possible factors that influence a self-driving car's expected conduct. We pointed at different priorities that can have an impact on the car's actions. These priorities are both based on and justified by the values that the car's manufacturer or society has decided to implement in it. For instance we may prefer minimizing harm to our property more than minimizing the harm to other people's property or we may prefer saving the people the car carries more than pedestrians etc. It is evident that the logical framework responsible for the self-driving car's ethical control should be able to explicitly express norms and priorities on them (see e.g. [7, 23]).

A need for ethical reasoning often appears in situations where there are a few options that have to be judged in the light of ethical values and preferences. It means that we need more than rules of the form of "if condition A, then do B". A "higher level" deontic action logic reasoning (like those found in [11]) is needed that will constitute deliberation layer where possible actions could be identified, evaluated and finally the best option is chosen.

Then ethical reasoning requires taking into account both basic actions and their social interpretation. In the case of self-driving cars by basic actions we mean "turn left", "go straight","stop" etc. and their social interpretation in the particular conditions could be: "kill", "hurt", "protect", etc. Moreover, normative evaluation of actions themselves (like if they obey traffic regulations) and their results (like harm they make to the environment) should be considered [18, 21].

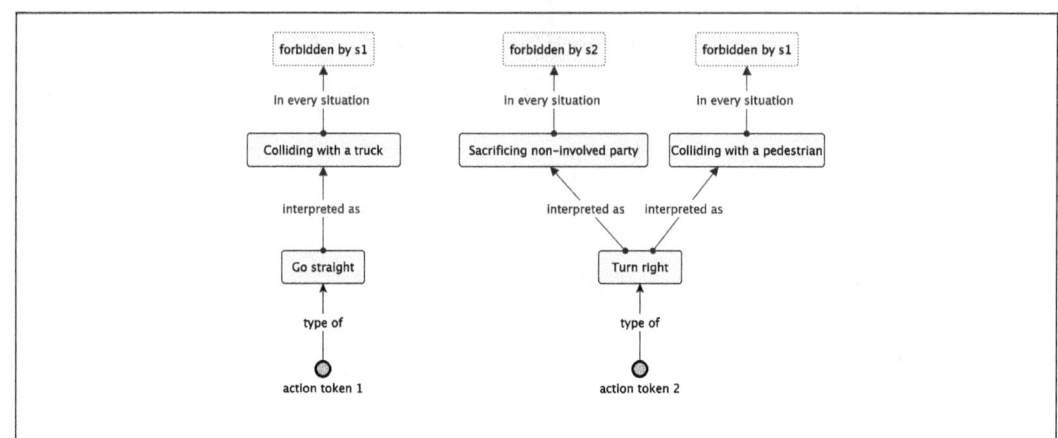

Figure 1: A situation where there are in force: a rule (from a normative source s1) that forbids collision with an object and a rule (from a normative source s2) stating that parties involved in the generation of mobility risks must not sacrifice non-involved parties. The self-driving car can carry out only two actions – action token 1 and action token 2.

For our example, if crash-optimization would be the main decision factor, then the choice between these options would be based on the evaluation of which object to hit in order to optimize the crash (the pedestrian has lesser weight than the car, contrary to the truck, so...). We could also discuss what would happen if the rule saying that "parties involved in the generation of mobility risks must not sacrifice non-involved parties" [4, p.11] would be implemented in the car. This rule could be of crucial importance and would forbid hitting the pedestrian that action would be recognized as "sacrificing a non-involved party". It is important here to interpret the car making the "turning right" action as making the pedestrian involved. The former is not forbidden, while the latter is.

Following [11] we propose a modelling of the situation as depicted in figure 1. If we assume that the source of norms s2 is preferred over s1, then the deontic qualifications coming from s1 will be somehow "removed" from the normative system as being less important. Then we can conclude that the car should go straight. If neither of norms is preferred we face a dilemma situation and the procedure chosen for such a situation should be applied – in [11] several approaches for such a scenario are discussed. An implementation is proposed in a paper under review[4], however the programs in said paper are available online at http://kpi.kul.pl/deonticmachine.

5 Conclusions

In this paper we have presented a justification for formal ethics for autonomous cars. The main point here is the need for transparency in car's behaviour that, as we believe, is a necessary precondition for the social acceptance of and the widespread introduction of this technology.

Here we have also discussed some foundational problems of ethics for self-driving cars: the question what ethics means in this context and how cars can be understood as normative agents.

Finally we have examined the possibility of a "if-then rules" based approach to the specification of the expected behaviour of an autonomous vehicle. This approach, while useful, is not satisfactory when more complex situations are considered. That has led us the conclusion that more powerful logical tools are needed, and we have provided a list of the basic requirements of such a logic. Developing the particular details of this logical approach to the issue is planned as future work based on the findings of the present present paper.

References

[1] Selmer Bringsjord and Atriya Sen. On creative self-driving cars: Hire the computational logicians, fast. *Applied Artificial Intelligence*, 30(8):758–786, 2016. https://doi.org/10.1080/08839514.2016.1229906.

[2] European Commision. The Report of the High Level Group on the Competitiveness and Sustainable Growth of the Automotive Industry in the European Union FINAL REPORT - 2017, October 2017.

[3] European Parliament Press Room. Robots: Legal Affairs Committee calls for EU-wide rules. Press Release, Jan 2017. http://www.europarl.europa.eu/news/en/press-room/20170110IPR57613/robots-legal-affairs-committee-calls-for-eu-wide-rules.

[4] Federal Ministry of Transport and Digital Infrastructure, Ethics Commission. Automated and Connected Driving, June 2017.

[5] Jan Gogoll and Julian F. Müller. Autonomous cars: In favor of a mandatory ethics setting. *Science and Engineering Ethics*, 23(3):681–700, 2017.

[6] Noah Goodall. Ethical Decision Making During Automated Vehicle Crashes. *Transportation Research Record: Journal of the Transportation Research Board*, 2424(58–65), 2014.

[7] Jörg Hansen. Deontic logics for prioritized imperatives. *Artif. Intell. Law*, 14(1-2):1–34, 2006. https://doi.org/10.1007/s10506-005-5081-x.

---

[4]Submitted to Fundamenta Informaticae, 2017-07-07

[8] Wesley Hohfeld. Some fundamental legal conceptions as applied in judicial reasoning. *Yale Law Journal*, 23, 1913. http://digitalcommons.law.yale.edu/ylj/vol23/iss1/4.

[9] Iyad Rahwan Jean-François Bonnefon, Azim Shariff. The social silemma of autonomous vehicles. *Science*, 352(6293):1573–1576, June 2016.

[10] Rolf Johansson and Jonas Nilsson. Disarming the Trolley Problem –Why Self-driving Cars do not Need to Choose Whom to Kill. In Matthieu Roy, editor, *Workshop CARS 2016 - Critical Automotive applications : Robustness & Safety*, CARS 2016 - Critical Automotive applications : Robustness & Safety, Göteborg, Sweden, September 2016. https://hal.archives-ouvertes.fr/hal-01375606.

[11] Piotr Kulicki and Robert Trypuz. Multivalued logics for conflicting norm. In *Deontic Logic and Normative Systems (DEON 2016)*, pages 123–138. College Publications, 2016.

[12] Patrick Lin. *Autonomous Driving: Technical, Legal, and Social Aspects*, chapter Why Ethics Matters for Autonomous Cars, pages 69 – 86. Springer.

[13] Hod Lipson and Melba Kurman. *Driverless: Intelligent Cars and the Road Ahead (MIT Press)*. The MIT Press, 2016.

[14] Hillel Steiner Matthew Kramer, N. E. Simmonds. *A Debate Over Rights: Philosophical Enquiries*. 2000.

[15] Neil McBride. The ethics of driverless cars. *SIGCAS Computers & Society*, 2015.

[16] Rapjael Orlove. Now mercedes says its driverless cars won't run over pedestrians, that would be illegal. Internet, October 2016. https://jalopnik.com/now-mercedes-says-its-driverless-cars-wont-run-over-ped-1787890432.

[17] Laurence F. White Samar Chopra. *A legal Theory for Autonomous Artificial Agents*. University of Michigan Press, 2011.

[18] Marek Sergot. Some examples formulated in a 'seeing to it that' logic: Illustrations, observations, problems. In *Outstanding Contributions to Logic*, pages 223–256. Springer International Publishing, 2014.

[19] Michael Taylor. Self-driving mercedes-benzes will prioritize occupant safety over pedestrians. Blog. https://blog.caranddriver.com/self-driving-mercedes-will-prioritize-occupant-safety-over-pedestrians/.

[20] Robert Trypuz. *Formal Ontology of Action*. Elpil, 2008.

[21] Robert Trypuz and Piotr Kulicki. Connecting actions and states in deontic logic. *Studia Logica*, 105(5):915–942, 2017.

[22] U.S. Department of Transportation, National Highway Traffic Safety Administration. *Federal Automated Vehicle Policy Accelerating the Next Revolution in Road Safety.* 2016. US Federal policy concerning AV.

[23] Johan van Benthem, Davide Grossi, and Fenrong Liu. Priority structures in deontic logic. *Theoria*, 80(2):116–152, aug 2013.

[24] Ota Weinberger. *Law, Institution and Legal Politics: Fundamental Problems of Legal Theory and Social Philosophy.* Law and Philosophy Library 14. Springer Netherlands, 1 edition, 1991.

[25] Lihua Zhao, Naoya Arakawa, Hiroaki Wagatsuma, and Ryutaro Ichise. An ontology based map converter for intelligent vehicles. In Takahiro Kawamura and Heiko Paulheim, editors, *Proceedings of the ISWC 2016 Posters & Demonstrations Track co-located with 15th International Semantic Web Conference (ISWC 2016), Kobe, Japan, Octo-*

ber 19, 2016., volume 1690 of *CEUR Workshop Proceedings*. CEUR-WS.org, 2016. http://ceur-ws.org/Vol-1690/paper44.pdf.

[26] Lihua Zhao, Ryutaro Ichise, Zheng Liu, Seiichi Mita, and Yutaka Sasaki. Ontology-based driving decision making: A feasibility study at uncontrolled intersections. *IEICE Transactions*, 100-D(7):1425–1439, 2017. http://search.ieice.org/bin/summary.php?id=e100-d_7_1425.

[27] Lihua Zhao, Ryutaro Ichise, Seiichi Mita, and Yutaka Sasaki. Core ontologies for safe autonomous driving. In Serena Villata, Jeff Z. Pan, and Mauro Dragoni, editors, *Proceedings of the ISWC 2015 Posters & Demonstrations Track co-located with the 14th International Semantic Web Conference (ISWC-2015), Bethlehem, PA, USA, October 11, 2015.*, volume 1486 of *CEUR Workshop Proceedings*. CEUR-WS.org, 2015. http://ceur-ws.org/Vol-1486/paper_9.pdf.

A Formalization of Kant's Second Formulation of the Categorical Imperative

Felix Lindner
Foundations of Artificial Intelligence, University of Freiburg, Germany
lindner@informatik.uni-freiburg.de

Martin Mose Bentzen
Management Engineering, Technical University of Denmark, Lyngby, Denmark
mmbe@dtu.dk

We present a formalization and computational implementation of the second formulation of Kant's categorical imperative. This ethical principle requires an agent to never treat someone merely as a means but always also as an end. Here we interpret this principle in terms of how persons are causally affected by actions. We introduce Kantian causal agency models in which moral patients, actions, goals, and causal influence are represented, and we show how to formalize several readings of Kant's categorical imperative that correspond to Kant's concept of strict and wide duties towards oneself and others. Stricter versions handle cases where an action directly causally affects oneself or others, whereas the wide version maximizes the number of persons being treated as an end. We discuss limitations of our formalization by pointing to one of Kant's cases that the machinery cannot handle in a satisfying way.

1 Introduction

It has been suggested that artificial agents, such as social robots and software bots, must be programmed in an ethical way in order to remain beneficial to human beings. One prominent ethical theory was proposed by Immanuel Kant [1]. Here, we propose a formalization and implementation of Kant's ethics with the purpose of guiding artificial agents that are to function ethically. In particular, the system will be able to judge whether actions are ethically permissible according to Kant's ethics. In order to accomplish this we focus on the second formulation of Kant's

---

We would like to thank the three anonymous reviewers for their valuable comments that helped us to improve the paper.

categorical imperative. Kant proposed three formulations of the categorical imperative. We formalize and implement the second formulation and do not take a stance on the interrelation of Kant's three formulations. The second formulation of Kant's categorical imperative reads:

> Act in such a way that you treat humanity, whether in your own person or in the person of any other, never merely as a means to an end, but always at the same time as an end. (Kant, 1785)

We take it to be the core of the second formulation of the categorical imperative that all rational beings affected by our actions must be considered as part of the goal of the action.

The paper is structured as follows: We first briefly review related work. Then, building upon our earlier work [2], we introduce an extension of Pearl-Halpern-style causal networks which we call Kantian causal agency models. These models serve as a formal apparatus to model the morally relevant aspects of situations. We then define an action's permissibility due to the categorical imperative, while considering two readings of *being treated as a means*. To deal with Kant's wider duties, we introduce an extra condition according to which an agent should maximize the number of persons being treated as an end. Finally, we briefly showcase the computational implementation of the categorical imperative within the HERA software library[1].

2 Related work

In machine ethics, several ethical theories have been formalized and implemented, e.g., utilitarianism, see [3, 4], the principle of double effect, see [5, 6], pareto permissibility, see [2], and Asimov's laws of robotics, see [7].

It has been suggested for some time that Kant's ethics could be formalized and implemented computationally, see [8, 9]. Powers [8] suggests three possible ways of formalizing Kant's first formulation of the categorical imperative, through deontic logic, non-monotonic logic, or belief revision. The first formulation of the categorical imperative states that you must be able to want that the subjective reasoning (or maxim) motivating your action becomes a universal law and as Kant claims that this in some cases is a purely formal matter, it should be possible to formalize it. However, Powers does not provide details of a formalization or a computational implementation, so the formalization of the first formulation in effect remains an open problem.

The work presented here differs in that we focus on the second formulation of the categorical imperative and in that we present a precise formal representation and computational implementation of the formal theory. Rather than taking a

---

[1] http://www.hera-project.com

starting point in one of the paradigms Powers suggests, we use formal semantics and causal agency modelling as this is fitting for the means-end reasoning central to the second formulation. Philosophically, our formalization is best seen as a rational reconstruction within this framework of what we take to be the central ideas of Kant's second formulation.

We think the second formulation has some intuitive appeal also to modern people, many people perceive that there is something morally wrong in using people (including yourself) without consideration of how it affects them. Ultimately, although we are sensitive to Kant's original text, the goal of our work is not to get close to a correct interpretation of Kant, but to show that our interpretation of Kant's ideas can contribute to the development of machine ethics. To meet this goal, our interpretation has to be detailed and explicit enough to provide a decision mechanism for the permissibility or not of specific actions in specific situations.

3   Kantian causal agency models

In order to formalize the second formulation of the categorical imperative, we assume some background theory. First, we assume that *actions* are performed by *agents*, and that actions and their *consequences* can *affect* a set of *moral patients*, i.e. persons who must be considered ethically in a situation. The agent itself is also one of the moral patients. The agent has available a set of actions which will have consequences given *background conditions*. Some of the action's consequences are the *goals* of the action. The actions and consequences that together cause my goal are the *means* of the action. Patients, who are affected by these means are *treated as a means*, and patients, who are affected by my goal are *treated as an end*. For example, I (agent) have the option available to press the light switch (action), and given that the light bulb is not broken (background condition), the light will go on (consequence), which leads to me being able to read my book (consequence). The last consequence was also my goal, and it affects me in a positive way. The action thus treats me as an end.

Within this informally characterized framework, we can reformulate the second formulation of the categorical imperative as follows:

> Act in such a way, that whoever is treated as a means through your action (positively or negatively and including yourself), must also be treated as an end of your action.

The purpose of what follows is to formalize these intuitions. As a first step, we now give the formal definition of the models we will be using in Definition 1. We call these models *Kantian causal agency models* to set them apart from the causal agency models we used in our earlier work [2], and which had no formal tools to consider moral patients affected by one's actions.

Definition 1 (Kantian Causal Agency Model). *A Kantian causal agency model M is a tuple (A, B, C, F, G, P, K, W), where A is the set of action variables, B is a set of background variables, C is a set of consequence variables, F is a set of modifiable boolean structural equations, $G = (Goal_{a_1}, \ldots, Goal_{a_n})$ is a list of sets of variables (one for each action), P is a set of moral patients (includes a name for the agent itself), K is the ternary affect relation $K \subseteq (A \cup C) \times P \times \{+, -\}$, and W is a set of interpretations (i.e., truth assignments) over $A \cup B$.*

A (actions), B (background variables) and C (consequences) are finite sets of boolean variables with B and C possibly empty. W is a set of boolean interpretations of $A \cup B$. Thus, the elements of W set the truth values of those variables that are determined externally, and thus specify the concrete situation. We require that all interpretations in W assign true to exactly one action $a \in A$. As a notational convention, by M, w_a and M, w_b we distinguish two situations that only differ in that in the first situation, action a is performed, and in the second situation, action b is performed.

Causal influence is determined by the set F of boolean-valued structural equations. Each variable $c_i \in C$ is associated with the function $f_i \in F$. This function will give c_i its value under an interpretation $w \in W$. An interpretation w is extended to the consequence variables as follows: For a variable $c_i \in C$, let $\{c_{i1}, \ldots, c_{im-1}\}$ be the variables of $C \setminus \{c_i\}$, $B = b_1, \ldots, b_k$, and $A = \{a_1, \ldots, a_n\}$ the action variables. The assignment of truth values to consequences is determined by:

$$w(c_i) = f_i(w(a_1), \ldots, w(a_n), w(b_1), \ldots, w(b_k), w(c_{i1}), \ldots, w(c_{im-1}))$$

To improve readability, we will use the notation $c := \phi$ to express that c is true if ϕ is true, where ϕ can be any boolean formula containing variables from $A \cup B \cup C$ and its negations. For instance, the boolean structural equations for the light-switch example will be written as $F = \{lightOn := press \land \neg bulbBroken, canReadBook := lightOn\}$.

In the general setting, it may be unfeasible to extend an interpretation from the action variables to the rest of the variables, because it is possible that the value of some variable depends on the value of another variable, and the value of the latter variable depends on the value of the former. Dependence is defined in Definition 2.

Definition 2 (Dependence). *Let $v_i \in C$, $v_j \in A \cup B \cup C$ be distinct variables. The variable v_i depends on variable v_j, if, for some vector of boolean values, $f_i(\ldots, v_j = 0, \ldots) \neq f_i(\ldots, v_j = 1, \ldots)$.*

Following Halpern [10], we restrict causal agency models to acyclic models, i.e., models in which no two variables are mutually dependent on each other. First, note that the values of action variables in set A and the values of background variables in set B are determined externally by the interpretations in W. Thus, the truth values of action variables and background variables do not depend on any other variables. Additionally, we require that the transitive closure, $\prec$, of the dependence relation is a partial order on the set of variables: $v_1 \prec v_2$ reads "v_1 is causally modified by v_2". This enforces absence of cycles. In case of acyclic models, the values of all consequence variables can be determined unambiguously: First, there will be consequence variables only causally modified by action and/or background variables, and whose truth value can thus be determined by the values set by the interpretation. Call these consequence variables *level one*. On *level two*, there will be consequence variables causally modified by action variable, background variables, and level-one consequence variables, and so on [5, 10].

Some of the definitions below will make use of causality. Thus, to take causation into account, Definition 3 defines the relation of y being a but-for cause of ϕ, see [10]. Definition 3 makes use of *external interventions* on models. An external interventions X consists of a set of literals (viz., action variables, consequence variables, background variables, and negations thereof). Applying an external intervention to a causal agency model results in a new causal agency model M_X. The truth of a variable $v \in A \cup B \cup C$ in M_X is determined in the following way: If $v \in X$, then v is true in M_X, if $\neg v \in X$, then v is false in M_X, and if neither $v \in X$ nor $\neg v \in X$, then the truth of v is determined according to its structural equation in M. External interventions thus override structural equations of the variables occurring in X.

Definition 3 (Actual but-for cause). *Let y be a literal and ϕ a formula. We say that y is an actual but-for cause of ϕ (notation: $y \rightsquigarrow \phi$) in the situation the agent choses option w_a in model M_X, if and only if $M_X, w_a \models y \wedge \phi$ and $M_{(X \setminus \{y\}) \cup \{\neg y\}}, w_a \models \neg \phi$.*

The first condition requires that both the cause and the effect must be actual. The second condition requires that if y had not been the case, then ϕ would have not occurred. Thus, in the chosen situation, y was necessary to bring about ϕ. Consider again the book reading situation M, w, such that $w(bulbBroken) = \bot, w(press) = \top$. Due to the structural equations (see above), we have both $M, w \models press$ and $M, w \models canReadBook$. Also, in the intervention where the agent does not press the light switch, the agent cannot read the book, $M_{\{\neg press\}} \models \neg canReadBook$. Therefore, in situation M, w, *press* is a but-for cause of *canReadBook*.

Generally, the definition of but-for cause allows to talk about individual actions and consequences and their causal effects on other individual consequences in the given situation, as well as counterfactual effects if the situation were different from the actual situation. The definition does not allow conjunctive or disjunctive causes. Consequently, this definition of causality does not cope with cases of preemption. For instance, consider the agent shoots at someone who is already about to die, because he was poisoned just a minute ago. In this case, the agent's shot is not a but-for cause for the patient's death—but the disjunction of the agent's shot and the patient being poisoned is. Our examples work with the simpler but-for causality, so we do not discuss more sophisticated definitions of causality (but see [10]).

Based on the concept of but-for cause the useful concept of *direct consequences* is introduced via Definition 4.

Definition 4 (Direct Consequence). *A consequence $c \in C$ is a direct consequence of $v \in A \cup B \cup C$ in the situation M_X, w_a iff $M_X, w_a \models v \rightsquigarrow c$.*

With regard to modeling moral patients affected by effects, we assert that persons can be affected by actions or consequences either in a positive or in a negative way. To represent that some action or consequence (knowingly) affects a person positively or negatively, we introduce the notations $\triangleright_+$ and $\triangleright_-$, respectively. Thus, $M_X, w_a \models c \triangleright_+ p$ holds iff $(a, c, +) \in K$, and $M_X, w_a \models c \triangleright_- p$ holds iff $(a, c, -) \in K$. We use $\triangleright$ in case the valence of affection is not relevant. As a means to refer to the goals of some action, we define $M_X, w_a \models Goal(c)$ iff $c \in Goal_a$, i.e., a consequence c is the goal in the agent's chosen situation w_a iff c is in the set of goals associated with action a (cf., Def. 1).

This finalizes the exposition of the background theory.

4 Categorical imperative defined

We now consider how to make permissibility judgments about actions as defined in the context of Kantian causal agency models using the categorical imperative. The second formulation of the categorical imperative requires an agent to never treat someone merely as a means but always also as an end. Thus, to formalize under which conditions an action is permitted by the categorical imperative, we first define the concept of someone being *treated as an end* (Definition 5). We then proceed to formalize two possible readings of the concept of someone being *treated as a means* (Definition 6 and Definition 7).

Definition 5 (Treated as an End). *A patient $p \in P$ is treated as an end by action a, written $M_X, w_a \models End(p)$, iff the following conditions hold:*

1. *Some goal g of a affects p positively.*
   $M_X, w_a \models \bigvee_g (Goal(g) \wedge g \triangleright_+ p)$.

2. *None of the goals of a affect p negatively.*
   $M_X, w_a \models \bigwedge_g (Goal(g) \rightarrow \neg(g \triangleright_- p))$

Thus, being treated as an end by some action means that some goal of the action affects one in a positive way. One could say that the agent of the action, by performing that action, considers those who benefit from her goal. Things are less clear regarding the concept 'being treated as a means'. As a first step, we define two versions of the concept which we refer to as *Reading 1* and *Reading 2*. Both readings make use of the causal consequences of actions. Reading 1 considers a person used as a means in case she is affected by some event that causally brings about some goal of the action.

$$A = \{pull\}$$
$$C = \{survive1, \ldots, survive6\}$$
$$P = \{person1, \ldots, person6\}$$
$$F = \{survive1 := pull, \ldots, survive5 := pull, survive6 := \neg pull\}$$
$$K = \{(survive1, person1, +), (\neg survive1, person1, -), \ldots\}$$
$$G = (Goal_{pull} = \{survive1, \ldots, survive5\})$$

Figure 1: Model of the classical trolley problem. Person 1 to person 5 are together on the one track, person 6 alone on the other track.

Definition 6 (Treated as a Means, Reading 1). *A patient $p \in P$ is treated as a means by action a (according to Reading 1), written $M_X, w_a \models Means_1(p)$, iff there is some $v \in A \cup C$, such that v affects p, and v is a cause of some goal g, i.e., $M_X, w_a \models \bigvee_v ((a \rightsquigarrow v \wedge v \triangleright p) \wedge \bigvee_g (v \rightsquigarrow g \wedge Goal(g)))$.*

As a consequence, negative side effects are permitted under Reading 1. Consider, for instance, the classical trolley dilemma, where the agent has the choice to either pull the lever to lead the tram onto the second track killing one person, or refraining from pulling letting the tram kill five persons on the first track (see Fig. 1). Under Reading 1, in case of pulling, the one agent—person 6 in Fig. 1—is not treated as a means: If, counterfactually, person 6 survived although the switch was pulled, then this would not deactivate any of the agent's goals. Therefore, the formula $\neg survive6 \rightsquigarrow survive1 \vee \ldots \vee \neg survive6 \rightsquigarrow survive5$ is not satisfied by the model of the classical trolley problem. Reading 1 is probably closest to what we informally mean by 'being treated as a means'.

Reading 2 requires that everybody affected by any direct consequence of the action is considered as a goal.

Definition 7 (Treated as a Means, Reading 2). *A patient $p \in P$ is treated as a means by action a (according to Reading 2), written $M_X, w_a \models Means_2(p)$, iff there is some direct consequence $v \in A \cup C$ of a, such that v affects p, i.e., $M_X, w_a \models \bigvee_v (a \rightsquigarrow v \wedge v \triangleright p)$.*

Hence, under Reading 2, also the person on the second track must be considered as a goal. Consequently, everyone treated as a means according to Reading 1 is also treated as a means according Reading 2, but Reading 2 may include additional patients. Reading 2 is further from the everyday understanding of means-end reasoning, but is probably closer to what some people expect of a Kantian ethics, viz., that everyone affected by the direct consequences of one's actions must be considered. We consider it a feature of a formal framework that this distinction can be formalized, but we leave it for the modeler to decide which one of the readings is more useful for a given application. One thing to note is that Kantian causal-agency models are meant to represent what an agent considers possible. Hence, the agent uses some patient as a means in case she knowingly affects that patient. Thus, the formalization does not require an agent to consider affected moral patients she was not aware of. For instance, if the reader of this paper feels affected by what she reads, then, of course, the authors are not using her as a means.

Having defined both *being treated as an end* and *being treated as a means*, the permissibility of actions according to the second formulation of the categorical imperative can now be defined in Definition 8. The formulation requires that no-one is merely used as a means, but always at the same time as an end.

Definition 8 (Categorical Imperative). *An action a is permitted according to the categorical imperative, iff for any $p \in P$, if p is treated as a means (according to Reading N) then p is treated as an end $M_X, w_a \models \bigwedge_{p \in P}(Means_N(p) \to End(p))$.*

There are thus two main reasons why an action is not permitted: Either a patient is treated as a means but is left out of consideration by the end of the action. Or, the action is done for an end that affects someone negatively.

5 Cases of strict duty

We will now provide examples that highlight aspects of the definition of the categorical imperative. Although these do not prove it correct in any formal sense they can be used to discuss its appeal as an ethical principle as an explication of Kant's ideas. First, we rephrase three cases that contain what Kant calls strict duties (and two of which Kant himself used to explain his ideas).

5.1 Example 1: Suicide

Bob wants to commit suicide, because he feels so much pain he wants to be relieved from. This case can be modeled by a causal agency model M_1 that contains one action variable *suicide* and a consequence variable *dead*, see Figure 2a. Death is the goal of the suicide action (as modeled by the set G), and the suicide affects Bob (as modeled by the set K). In this case, it does not make a difference whether the suicide action affects Bob positively or negatively.

The model assumes that the suicide affects no-one other than Bob, because Kant's argument is not about the effect of suicide on other people but about the lack of respect of the person committing suicide. The reason why Bob's suicide is not permitted is that the person affected by the suicide, viz., Bob, does not benefit from the goal, because he is destroyed and thus cannot be affected positively by it. He is thus treated as a means to his own annihilation from which he receives no advancement. Therefore, the first condition of the categorical imperative (Definition 8) is violated

$$A = \{suicide\}$$
$$C = \{dead\}$$
$$P = \{Bob\}$$
$$F = \{dead := suicide\}$$
$$K = \{(suicide, Bob, +)\}$$
$$G = (Goal_{suicide} = \{dead\})$$

(a) Model M_1

$$A = \{amputate\}$$
$$C = \{survives\}$$
$$P = \{Bob\}$$
$$F = \{survives := amputate\}$$
$$K = \{(amputate, Bob, -),$$
$$(survives, Bob, +)\}$$
$$G = (Goal_{amputate} = \{survives\})$$

(b) Model M_1^*

Figure 2: Kantian Causal Agency Models yielding the impermissibility of Suicide (M_1) and the permissibility of Amputation (M_1^*).

according to both readings (1 and 2), because $M_1, w_{suicide} \models Means_{\{1,2\}}(Bob)$ holds but $M_1, w_{suicide} \models End(Bob)$ does not.

As noted above, it could also be said that the suicide affects Bob negatively, and the action would also be impermissible. The reason for the impermissibility of suicide also in this case is not due to the fact that Bob does something harmful towards himself. As Kant also remarks, other harmful actions would be allowed, e.g., risking your life or amputating a leg to survive. To see this, consider Fig. 2b, where M_1 has been be slightly modified to M_1^*: Rename *suicide* to *amputate* and *dead* to *survives*. Moreover, add $(amputate, Bob, -)$ to K. In this case, Bob is positively affected by the goal, and thus the act of amputation is permitted. The modified example also shows that in some cases, the categorical imperative is more permissive than the principle of double effect, which strictly speaking never allows negative means to an end (cf., [5]).

5.2 Example 2: Giving flowers

The fact that an action can be judged as impermissible by the categorical imperative although no-one is negatively affected is a property of the categorical imperative that inheres in no other moral principles formalized so far. The following example showcases another situation to highlight this property: Bob gives Alice flowers in order to make Celia happy when she sees that Alice is thrilled about the flowers. Alice being happy is not part of the goal of Bob's action. We model this case by considering the Kantian causal agency model M_2 shown in Figure 3a.

In the model M_2, the action *give_flowers* is not permitted according to the categorical imperative, because Bob is using Alice as a means to make Celia happy, but not considering her as part of the goal of the action. This action is immoral,

$$
\begin{aligned}
A &= \{give\_flowers\} \\
C &= \{alice\_happy, celia\_happy\} \\
P &= \{Bob, Alice, Celia\} \\
F &= \{alice\_happy := give\_flowers \\
&\quad celia\_happy := alice\_happy\} \\
K &= \{(alice\_happy, Alice, +), \\
&\quad (celia\_happy, Celia, +)\} \\
G &= (Goal_{give\_flowers} = \\
&\quad \{celia\_happy\})
\end{aligned}
$$

(a) Model M_2

$$
\begin{aligned}
A &= \{give\_flowers\} \\
C &= \{alice\_happy, celia\_happy\} \\
P &= \{Bob, Alice, Celia\} \\
F &= \{alice\_happy := give\_flowers \\
&\quad celia\_happy := alice\_happy\} \\
K &= \{(alice\_happy, Alice, +), \\
&\quad (celia\_happy, Celia, +)\} \\
G &= (Goal_{give\_flowers} = \\
&\quad \{celia\_happy, alice\_happy\})
\end{aligned}
$$

(b) Model M_2^*

Figure 3: Kantian Causal Agency Models yielding the impermissibility of giving flowers to Alice to make Celia happy (M_2) and the permissibility of doing so if making Celia happy is a goal as well (M_2^*).

even though the action has positive consequences for all, and no bad consequence are used to obtain a good one. Again, this example shows how the Kantian principle differs from other ethical principles such as utilitarianism and the principle of double effect, because these principles would permit the action.

The model M_2 can be extended to model M_2^* shown in Figure 3b. In model M_2^*, Bob's action is permitted by the Kantian principle. The only thing in which M_2^* differs from M_2 is that the variable $alice\_happy$ is added to the set $Goal_{give\_flowers}$. In this case, Alice is both treated as a means and treated as an end, which is permitted by the categorical imperative.

The flower example demonstrates how demanding the categorical imperative is, because the principles requires that everybody affected by ones' action must be treated as a goal: This includes the taxi driver that drives you to your destination, as well as the potential murderer you defend yourself against. In these examples, the ethical principle requires one to, e.g., have the taxi driver's earning money among one's goals, and the murderer's not going to jail.

5.3 Example 3: False promise

We return to a case mentioned by Kant himself. Consider that Bob makes a false promise to Alice. Bob borrows one 100 Dollars from Alice with the goal of keeping the money forever. He knows that it is an inevitable consequence of borrowing the

$$A = \{borrow\}$$
$$C = \{bob\_keeps\_100Dollar\_forever\}$$
$$P = \{Alice\}$$
$$F = \{bob\_keeps\_100Dollar\_forever := borrow\}$$
$$K = \{(borrow, Bob, +), (borrow, Alice, -),$$
$$(bob\_keeps\_100Dollar\_forever, Bob, +),$$
$$(bob\_keeps\_100Dollar\_forever, Alice, -)\}$$
$$G = (Goal_{borrow} = \{bob\_keeps\_100Dollar\_forever\})$$

Figure 4: Model M_3 for the case of Bob making a false promise to Alice.

money that he will never pay it back. Figure 4 shows the model of this situation, M_3. The action is impermissible, because Alice is treated as a means (by both Reading 1, Definition 6, and Reading 2, Definition 7). However, none of the two conditions for 'being treated as an end' (Definition 5) are met: None of the goals affects Alice positively, and Bob's goal affects her negatively.

6 Cases of wide duty

Examples 1, 2 and 3 are instances of what Kant calls necessary, strict, narrower duties to oneself and to others, and it seems obvious they involve using a person as a means. Kant also presents two other examples to which we now turn in this section. These involve what Kant calls contingent, meritorious, or wider duties. His arguments for these appear more vague and at least from our perspective harder to handle. We now turn to wide duties and discuss, through an example, how actions that indirectly affect others by refraining from preventing harmful consequences could be handled in the formal framework. Another example will demonstrate where the limitations of our formalization attempt are.

6.1 Example 4: Not helping others

Bob who has everything he needs, does not want to help Alice who is in need. Let us assume she is drowning and Bob is refraining from saving her live. Formally, the situation in the example can be represented with a causal agency model M_4 that contains one background variable *accident* representing the circumstances that led to Alice being in dire straits, two action variables *rescue* and *refrain* and a consequence variable *drown*. Moreover, ¬*drown* is the goal of *rescue*. See Figure 5 for the specification of the model.

$$\begin{aligned}
A &= \{rescue, refrain\} \\
B &= \{accident\} \\
C &= \{drown\} \\
P &= \{Alice, Bob\} \\
F &= \{drown := accident \wedge \neg rescue\} \\
K &= \{(drown, Alice, -), (\neg drown, Alice, +)\} \\
G &= (Goal_{rescue} = \{\neg drown\}, Goal_{refrain} = \emptyset)
\end{aligned}$$

Figure 5: Model M_4 for the impermissibility of not helping others.

According to the categorical imperative using Readings 1 and 2 of 'being treated as a means' both *rescue* and *refrain* are permitted. Bob is strictly speaking not using Alice as a means by going about his business. Kant gives us a clue of how to formalize an argument against refraining in that he says we have to make other people's ends our own as far as possible. Kant writes that 'For a positive harmony with humanity as an end in itself, what is required is that everyone positively tries to further the ends of others as far as he can.' One way of understanding this is as an additional requirement on top of the categorical imperative of choosing an action whose goals affect most people positively. This understanding is captured in Definition 9.

Definition 9 (Meritorious principle). *Among actions permitted by the categorical imperative, choose one whose goals affect most patients positively.*

The meritorious principle thus goes beyond simply avoiding to treat others as means by actively helping them. As formulated here, the principle is compatible with the categorical imperative. In our example, it requires of the agent to choose saving Alice, because the goal advances her. There may be several actions advancing the same number of agents, in which case the agent can choose freely (or randomly) amongst them. One could also take Kant to imply a second condition to the meritorious principle, to prevent as many people being negatively affected by circumstances as possible. In the current example, both conditions would lead to the same result.

6.2 Unhandled case: Not using your talent

As a final example, consider the following situation: Bob has the talent to become a great artist. However, he wonders whether it is permissible to just be lazy and enjoy life instead of working hard to improve himself. Strictly speaking Bob is not working to anyone's disadvantage by being lazy and thus the definitions of 'being treated as a means' advanced above will not cover this example. As the goal of enjoying life and the goal of making art both benefit Bob, the meritorious principle also cannot be used to make the distinction. What Kant says is that laziness could be consistent with the preservation of humanity but does not harmonize with its advancement. He also writes that a rational being necessarily wills that all his capacities are developed. However, it is not clear to us what constitutes the advancement of humanity beyond the sheer feeling of happiness. The example is further complicated by the fact that Kant says that this is a duty one has towards oneself, not others. Therefore, it would be inappropriate to solve this case by introducing others into the model that would benefit from Bob becoming an artist.

In the current formalization, we have no means to represent the relevant aspects that render laziness impermissible and becoming an artist permissible for the right reasons. We thus take this example to showcase a limitation of our treatment of Kant's ethics, and leave a formalization that could capture this last example for further research.

7 Implementation within the HERA framework

The formalization of Kant's second formulation of the categorical imperative has been implemented within the Hybrid Ethical Reasoning Agent software library (short: HERA).[2] The general goal of the HERA project is to provide theoretically well-founded and practically usable logic-based machine ethics tools for implementation in artificial agents, such as companion robots with moral competence [11]. The core of HERA consists of a model checker for (Kantian) causal agency models. Thus, the situations the agent can reason about are represented in terms of models, and ethical principles like the categorical imperative are implemented as (sets of) logical formulae. To showcase the use the categorical imperative from a Python program, Listing 1 reconsiders a representation of the suicide case.

```
{
"actions": ["suicide"],
"background": [],
"consequences": ["dead"],
"patients": ["Bob"],
"mechanisms": {"dead": "suicide"},
"affects": {"suicide": [["Bob", "+"]],
"dead": []},
"goals": {"suicide": ["dead"]}
}
```

Listing 1: A sample JSON encoding of the suicide case.

---

[2]The HERA software is available from http://www.hera-project.com. It is fully implemented in Python and can be installed via the PyPI repository (package name: ethics).

The workflow for using HERA requires to first generate a causal agency model like the one in Listing 1. Given such a model, arbitrary logical formulae can be checked for being satisfied or not by this model. This way, the conditions of ethical principles like the Kantian categorical imperative as defined in Definition 8 can be checked for satisfaction.

To support the usage of the HERA library, the logical formulae to be checked for ethical principles already included in HERA are encapsulated into prepared classes. Listing 2 shows a sample interaction. The first three commands load the implementations of two syntactical entities of the logical language (the predicates **Means** and **End**), the causal agency model from the **semantics** package, and the categorical imperative using Reading 1 of 'being treated as a means' from the **principles** package. The third command loads the suicide example and sets the external variable *suicide* to the value *True*. This way, the *suicide* action is chosen in the situation, and the truth values of the consequence variables can be evaluated the way explained in Section 3. In the concrete case, *True* will be assigned to the variable *dead*. The fourth command asks whether, in the resulting situation, Bob is used as a means according to Reading 1 (see Definition 6). The answer is *True*, because Bob is affected by the action (*suicide*) and the action is a but-for cause of Bob's goal (*dead*). The fifth command asks if Bob is used as an end. This query returns *False*, because Bob is not affected by the goal (see Section Example 1: Suicide). All in all, the action is not permissible according to the categorical imperative, and the output of the last command is accordingly.

```
from ethics.language import Means, End
from ethics.semantics import CausalModel as cm
from ethics.principles import KantianHumanityPrinciple as ci
m = cm("suicide.json", {"suicide": True})
m.models(Means("Reading-1", "Bob"))
```
output: *True*
```
m.models(End("Bob"))
```
output: *False*
```
m.evaluate(ci)
```
output: *False*

Listing 2: A sample interaction with the Python package **ethics**, which we develop and maintain as the standard implementation of HERA.

8 Conclusion

We have shown proof of principle how Kant's second formulation of the categorical imperative can be formalized and implemented computationally. The strict duties towards yourself and others are defined, given goals, structural equations, and the affects relation. To define permissibility according the categorical imperative, we have defined 'being treated as an end', and we formalized two readings of 'being treated as a means' that meet different intuitions about this concept. The formalization deals well with Kant's own examples of strict duties. We were also able to partly deal with Kant's wide duties by defining an additional condition that requires agents to maximize the number of persons being treated as an end.

We envision that the theory will be used as a tool for the comparison of morally relevant aspects of different views on morally delicate cases, thus helping people to have moral discussions. Moreover, we aim at allowing automatic moral judgments in line with Kant in robots such as self-driving cars, care robots, robot companions, and robotic tutors. Our current research investigates whether and under which circumstances Kantian reasoning the way it is presented here is perceived as appropriate for social robots as compared to other types of moral reasoning already defined within HERA.

References

[1] Kant, I. 1785. *Grundlegung zur Metaphysik der Sitten*.

[2] Lindner, F., Bentzen, M., and Nebel, B. 2017. The HERA approach to morally competent robots. In *Proceedings of the 2017 IEEE/RSJ International Conference on Intelligent Robots and Systems (IROS)*.

[3] Horty, J. F. 2001. *Agency and Deontic Logic*. Oxford University Press.

[4] Arkoudas, K.; Bringsjord, S.; and Bello, P. 2005. Toward ethical robots via mechanized deontic logic. Technical Report, AAAI Fall Symposium on Machine Ethics, AAAI.

[5] Bentzen, M. 2016. The principle of double effect applied to ethical dilemmas of social robots. In *Robophilosophy 2016/TRANSOR 2016: What Social Robots Can and Should Do*. IOS Press. 268–279.

[6] Govindarajuli, N. S., and Bringsjord, S. 2017. On automating the doctrine of double effect. In *Proceedings of the Twenty-Sixth International Joint Conference on Artificial Intelligence (IJCAI)*. 4722–4730.

[7] Winfield, A. F.; Blum, C.; and Liu, W. 2014. Towards an ethical robot: internal models, consequences and ethical action selection. In Mistry, M.; Leonardis, A.; M.Witkowski; and Melhuish, C., eds., *Advances in Autonomous Robotics Systems*. Springer. 85–96.

[8] Powers, T. M. 2006. Prospects for a kantian machine. *IEEE Intelligent Systems* 21(4):46–51.

[9] Abney, K. 2012. Robotics, ethical theory, and metaethics: A guide for the perplexed. In Lina, P.; Abney, K.; and Bekey, G. A., eds., *Robot Ethics: The Ethical and Social Implications of Robotics*. MIT Press. 35–52.

[10] Halpern, J. Y. 2016. *Actual Causality*. The MIT press.

[11] Lindner, F., and Bentzen, M. 2017. The hybrid ethical reasoning agent IMMANUEL. In *Proceedings of the Companion 2017 Conference on Human-Robot Interaction (HRI)*. 187–188.

You Must! Maybe You Won't.

Matthew Mandelkern
All Souls College, Oxford
matthew.mandelkern@all-souls.ox.ac.uk

I introduce and discuss what I call *practical Moore sentences*: sentences which combine a command with an admission that the speaker does not know whether she will be obeyed. Practical Moore sentences are infelicitous. I argue that this infelicity is surprising, and can be best explained by adopting a striking hypothesis about human psychology: when giving an order, we must act as if we believe we will be obeyed.

1 Introduction

Moore [17] observed that there is something peculiar about a sentence like (1):

(1) #It's raining but I don't believe it's raining.

This is at least *prima facie* puzzling, because (1) can certainly be *true*. The lesson of Moore sentences, many have thought, is that you must stand in a certain cognitive relation to what you assert—you must know, or at least believe, what you assert. And given plausible assumptions about the logics of knowledge and belief, it is impossible to know or believe the content of a Moore sentence. Suppose, for instance, that you knew (1). Then, assuming knowledge distributes over conjuncts, you would know that you don't believe it's raining. Intuitively, it is then impossible for you to know that it's raining (this intuition is easy to encode in the logic of belief and knowledge). But then you don't know the first conjunct, contrary to assumption.[1]

In this paper I discuss a parallel to Moore sentences in the practical domain. *Practical Moore sentences* are sentences like (2) (and close variants): sentences in which one simultaneously gives a command to do something while expressing that one is not certain one's command will be fulfilled.

(2) #You must close the door, but I don't know whether you will.

Many thanks to three anonymous referees for *DEON* 2018, Agnes Callard, Fabrizio Cariani, Hasan Dindjer, Cian Dorr, Jeremy Goodman, Daniel Harris, Brendan de Kenessey, Dilip Ninan, Jonathan Phillips, Daniel Rothschild, and Zoltán Szabó for very valuable comments and discussion.

[1] See [13]. Wittgenstein's variant with 'might' in [31, II.x.109] introduces some interesting complications, but these complications are not relevant for our purposes for reasons I discuss in §4.3 below.

I argue that the infelicity of practical Moore sentences is surprising, because there are many circumstances in which both conjuncts of a sentence like (2) seem to be true. Indeed, the situation is even more puzzling than for Moore sentences, because (unlike for Moore sentences) there are many situations in which both conjuncts of (2) are knowable and believable; and, indeed, many situations in which each conjunct on its own is assertable—despite the fact that the whole conjunction is not assertable in those situations. I argue that, disanalogies aside, practical Moore sentences, like Moore sentences, teach us something new about attitudes, though the lesson is surprising: in giving an order, you must act towards your addressee as if you believe your order will be obeyed.

2 Practical Moore sentences

Suppose Sue wants to tell her student Liz to turn in her final paper on time. She might do so by saying one of the following:

(3)   You must turn in your final paper on time.

(4)   Turn in your final paper on time.

(5)   I [order/command] you to turn in your final paper on time.

Now, Sue is an experienced teacher. She knows that students don't always get their papers in on time. She knows that Liz in particular is a chronic procrastinator. She wants Liz to know that she knows this, too. She might communicate this with any of the following:

(6)   You might not turn your paper in on time.

(7)   I don't know if you will turn your paper in on time.

(8)   For all I know, you won't turn your paper in on time.

This is all perfectly quotidian. The observation at the heart of this paper is that there is something wrong with asserting the *conjunction* of any of the sentences in (3)–(5) together with any of the sentences in (6)–(8). (The part of the observation concerned with deontic modals was first made, as far as I know, in [27, 28]; the corresponding observation concerning imperatives and performatives has not to my knowledge been discussed.)[2] Here is a sampling of possible combinations:[3]

(9) #You must turn in your final paper on time, but you might not.

(10) #You must turn in your final paper on time, but I don't know if you will.

(11) #You must turn in your final paper on time, but for all I know you won't.

(12) #You might not turn your final paper on time, but turn in your final paper on time.

(13) #I don't know if you will turn in your final paper on time, but turn in your final paper on time.

(14) #I order you to turn in your final paper on time, but you might not.

(15) #I order you to turn in your final paper on time, but I don't know if you will.

(16) #You might not turn your paper in on time, but you must turn it in on time.

(17) #I don't know if you will turn in your final paper on time, but I order you to.

Sentence (3)–(5) are all order-giving sentences: sentences in which the speaker orders the addressee to make true some proposition φ. Let us use 'ORDER(φ)' as a schema which ranges over sentences which are used as orders to make φ the case:[4] this includes sentences containing strong deontic necessity modals intended as commands (⌜You must do φ⌝, ⌜You have to do φ⌝, ⌜You can't do φ⌝; though not when these are restricted in a way which makes it clear that they are not intended to be orders, as in 'According to certain morally bad laws, you must kill your grandmother'); imperatives which are intended as commands (⌜Do φ!⌝; though not in permissive uses, as in 'Close the door! Leave it open! I don't care!'); and performative commands (⌜I order you to do φ!⌝). Sentences (6)–(8) are all proposals to leave it open that some state of affairs obtains. Let us use 'OPEN(φ)' as a schema which ranges over such sentences: these include epistemic possibility modals (⌜Might φ⌝, ⌜Could φ⌝), as well as avowals of ignorance (⌜I don't know whether φ⌝, ⌜For all I know, φ⌝). I propose that from the infelicity of sentences like (9)–(17) we should conclude that

---

[2][27]'s example is 'I'm not sure if Alice will come clean, but she must'. Silk's discussion is brief, but, as will become clear, I am in agreement with the basic idea: 'Uttering ⌜Must φ⌝ not only conveys that ⌜¬φ⌝-possibilities are unacceptable; it suggests that ⌜¬φ⌝-possibilities aren't even on the table for consideration. Of course obligations can go unfulfilled. What is interesting is that speakers appear to assume otherwise, at least for the purposes of conversation, when expressing obligations with 'must'.' [18] discusses similar but importantly different sentences; see §4.4 for discussion. [6, 26, 14, 2, 11] discuss 'deontic Moore paradoxes', which are conjunctions like 'Call him at home! I don't want you to'; these are very interesting, but are obviously quite different from our practical Moore sentences.

[3]I assume that 'but' and 'and' both have the core semantics of conjunction, and use '∧' to range over both.

[4]I will sometimes sloppily speak of 'doing φ'. Since many sentences which can be used as an order can also be used to other effects, ⌜ORDER(φ)⌝ should really range over particular uses of sentences.

sentences of the form $\ulcorner\text{ORDER}(\varphi) \land \text{OPEN}(\neg\varphi)\urcorner$ or $\ulcorner\text{OPEN}(\neg\varphi) \land \text{ORDER}(\varphi)\urcorner$ are typically infelicitous.

This generalization explains the infelicity of (9)–(17), since all those sentences have this form. Moreover, it makes the right predictions about further data in the neighborhood. First, note that further permutations which have the form of practical Moore sentences are all, as far as I can tell, infelicitous. Thus to get one more case on the table, suppose Liz gets up to storm out of Sue's office. Sue wants Liz to close the door behind her, but knows that Liz might not obey her. The following all sound quite weird in this situation:

(18) #You might not close the door, but close it!

(19) #You have to close the door after you, but you might not!

(20) #I don't know if you'll do so, but I order you to close the door!

And so on. Second, nearby permutations on practical Moore sentences which do *not* amount to orders are *not* infelicitous. Consider variants in which we replace $\ulcorner\text{ORDER}(\varphi)\urcorner$ with a construction which tells one's addressee what she *should* or *ought* to do, or what the speaker would *like* her to do, without rising to the level of giving an order:

(21) You should open the door.

(22) You ought to open the door.

(23) I want you to open the door.

I'll schematize claims with this form with 'SHOULD(φ)'. Conjunctions with the form $\ulcorner\text{SHOULD}(\varphi) \land \text{OPEN}(\neg\varphi)\urcorner$ and of the form $\ulcorner\text{OPEN}(\neg\varphi) \land \text{SHOULD}(\varphi)\urcorner$ do not have the same infelicity as practical Moore sentences. Sue can say any of the following:

(24) You should close the door, but you might not.

(25) You might not close the door, but you should.

(26) I want you to close the door, but I don't know whether you will.

(27) You ought to close the door, but for all I know, you won't.

And so on. These are structurally very much like practical Moore sentences, with the one striking difference that sentences of the form $\ulcorner\text{SHOULD}(\varphi)\urcorner$ clearly do not amount to *orders* to do φ (as evidenced by the felicity of conjunctions like 'You should close the door, but you don't have to', or 'I want you to close the door, but I'm not ordering you too').[5] This provides confirmation, again, that our generalization above is correct: the infelicity of (9)–(17) comes from the fact that they conjoin an *order* (not just advice) to φ with a claim that the speaker leaves it open that φ is false.

---

[5] See discussion in [18, 5] and citations therein.

Further support comes from deontic necessity claims and imperatives which do not count as commands, like the following:

(28) According to local custom, you have to take two lumps of sugar in your coffee. But you should not feel bound by local custom!

(29) Open the window! Don't open the window! I don't care at all.

Our generalization predicts that, since these are not orders, they should be felicitous when conjoined with the corresponding OPEN sentences. And this seems to be correct:

(30) You should not feel bound by it, but according to local custom, you have to take two lumps of sugar in your coffee. You might take only one lump, though, since you're not from around here!

(31) Open the window! Don't open the window! I don't know or care what you're going to do.

Our generalization is about a certain kind of *speech act*: one which combines an order to do something with an avowal of ignorance about whether it will get done. The same form of words, in different contexts, can be used, not to given an order, but rather to describe a set of rules or to give permissions. Our generalization predicts that in those cases, the relevant conjunctions will be felicitous; (30) and (31) suggest this is correct.

It is worth noting, finally, that there is nothing special about *conjunction* here: immediately consecutive assertions of ⌜ORDER(φ)⌝ followed by ⌜OPEN($\neg\varphi$)⌝, or the reverse direction, are equally infelicitous:

(32) #You must close the door. (And) I don't know if you will.

(33) #Close the door! (But) You might not do so.

For simplicity, I will continue to focus on the single-sentence variants, but everything I say applies to these sequential versions as well.

3 Why this is so weird

What is so weird about practical Moore sentences is that it looks like there are many situations in which practical Moore sentences can be true; in which they can be believed and known; and in which either conjunct, on its own, can be felicitously asserted—despite the fact that the conjunctions themselves are unacceptable.

Taking each of these in turn: first, practical Moore sentences can be true (at least setting aside the imperative variants, since imperatives are not obviously truth evaluable). In the situation above, (34) seems true:

(34)   Liz must turn her final paper in on time.

(34) is simply a true description of the requirements that bind Liz with respect to her paper. And the following also seem true in the situation above:

(35)   Liz might not turn her paper in on time.

(36)   I don't know if you will turn your paper in on time.

(35) is true because of facts about Liz: she is a procrastinator, chronically turning in papers late. (36), as spoken by Sue to Liz, is true because of Sue's mental state: she simply doesn't know whether Liz will turn her paper in on time, given Liz's tendency towards procrastination. Assuming that truth of both conjuncts suffices for the truth of a conjunction, then, the conjunction of (34) with either (35) or (36) will be true in the situation described above. So practical Moore sentences can be true.

Moreover, practical Moore sentences—unlike Moore sentences—seem perfectly knowable (setting aside, again, the imperative variants). Sue knows her course requirements. So nothing stops her from knowing (34). And Sue knows that Liz procrastinates. So nothing stops Sue from knowing both (35) and (36). Then, assuming that knowledge of both conjuncts suffices for knowledge of a conjunction, it follows that practical Moore sentences can be known (and thus believed). This makes their infelicity even more puzzling than that of Moore sentences: not only are practical Moore sentences consistent (like Moore sentences); they are also (unlike Moore sentences) knowable, meaning that we cannot explain their infelicity in the same way that we account for the infelicity of Moore sentences.

Turning to assertability, things look even more puzzling. As we saw above, Sue is entirely within her rights to order Liz to turn her paper in on time: in the situation above, she has the right kind of authority, knowledge, etc. to tell Liz when she must turn in her paper by. And she is also entirely within her rights to inform Liz that Liz might not turn her paper in on time: Sue knows this is so, and may have good reason to let Liz know that Sue knows it is so. So there are cases in which each conjunct of a practical Moore sentence is assertable—despite the fact that there do not seem to be situations in which their conjunction is assertable.

4   What practical Moore sentences are not

This should give a sense of why practical Moore sentences are so puzzling. It's perfectly alright to give an order to do φ, even when you leave it open that (even after that order) your addressee may not do φ; in the same situation, it's perfectly alright to assert that you leave it open that your addressee won't do φ; but it does not seem permissible to do both of these things together. Before offering a tentative account of the infelicity of practical Moore sentences in the next section, I will discuss and dismiss a number of *prima facie* attractive, but unsuccessful accounts.

4.1 Rhetoric

A natural thought at this point is that there is something rhetorically strange about our conjunctions. Why highlight that you don't know whether your addressee will obey you right at the moment when you are telling them what they should do? There is just something strange about putting these two speech acts together.

I think there is much to this thought, and a precisification of it will form the core of my account below. But we cannot help ourselves to this line in a straightforward way, because the SHOULD variants on practical Moore sentences which we saw above are not infelicitous in the same way, despite likewise putting together two thoughts which do not naturally coexist. The fact that sentences with this form do not have the infelicity of practical Moore sentences shows that the infelicity of the latter thus does not arise simply from the fact that they are conjunctions which put together two thoughts which somehow pull in opposing directions.

4.2 Common ground

A different thought is that, though a speaker can know a practical Moore sentence, the addressee cannot know that the speaker knows it. In other words, it may be fine to order someone to do φ when you think they might not obey you, but not when *they* know this, or not when this fact is common ground between you. But, again, things can't be this simple: it's perfectly felicitous to order someone to do something even when it's common ground that they might disobey you. To illustrate, suppose that Mark is being kidnapped by some particularly vicious-seeming kidnappers. As he is being carried away over his kidnapper's back, he knows it is extraordinarily unlikely that they will let him go; and it is surely plausible that this is common ground in the scenario (i.e. the kidnappers know it too, know that he knows it, and so on). Still, it seems perfectly felicitous for Mark to cry out 'Let me go!', 'You have to let me go!', and so on. So an order can be felicitous even when it is common ground that the addressee might disobey. Even in such situations, however, the conjunction of the order with the claim that his addressee might disobey yields striking infelicity:

(37) #You might not let me go, but let me go!

(38) #You have to let me go right now, even though I don't know if you will!

4.3 Modal semantics

A different thought, which is very tempting given recent developments in the literature on epistemic modals, would be that our phenomenon has something to do with the details of the semantics of modals. I do not think this is right, though. Practical Moore sentences bear a superficial similarity to Wittgenstein's variation on Moore sentences: sentences of the form $\ulcorner \varphi \wedge \text{MIGHT}(\neg \varphi) \urcorner$ and $\ulcorner \text{MIGHT}(\neg \varphi) \wedge \varphi \urcorner$, like 'It might be raining and it's not'. Sentences like this have played an important role in recent work on epistemic modals (see e.g. [8, 32, 16]). But this resemblance is, as far as I can tell, superficial. The infelicity of unembedded sentences like these is straightforward to explain, along just the same lines as Moore sentences; but as we just saw, nothing similar extends to practical Moore sentences. The discussion of Wittgenstein's sentences in the recent literature has focused on how to account for the infelicity of these sentences in *embedded* environments. None of the proposals in that literature helps account for the infelicity of unembedded practical Moore sentences. And, moreover, as we will see in §6, practical Moore sentences do not embed like contradictions, unlike Wittgenstein sentences. A final obstacle for an approach along these lines is that practical Moore sentences sound just as bad when the OPEN conjunct is of the form $\ulcorner$I don't know whether $\varphi \urcorner$, $\ulcorner$For all we know, not $\varphi \urcorner$, and so on. These are not epistemic modal constructions, and, as is now well known in the literature on epistemic modals, do not pattern like epistemic modal constructions in their embedding behavior. So the details of the semantics of epistemic modals will not help with the present puzzle.

A different thought is that the infelicity of practical Moore sentences follows from the *impurity* of modals. [15] argues that many uses of modals mix "flavors" (see also [20]). One possible explanation of our data, inspired by this line, goes as follows. When we process a sentence like 'You must open the door, even though you might not', the two modals in the sentence appear to be interpreted differently: the first is deontic, the second epistemic. But if modals are really impure, perhaps both modals are interpreted on more of an intermediate basis, which combines both epistemic and deontic flavors. Indeed, perhaps both are interpreted relative to the *same* accessibility relation, in which case—assuming that 'might' and 'must' are duals—it would be impossible for both to be true.

While I am sympathetic to the general idea that there are impure uses of modals, this particular line seems to me to take it too far. It also fails to explain the full range of data. First, as we just discussed, not all of our sentences contain modals in the first place. Defenders of the impure line might argue that, although imperatives, performatives, and attitude ascriptions are not literally modals, they have a core modal meaning which suffices to make them prone to impure modal readings. That these have a core modal meaning is quite plausible. But that they are subject to this degree of 'impurity' does not, particularly for attitude ascriptions—I do not know of any independent evidence to that effect. Another argument against this approach is that it does not predict the felicity of variants on practical Moore sentences which contain strong necessity modals or imperatives but which are not used to give orders. This is because this impure modal account does not say anything about *ordering* in particular; and so it seems to entirely miss the crucial generalization that it is the combination of *orders* in particular with OPEN sentences which strike us as so weird, not the combination of strong necessity modals/imperatives in general with OPEN sentences. Finally, again, practical Moore sentences do not embed like contradictions, contrary to what seem to me to be the predictions of this approach.

4.4 Ninan 2005

Let me finally compare practical Moore sentences with similar sentences discussed by [18] (see also [19, 26, 14]), which have the form ⌜ORDER(φ) ∧ ¬φ⌝:[6]

(39) #You must go to confession, but you're not going to.

(40) #Go to confession! You're not going to go to confession.

Practical Moore sentences are much like Ninan's sentences. Indeed, any instantiation of Ninan's sentence schema will *entail* many corresponding practical Moore sentences, since, schematically, φ entails many instances of ⌜OPEN(φ)⌝. For instance, the factivity of knowledge guarantees that φ entails ⌜For all we know, φ⌝; and epistemic modals are standardly associated with reflexive accessibility relations, guaranteeing that φ entails ⌜Might φ⌝. Thus Ninan's data are strictly stronger than mine.

This makes it tempting to look for a common explanation. I will ultimately propose such an explanation. But unfortunately Ninan's own explanation won't do the trick. The problem, at a high level, is that Ninan's explanation crucially exploits the fact that in asserting ⌜ORDER(φ) ∧ ¬φ⌝, the speaker commits herself to the falsity of φ. But there is obviously no such commitment in the case of practical Moore sentences. In more detail, Ninan's explanation runs as follows. Ninan takes as background a widely accepted assumption of action theory [1]: namely, that if you try to bring it about that φ, you must leave it open that φ will happen. So, if you're trying to make someone do φ, you must leave it open that they will do φ; i.e., you can't reject the possibility that they will do φ. This accounts for the infelicity of ⌜ORDER(φ) ∧ ¬φ⌝: such a conjunction is an attempt to get someone to do something, made while rejecting the possibility that they will do it.

This strikes me as a good explanation of Ninan's data. But it is clear that it won't extend to account for our data. The issue, again, is that Ninan's explanation depends on the key fact that an assertion of ⌜ORDER(φ) ∧ ¬ϕ⌝ communicates (in part) that the speaker believes that φ is false. By contrast, one does *not* have to believe φ is false in order to assert a practical Moore sentence, ⌜ORDER(φ) ∧ OPEN(¬φ)⌝: you need only believe that there is a *possibility* that φ is false. Thus asserting a practical Moore sentence requires only that you leave open the possibility that what you are ordering your interlocutor to do will not happen. But *that* is not ruled out by the action theoretic principle which Ninan assumes above, which entails that you should not order someone to do something you think *will* not happen, but says nothing to

---

[6]Plus order variants; I will stop appending that caveat, but it applies throughout the remainder of the paper. Ninan is also concerned with the distribution of deontic versus epistemic readings across different tenses; I will not address that second issue.

rule out ordering someone to do something that you think *might* not happen.

A natural idea is just to strengthen Ninan's principle so that it does cover our cases. Instead of saying that, if you're trying to get someone to do φ, you must leave it open that they will φ, we could say that if you're trying to get someone to do φ, you have to believe that they will do φ. A principle like this would strengthen Ninan's account just enough to cover our data. The problem with this principle is that it is obviously false: cases like the kidnapper case above show that there is nothing wrong with trying to get someone to do φ even when you are not sure they will obey you (and indeed, even when you think there is a rather substantial chance that they will not).

5 Posturing

Ninan's key principle, again, says that there is something wrong with giving a command to φ if one believes that φ won't obtain. Although, as we have just seen, the natural extension of this to account for our data is false, there is a more plausible extension which also relates the performance of certain speech acts to certain attitudes, but in a more subtle way, as follows:

> *Posturing*: When you order someone to φ, you must act towards them as if you believe that they will φ.

I argue that the lesson of practical Moore sentences is that *Posturing* is true. When you order someone to φ, you need not believe that they will φ, but you must act towards them *as if* you believe they will φ. You cannot acknowledge that your order might not be obeyed: you must assume an authoritative position with respect to your command.

Let's see first how *Posturing* accounts for our data. When you assert a practical Moore sentence, you are ordering your addressee to φ. You are also making it clear that you are not confident that φ will happen: you are making it manifest that ⌜$\neg\varphi$⌝ remains a live possibility for you, by conjoining your order with a claim of the form ⌜OPEN($\neg\varphi$)⌝. So you are failing to act as if you believe φ will obtain. If *Posturing* is true, it follows that there is something wrong with your speech act.

Posturing also accounts for the contrast between practical Moore sentences and the SHOULD variants. Following the standard line in the literature, again, I assume that claims of the form ⌜SHOULD(φ)⌝ do not amount to commands; they are something weaker, expressions of preference or commitment or advice. Then *Posturing* does not predict there to be anything wrong with an assertion of the form ⌜SHOULD(φ) ∧ OPEN($\neg\varphi$)⌝, matching observation, and capturing the crucial contrast between practical Moore sentences and SHOULD variants. *Posturing* likewise accounts, in a parallel way, for the contrast observed above between practical Moore sentences and variants which are used in contexts where the deontic modal or imperative is not interpreted as an order.

Finally, *Posturing* extends immediately to account for Ninan's data. If you assert something of the form ⌜ORDER(φ) $\wedge$ $\neg\varphi$⌝, you are ordering your interlocutor to do φ while manifestly not acting as if they will (this follows from the fact that ⌜$\neg\varphi$⌝ entails ⌜OPEN($\neg\varphi$)⌝). And so, as desired, we have a unified explanation of both our data and of Ninan's. (This explanation of Ninan's data is of course consistent with Ninan's; it may well be that the infelicity of Ninan's data is overdetermined.)

Why is *Posturing* plausible? As I see it, *Posturing* is a reasonable way of spelling out the broadly rhetorical strategy mooted in §4.1. The idea there was that what is wrong with our infelicitous conjunctions is broadly rhetorical: why talk about your uncertainty about whether someone will do something when you're telling them that they ought to do it? As we saw above, this formulation of the rhetorical strategy is too broad: it does not predict the observed contrast between our order-giving practical Moore sentences and the advice-giving variants with SHOULD. *Posturing* makes the rhetorical strategy more discriminating: it is not that there is something wrong with giving advice while at the same time expressing doubt at whether it will be followed; what is bad is rather specifically *commanding* your interlocutor to do φ without acting towards them as though you are confident they will do φ. To acknowledge that your interlocutor might not obey you is to acknowledge that you are in a position of some weakness with respect to your act of ordering them to do something. And, it seems, we are disinclined to acknowledge such weakness, even when it is manifestly present—indeed, even when it is common ground, as in the kidnapping case. We are obligated, as it were, to put a good face on things, and act as though we are in full control of the situation—and our interlocutor—even when we are not.

So I think that we should accept *Posturing*, based on the fact that it provides a plausible explanation of the infelicity of practical Moore sentences. *Posturing* is fascinating as a thesis about human psychology, and there is much work to do in situating it within a broader philosophical and psychological theory of command and authority. I will not try to give an account here; my main goal is rather to explore the phenomenon of practical Moore sentences, and provide a first-pass account of them, and I do not have space for an adequate discussion of the subtle action theoretic principles that might further explain *Posturing*. But I suspect that a full account of *Posturing* will connect it to the way that we think about authority in general, and I will say just a bit about that connection. To see the kind of account I have in mind, consider the following possible explanation of *Posturing*. In the philosophical literature on authority, it has been suggested that to fully exercise authority over someone is to give that person practical reasons for action: 'The commander characteristically intends his hearer to take the commander's will instead of his own as a guide to action' [12].[7] And it has been proposed in the action theoretic literature that when you practically commit to do something—i.e. when you intend to do it—you must believe you will do it.[8] Putting these thoughts together, it follows that for A to fully exercise authority over B in ordering B to φ, A is, essentially, intending that φ happen *via* B's agency, and thus A must believe that φ will happen. Now we have seen that it is implausible to say that, when you give a command, you must believe it will be obeyed. But a more plausible thought is in the neighborhood: we could say that one can give a command without having full authority over one's addressee in the sense just sketched; but that, in giving a command, one must act as if one has full authority over one's addressee. Putting all this together, we arrive at *Posturing*: in giving a command to φ, you must act as if you have full authority over your addressee, which in turn requires acting (towards them) as though you believe they will actualize your intention to φ.[9]

Every step of this story is controversial, and I will not commit myself to any of it here (I am not sure how much of it I myself believe). My goal with this sketch is not to give an explanation of *Posturing*, but rather to give a sense of the kind of explanation we might find of *Posturing* by way of a more careful exploration of broad principles in action theory and the theory of authority. Such considerations will help us understand *Posturing*; and, of course, *Posturing* may well help us decide between rival theories of authority and commands, since, if my main argument in this paper is correct, our theory of commands will have to validate *Posturing*.

Let me close this section with a final high-level remark. If *Posturing* is correct, it shows that certain rhetorical rules which might seem entirely elective are, in fact, cognitively hardwired. In other words, the advice given in *Posturing* might seem like reasonable advice to give someone about how to appear more authoritative; we might even imagine it being backed up by social psychological research which shows that people will be more likely to follow your orders if you act towards them as if you are confident that they will. But what seems quite surprising to me is that something like *Posturing* should be hardwired as a normative requirement on speech acts: i.e. that violating it leads not only to the judgment that the speaker is doing something odd or inefficient (as we might judge about the $\ulcorner \text{SHOULD}(\varphi) \wedge \text{OPEN}(\neg\varphi) \urcorner$ sentences above), but that they have done something *wrong*, something which results in a distinctive intuition of linguistic infelicity. In other words, what we seem to find here is something which *prima facie* looks like nothing more than a good rhetorical rule of thumb, but which turns out to be encoded as linguistic knowledge (or something like it). I find this situation quite surprising, and perhaps that means that *Posturing* is not the right explanation of our data. On the other hand, it is the best explanation I can think of, and so for the present I am inclined to accept it, surprising as it is,

---

[7] See e.g. [22, 23, 24, 25, 9] for related discussion.

[8] Ninan discusses this principle, though remains non-committal on it; see [7, 10, 30, 21] for arguments in favor and [3, 1] for arguments against.

[9] Thanks to Hasan Dindjer and Brendan de Kenessey for very helpful discussion here.

and draw from it the lesson that rhetorical rules like *Posturing* sometimes are, to a surprising degree, built into our understanding of language.

6 Objections and replies

Before concluding, let me consider a few potential objections to my account. The first comes from third-personal variants on practical Moore sentences like (41):

(41)   Liz must turn in her paper on time, but she might not.

Note first that there are contexts in which (41) sounds felicitous. Those are contexts where it is clear that the first conjunct in (41) is not a command, but simply a description of the requirements that bind Liz. Suppose that two TAs are speculating about whether Liz will pass the class. One says:

(42)   Well, I'm not sure she will. She has to turn her paper in on time, but she might not get it done. Then she would fail.

Here it is clear that the TA is not giving a command to Liz of any kind; she is simply describing the constraints that bind Liz. And in this case, the conjunction in (42) sounds perfectly felicitous, in line with our observation at the outset that the infelicity of practical Moore sentences arises from the fact that they conjoin OPEN sentences with *orders*.

But there are other contexts where (41) sounds infelicitous, and these contexts constitute a *prima facie* challenge for our approach. Suppose Sue says (41) to her TA, who has asked Sue whether Liz, a chronic procrastinator, can have an extension on her final paper. In that context, (41) sounds very odd. If there were no such thing as third-personal commands, then our account would not predict the infelicity of (41) in this case. But I think that in this case (41) *is* a command: although Liz is not present when (41) is asserted, it is still reasonable to interpret (41) as a command which requires Liz to turn her paper in on time, and thus requires Sue's TA to not grant Liz an extension.[10] The infelicity of (41) in this case would be a problem for our account if (41) were not an order, since our account makes crucial reference to ordering. But if in this context (41) is an order, as I believe it is, then we rightly predict it will be infelicitous; that this is the right prediction is, again, brought out by the contrast with the first context, where (41) is clearly not an order, and thus sounds felicitous, as we predict.

Second, the account I have given relies essentially on the *speech act* properties of practical Moore sentences: properties that adhere to *unembedded* practical Moore sentences. Thus if the infelicity of practical Moore sentences persists in embedding

---

[10][18] advocates a similar reply in response to the parallel objection to his account. The plausibility of the claim that there are third-personal commands of this kind seems to me to be increased by the existence of languages which have imperatives in the third person.

environments which cancel entailments, then my story is the wrong one; conversely, if the infelicity goes away, this provides support for my story. To see this point, it may be useful to compare the parallel situation for Moore sentences: if the standard account of Moore sentences is the correct one (they are consistent, just not assertable), then the infelicity of Moore sentences will not percolate up to sentences which embed them in entailment-canceling environments. This turns out to be exactly how things work for Moore sentences.[11] Thus e.g. the following is felicitous:

(43) Mark thinks that it's raining and that I don't know it's raining.

When we explore similar embedding constructions with practical Moore sentences, we find that, just as for Moore sentences, the infelicity of practical Moore sentences does not percolate up to sentences which embed them in entailment-canceling environments. This is exactly in line with the predictions of our account.[12] Thus for instance the following sound fine:

(44) If Liz has to turn in her paper on time but might not do so, we really need to tell her how seriously this is taken here.

(45) If we don't know whether Liz will turn in her paper on time but she has to do so, we really need to tell her how seriously this is taken here.

(46) Suppose that Liz must turn in her paper on time, and that we don't know whether she will do so.

(47) Suppose I command you to let me go, but I don't know whether you will.

(48) Mark thinks that Liz has to turn in her paper on time, but he also thinks that she might not do it.

The felicity of sentences like these suggests that—just as for Moore sentences—we are right to look for a speech-act level explanation of the infelicity of practical Moore sentences, which righly predicts that their infelicity does not extend to embeddings like this, since, in these embeddings, they are no longer being used to simultaneously order someone to do φ while leaving it open that φ won't happen.

The third objection, suggested to me by Daniel Rothschild,[13] comes from the interplay of orders and conditionals. Consider a sentence like (49) or (50), as asserted by Sue to Liz:

(49) You must turn your paper in on time, and if you don't turn it in on time, you'll fail the class.

---

[11]Though not for the variants on Moore sentences with 'might'. This provides another striking divergence between practical Moore sentences and those variants, providing further evidence that we should not look for a unified account of the two phenomena.

[12]The difficulty of embedding imperatives makes it hard to test this for the imperative variants.

[13][28, Fn. 36] discusses similar data.

(50) Turn your paper in on time! If you don't, you'll fail the class.

(49) and (50) are felicitous. This is somewhat surprising from the point of view of *Posturing* given standard assumptions about indicative conditionals. Indicative conditionals are often taken to be felicitous only when the speaker leaves open the possibility that the antecedent is true (see [29, 4] for discussion). On a natural way of spelling out this line, ⌜If φ, ψ⌝ entails (or presupposes) ⌜OPEN(φ)⌝, which means that (49) and (50) would entail practical Moore sentences: they would entail 'You must turn your paper in on time, but you might not', and 'Turn your paper in on time! You might not', respectively. But then we would expect (49) and (50) to be infelicitous. Put differently, if this thought about indicative conditionals were correct, then (49) and (50), like practical Moore sentences, would both make manifest that the speaker is not confident that she will be obeyed. Since they are also both orders, this should be ruled out by *Posturing*: *Posturing* seems to wrongly predict that (49) and (50) will be infelicitous.[14]

The underlying issues here are quite complex, but there is a simple way to defuse this objection: whatever the compatibility constraint on the antecedents of indicative conditionals amounts to, we can show that it falls short of entailing ⌜OPEN(φ)⌝. To see this, note the felicity of (51):

(51) Liz will turn her paper in on time. [And] if she doesn't, I'll be furious.

If ⌜If φ, ψ⌝ entailed ⌜OPEN(φ)⌝, then 'If she doesn't [turn her paper in on time], I'll be furious' would entail 'Liz might not turn her paper in on time' and 'For all I know, Liz won't turn her paper in on time'. Then (51) would entail (52) and (53):

(52) #Liz will turn her paper in on time. [And] she might not.

(53) #Liz will turn her paper in on time. [And] for all I know, she won't.

These of course both sound awful; they are, in fact, just Moorean sentences. So ⌜If φ, ψ⌝ does not, after all, seem to entail ⌜OPEN(φ)⌝.[15] I will not make a positive

---

[14] An interesting feature of these conjunctions/sequences is that they are sensitive to order; thus while (50) sounds fine, the reverse order in (i) does not:

(i) #If you don't turn your paper in, you'll fail the class. Turn your paper in on time!

This may suggest that the felicity of the conjunctions in (49) and (50) is due to something like a mid-sequence context change. But I think this would be mistaken. First, sequences like (51) sound fine in either order. Second, as Daniel Rothschild has pointed out to me, there are similar order effects quite generally when we put an order together with some kind of inducement to perform the order; thus compare 'Turn your paper in on time! There's a reward', which is felicitous, versus 'There's a reward for turning your paper in on time! Turn it in on time!', which is quite weird. So I suspect that whatever explains the infelicity of this latter variant also explains the infelicity of (i), and that it does not have to do with *Posturing*.

[15] As Cian Dorr and an anonymous referee have helpfully pointed out, perhaps ⌜If φ, ψ⌝ still

proposal about the compatibility requirements of indicative conditionals here; as these brief considerations suggest, this is clearly a very tricky topic. One possibility that all this leaves open, as an anonymous referee helpfully points out, is that ⌜If φ, ψ⌝ does communicate that the possibility of φ is somehow open, but in a very weak way—weaker than ⌜Might φ⌝ or ⌜For all I know, φ⌝. But this way of leaving open that possibility that φ is so weak that it seems to me compatible with acting as if φ is false; I would intuitively describe the speaker of (51) as leaving open, in a very weak way, the possibility that Liz won't turn her paper in on time, but nonetheless acting as if Liz will. If we do want to go this way, then the upshot would thus be that cases like (49) show that, in giving an order, you *can* leave open the possibility that you won't be obeyed, but only in an extremely weak way, and one which is consistent with the requirement of *Posturing* that you act as if you will be obeyed.

In short, then, the present data make it clear that indicative conditionals do not generally entail anything as strong as the corresponding OPEN sentences. This suffices to defuse the present objection: if indicative conditionals do not generally entail the corresponding OPEN sentences, then *Posturing* does not predict that (49) and (50) will be infelicitous.[16]

7 Conclusion

I have explored a practical analogue of Moore sentences, namely sentences in which one gives an order, while acknowledging that one is less than certain that it will be obeyed. The infelicity of these practical Moore sentences is surprising, because there are situations in which both conjuncts are true, knowable, and (individually) assertable, but no situations in which the whole conjunction is assertable.

On a parallel with the standard account of the infelicity of Moore sentences, I have offered a preliminary account of the infelicity of practical Moore sentences on the basis of general facts about human cognition. The standard account of Moore sentences goes by way of the fact that the speaker of a Moore sentence cannot know (or believe) both of its conjuncts. Something exactly parallel is not plausible for practical Moore sentences. Instead, I have proposed that the lesson of practical

---

presupposes ⌜OPEN(φ)⌝, and the difference between (51) versus (52) and (53) can be explained in terms of the pragmatic difference between entailment and presupposition. I am not sure exactly how such a story would go, but if an account like this could be spelled out, then it might well be that the same distinction also lets us defuse this point as an objection to *Posturing*.

[16] A similar sort of worry, raised by Hasan Dindjer, comes from orders combined with questions, as in 'You must close the door! Will you please do so?' If questions are appropriate only when the speaker does not know the answer, these pose a problem for *Posturing*. I am inclined to think, however, that in this case, the question 'Will you do so?' does not express ignorance but rather is seeking verbal consent to the order. After all, if the question expressed ignorance, then this sequence would entail 'You must close the door! I don't know if you will', which of course is infelicitous. (Compare: 'Liz will turn in her paper on time. *Won't* she, Liz?' Passive aggression aside, this is perfectly felicitous.)

Moore sentences is that when you order someone to do something, you must adopt a posture of confidence towards them: you must behave towards your interlocutor as though you are confident that they will comply. This explains the infelicity of practical Moore sentences, while accounting for the fact that close variants—with SHOULD, or with strong deontic necessity modals or imperatives not used to give orders—do not exhibit the same infelicity. It is, however, a surprising conclusion: *Posturing* sounds plausible enough as something like rhetorical good sense, but does not sound like the sort of thing that would be hard-wired psychologically to the extent that we experience its violation as something like linguistic infelicity. If my account is right, then, it reveals both a surprising specific fact about human psychology—that *Posturing* is true—as well as a surprising structural fact about human psychology—that rules like *Posturing* can be encoded in a way that we experience as linguistic infelicity.

My discussion here has been necessarily preliminary. Further exploration of the data; of alternate possible solutions; and of my proposed solution are certainly in order. In particular, we should explore the philosophical and psychological foundations of *Posturing*, as well as the possibility of finding evidence for *Posturing* in other domains, linguistic and non-linguistic (if *Posturing* is correct, there's no reason to think that evidence for it will be limited to the linguistic domain).

We should, finally, explore the ramifications of *Posturing* for the broader question of what norms govern speech acts in general. To give a sense of the kind of light *Posturing* might shed on this topic as a whole, consider the following. It is commonly accepted that one should know, or at least believe, what one asserts. A norm of this kind can be motivated on the basis of Moore sentences like those we saw at the outset, which are usually explained by way of such a norm. But there are variants on Moore sentences which sound equally bad, but are not explained by such norms, like (54):

(54) #It's raining, but I'm not absolutely certain that it's raining.

(54) is not explained straightforwardly by a knowledge or belief norm on assertion, since a speaker can believe, and indeed know, (54) (provided, as on most accounts, that knowledge does not require absolute certainty). (54) *would* be explained by a norm which said one must *know with absolutely certainty* whatever one asserts. But it is pretty clear there is no such norm: one can assert in the absence of absolute certainty. *Posturing* may offer a helpful lens through which to understand this situation: perhaps what is required here is not that you must know with certainty what you assert, but rather—far more plausibly—that you must *act as if* you do. There thus may be a whole bevy of norms which, like *Posturing*, require not that one have a certain cognitive relation towards a content in order to perform some speech act with that content, but rather that one act *as if* one has that cognitive relation. Further exploration of *Posturing* thus promises to illuminate not just practical speech

acts, but also more epistemic speech acts—which may turn out to be less different from practical speech acts than they first appear.[17]

References

[1] Michael Bratman. *Intention, Plans, and Practical Reason*. Harvard University Press, 1987.

[2] Cleo Condoravdi and Sven Lauer. Imperatives: meaning and illocutionary force. In Christopher Piñón, editor, *Empirical Issues in Syntax and Semantics*, volume 9, pages 37–58, 2012.

[3] Donald Davidson. Intending. In *Essays on Actions and Events*. Oxford University Press, 1980.

[4] Kai von Fintel. The presupposition of subjunctive conditionals. In Orin Percus and Uli Sauerland, editors, *The Interpretive Tract(25)*, pages 29–44. MIT, Department of Linguistics, Cambridge, Massachusetts, 1998.

[5] Kai von Fintel and Sabine Iatridou. How to say 'ought' in foreign: The composition of weak necessity modals. In J. Guéron and J. Lecarme, editors, *Time and Modality*, pages 115–141. Springer, 2008.

[6] Annette Frank. *Context Dependence in Modal Constructions*. PhD thesis, Universität Stuttgart, 1996.

[7] H. Paul Grice. Intention and uncertainty. In *Proceedings of the British Academy*, volume LVII, 1972.

[8] Jeroen Groenendijk, Martin Stokhof, and Frank Veltman. Coreference and modality. In *Handbook of Contemporary Semantic Theory*, pages 179–216. Oxford: Blackwell, 1996.

[9] Matthew Hanser. Doing another's bidding. In George Pavlokos and Veronica Rodriguez-Blanco, editors, *Reasons and Intentions in Law and Practical Agency*, pages 95–120. Cambridge University Press, 2015.

[10] Gilbert Harman. *Change in View*. MIT Press, Cambridge, MA, 1986.

[11] Daniel Harris. Imperatives and intention-based semantics. Manuscript, Hunter College, July 2017.

[12] H.L.A. Hart. *Essays on Bentham: Jurisprudence and Political Philosophy*. Oxford University Press, 1982.

[13] Jaakko Hintikka. *Knowledge and Belief: An Introduction to the Logic of Two Notions*. Cornell University Press, Ithaca, NY, 1962.

[14] Magdalena Kaufmann. Interpreting imperatives. In *Studies in Linguistics and Philosophy*, volume 88, New York, 2012. Springer.

[15] Joshua Knobe and Zoltán Gendler Szabó. Modals with a Taste of the Deontic. *Semantics and Pragmatics*, 6(1):1–42, 2013.

[16] Matthew Mandelkern. Bounded modality. *The Philosophical Review*, Forthcoming.

[17] G.E. Moore. A reply to my critics. In P.A. Schlipp, editor, *The Philosophy of G.E. Moore*. Northwestern University, Evanston, IL, 1942.

---

[17]Thanks to Daniel Rothschild for suggesting this possibility and for very helpful discussion.

[18] Dilip Ninan. Two puzzles about deontic necessity. In J. Gajewski, V. Hacquard, B. Nickel, and S. Yalcin, editors, *New Work on Modality*, volume 51, pages 149–178. MIT Working Papers in Linguistics, 2005.

[19] F. R. Palmer. *Modality and the English Modals*. Longman, New York, 1990.

[20] Jonathan Phillips and Joshua Knobe. The psychological representation of modality. *Mind and Language*, 33:69–94, 2018.

[21] Douglas W. Portmore. Opting for the best. Unpublished manuscript, Arizona State University, August 2017.

[22] Joseph Raz. *Practical Reason and Norms*. Oxford University Press, 1975.

[23] Joseph Raz. Reasons for action, decisions and norms. *Mind*, 84(336):481–499, 1975.

[24] Joseph Raz. The justification of authority. In *The Morality of Freedom*. Oxford University Press, 1988.

[25] Abraham Sesshu Roth. Shared agency and contralateral commitments. *The Philosophical Review*, 113(3):359–410, 2004.

[26] Magdalena Schwager. *Interpreting imperatives*. PhD thesis, Johann Wolfgang Goethe-Universität, Frankfurt am Main, 2006.

[27] Alex Silk. What normative terms mean and why it matters for ethical theory. In M. Timmons, editor, *Oxford Studies in Normative Ethics*, volume 5, pages 296–325. Oxford University Press, 2015.

[28] Alex Silk. Weak and strong necessity modals. In B. Dunaway and D. Plunkett, editors, *Meaning, Decision, and Norms: Themes from the Work of Allan Gibbard*. Forthcoming.

[29] Robert Stalnaker. Indicative conditionals. *Philosophia*, 5(3):269–86, 1975.

[30] David Velleman. *Practical Reflection*. Princeton University Press, 1989.

[31] Ludwig Wittgenstein. *Philosophical Investigations*. Wiley-Blackwell, 3rd edition, 1991/1953.

[32] Seth Yalcin. Epistemic modals. *Mind*, 116(464):983–1026, 2007.

Contraction of Combined Normative Sets

Juliano Maranhão*
University of São Paulo Law School
julianomaranhao@usp.br

Edelcio Gonçalves de Souza
University of São Paulo, Department of Philosophy-FFLCH
edelcio.souza@usp.br

We adopt a view of normative and legal interpretation as a dynamic of mutual adjustments and coherent revisions of meaning ascriptions and the content of rules. Within this activity, consistent sets of regulations may become incoherent in virtue of the ascriptions of meaning to its rules. In order to model it, using the framework of i/o-logics (Makinson and van der Torre [10]), we propose and characterize an operator of kernel contraction for single normative sets and three operators of kernel contraction for combined normative sets, called, respectively, constitutive, regulative and combined contraction. The constitutive contraction changes the set of meaning ascriptions, while the constitutive contraction changes the set of rules. The combined contraction is a general operator that may change, coherently, both meaning ascriptions and rules.

Keywords: Normative systems, interpretation, input/output logics, belief revision, combination of logics.

1 Introduction

Conflicts of norms with logically independent conditions of application represented once a watershed problem in deontic logic.

It seems intuitive that issuing opposite commands to different and independent contexts $O(x/a)$ ("it is obligatory to x in condition a") and $O(\neg x/b)$ should not be a problem (indeed should be usual) to norm-giving activity, while opposite demands for the same condition, i.e. $O(x/a)$ and $O(\neg x/a)$, should be a genuine conflict. On the other hand, it seems plausible to assume that the instantiation of the antecedent

---

We are grateful to Leon van der Torre for several insights and discussions on the subject of this paper and to Guilherme Kenzo. We are also indebted to the criticism by an anonymous referee, which led to significant improvements.

*I am grateful to CAPES and the *Alexander von Humboldt Foundation* for supporting this research project (Project nr.99999.000073/2016-04).

of a conditional norm should be sufficient to detach its normative consequent, given that the choice of such conditions by the lawgiver divides authoritatively what is and what is not relevant to the application of the deontic solution provided by the rule.

However, both intuitions collapse, given that, by the principle of *strengthening the antecedent*, one derives both $O(x/a \wedge b)$ and $O(\neg x/a \wedge b)$ from $O(x/a) \wedge O(\neg x/b)$. The clash of these intuitions provoked a schism in deontic logic, between the so-called Hansson-Lewis tradition, of defeasible deontic logics, and the so-called von-Wright-Alchourrón tradition, which drops the principle of non-contradiction for conditional norms. As it is well known, within this schism, Alchourrón has opposed defeasible deontic logics as philosophically unsound and defended that dyadic normative conflicts should be dealt with by normative change [1] not by weakening the inferential power of deontic logic.

A similar problem appears when the interpreter of conditional rules ascribes meanings to its conditions of application. Such meaning ascriptions may be the source of genuine normative conflicts even if the normative order is consistent.

Consider, for instance, an interpretive problem that appeared in many jurisdictions regarding the empowerment of police officers to access the message exchange stored in a cell phone collected in a search and seizure procedure. In some sense, such message exchange is simply stored data and "data" is an object of property rights. But, in another sense, such message exchange is also a sort of ongoing conversation and therefore it may be considered "communication". As it happens in different countries, the police is empowered by a search and seizure order to collect property and access any data and information found in the premises, but it is forbidden to intercept communication without an specific order, provided that freedom of communication is a fundamental right. Hence considering the rules *"it is forbidden to access communication"* and *"it is permitted to access property"*, ascriptions of meaning to the effect that *"message exchange counts as data"*, *"data counts as property"* and *"message exchange counts as communication"* would imply that police officers are both forbidden and permitted to access the content of message exchange stored in a cell phone collected at the searched premises.

In order to solve such conflicts, the interpreter or the legislator faces a dilemma between changing the meaning ascriptions previously assumed or changing the rules that are triggered by those meaning ascriptions, in order to keep coherence. There may be less flexibility to change authoritative rules of conduct, but meaning ascriptions may be also the content of positive law. Besides, changing or abandoning a meaning ascription may provoke a greater loss of normative consequences, since the same legal concept may trigger the application of several different and important rules of conduct.

This dilemma between changing legal concepts or changing legal rules is linked

---

[1] See Maranhão [14] for a discussion of Alchourrón's criticisms to non-monotonic logics

to the challenge of keeping coherence of the normative system, while losing less conditional norms or normative consequences as possible. Therefore, it has a close affinity with the main principles of belief revision (consistency and minimal change). The purpose of this paper is to present and characterize contraction operators that capture this interplay between meaning ascriptions and rules of conduct.

2 Normative interpretation as revision

Investigations into the logic of meaning ascriptions have been developed within the field of deontic logic in terms of so called "constitutive rules", a concept derived from Searle's broader theory of "institutional facts" [17]. One should then distinguish *regulatory rules*, that is, rules obliging, permitting or prohibiting actions or states of affairs, from *constitutive rules*, i.e. rules that authoritatively define or conceptualize actions or states of affairs in the context of an institution. Grossi et.al.[7] have proposed a modal formalism to represent constitutive rules as statements of the form "*X counts as Y in the context/institution C*". The "counts as" relation is simply a strict conditional, which holds in a restricted set of possible worlds (the context).

There are basically two approaches to combine constitutive with regulative rules to model legal interpretation. The modal defeasible approach and the norm based revisionist approach.

On the modal approach, Governatori and Rotolo [5] have argued that constitutive rules are intrinsically defeasible. Accordingly, they have offered an extension of default logic in order to represent institutional agency.

Soon it was noted that constitutive rules may be seen as statements that interpret concepts in regulative rules. So legal interpretation may be modeled as changes in constitutive rules. Governatori and Rotolo [6] have reinterpreted normative change in a framework of defeasible logics. The defeasible logic employed by them to model normative revision includes *defaults* and *defeaters*, which are defaults that do not derive any conclusion but may block the use of conclusions of other defaults. Contraction is interpreted as the addition of defeaters while expansion is interpreted as the conversion of a defeater into a default. Boella et.al. [5] added to that defeasible framework a function that connects each default rule to a value, in order to represent extensions and restrictions to legal concepts to make them cohere with values.

Boella and van der Torre [2] propose a different approach to constitutive rules within an alternative setting of deontic logic, which shifts from the modal semantics to "norm-based semantics", where truths are defined not with reference to morally best possible worlds but to a given set of commands and permissions. Boella and van der Torre represent constitutive rules directly as an explicit set of conditional rules "*X counts as Y*", abstracting from context, and then combine constitutive and regulatory rules within the framework of i/o-logics [10]. [2] The combination

---

[2]Lindahl and Odelstad [9] explore a similar idea from an algebraic perspective, where they study

or architecture makes the output of constitutive rules the input of regulative rules. Sun and van der Torre [18] explore different forms to combine constitutive with regulative rules in aggregative i/o-logics [16].

This norm-based approach has been combined with AGM-like revision [1]. Maranhão [13] and [14] has proposed a revision operator called "refinement" to represent the introduction of exceptions to rules in legal interpretation. More recently, Boella, Pigozzi and van der Torre [3] interpreted the AGM model in the framework of i/o-logics.

The models that represent legal interpretation as change have focused their attention into the modification of constitutive rules only. However, as Maranhão has remarked [15], legal interpretation is more flexible and may change not only constitutive rules, but may also modify or refine regulative rules and change the hierarchy of underlying values. Accordingly, he has suggested revision operators in i/o-logics to model coherent adjustments of rules, values and meaning ascriptions in legal interpretation.

The present paper contributes to this background discussion by providing and characterizing, within the framework of i/o-logics, contraction operators on normative systems combining constitutive and regulative rules, where both the constitutive set and the regulative set may be subject to change in order to restore coherence.

It will proceed as follows. In section 3, we briefly present i/o-logics, show how normative sets may be combined using those logics and define inconsistence and incoherence of normative combinations of constitutive and regulative rules. In section 4, we provide a construction and characterization of a Hansson-like kernel contraction operator on sets of conditional norms. In section 5, we provide a general construction and the characterization of a general contraction operator for the combination of constitutive and regulative rules, called *combined contraction*, as well as the characterization of two alternative contraction operators, one for the contraction of the set of constitutive rules with respect to the set of regulatory rules (called *constitutive contraction*), and the other for the contraction of regulative rules with respect to constitutive rules (called *regulative contraction*). In section 7, we present some examples of contraction in order to raise possible criteria for prefering a . Section 8 concludes and proposes challenges for future research.

3 Combining normative sets in input/output logics

Let L be a standard propositional language with propositional variables and logical connectives: $\neg, \wedge, \vee, \rightarrow, \bot, \top$. We call $N \subseteq L \times L$ a *normative set*, composed by *conditional norms*, which are here represented as pairs of propositional sentences. We shall employ the usual notation from set theory, with proper inclusion denoted

---

the link between constitutive and regulative rules from an abstract point of view by investigating the role of intermediary legal concepts.

by $\subseteq$ and difference between sets A and B by $A \backslash B$. For any $A \subseteq L$, $N(A)$ is the image of N under A, that is $N(A) = \{x : (a,x) \in N,$ for some $a \in A\}$. We write simply $N(a)$ to abbreviate $N(\{a\})$ and call it the image of N in context a. For any normative set N we define $bodyN = \{a : (a,x) \in N\}$ and $headN = \{x : (a,x) \in N\}$.

The main constructions and results of this paper on contraction functions are based on consequence relations that respect the structural features of a *Tarskian consequence operator Cn*:
(i) $A \subseteq Cn(A)$ (inclusion)
(ii) $Cn(Cn(A)) \subseteq Cn(A)$ (idempotence)
(iii) $Cn(A) = \bigcup\{Cn(A') : A' \subseteq A,$ finite$\}$ (finiteness)
(iv) $Cn(A) \subseteq Cn(A \cup B)$ (monotony).

Definition 1. (Makinson and van der Torre [10]) Let N be a normative set, $a \in L$ and V the set of all maximal consistent sets v in classical propositional logic[3]. Then we define the following output operators, where Cl is the operator of classical consequence:
(i) *simple minded*: $out_1(N, a) = Cl(N(Cl(a)))$
(ii) *basic*: $out_2(N, a) = \bigcap\{out_1(N, v) : a \in v,$ for $v \in V$ or $v = L\}$
(iii) *reusable*: $out_3(N, a) = \bigcap\{out_1(N, B) : a \in B$ and $out_1(N, B) \subseteq B = Cl(B)\}$
(iv) *basic reusable*: $out_4(N, a) = \bigcap\{out_1(N, v) : a \in v$ and $out_1(N, v) \subseteq v,$ for $v \in V$ or $v = L\}$
A conditional norm (a, x) is *implied* by a normative set N, i.e., $(a, x) \in out_i(N)$ if, and only if, $x \in out_i(N, a)$, for $i \in \{1, 2, 3, 4\}$.

It is easy to see that out_i is a consequence operator on a normative set.

A negative permission to x in context a in the sense of absence of prohibition to x in that context is defined as $(a, x) \in negperm_i(N)$ if, and only if, $(a, \neg x) \notin out_i(N)$. Permissions may also be independently and explicitly ruled. Positive permissions are defined below:

Definition 2. (Makinson and van der Torre [12]) Let N be a normative set and $P \subseteq (L \times L)$ a set of explicit permissions. Then, $(a, x) \in perm_i(P, N)$ if, and only if, $(a, x) \in out_i(N \cup Q)$, for some singleton or empty $Q \subseteq P$.

We shall focus on the combination between two kinds of normative sets: a set of constitutive rules (called the *constitutive set*) and a set of regulative rules (called the the *regulative set*).

Definition 3. Let $A \subseteq L$, out_i be an output operator, C and R, respectively, a constitutive and a regulative set. Then the combined output of C and R is defined as:

$$out_{ij}(C, R, A) = out_i(R, out_j(C, A))$$

---
[3]The characteristic function of an element $v \in V$ is a boolean valuation for L.

For $A \subseteq L$, out_i and out_j output operators, C a constitutive set and R a regulative set, we shall use the following conventions:
(i) $out_i(C, R, A)$ if $i = j$
(ii) $out_{ij}(C, R, a)$ denoting $out_{ij}(C, R, \{a\})$
(iii) $(a, x) \in out_{ij}(C, R)$ if $x \in out_{ij}(C, R, a)$
We call the pair of normative sets (C, R) the combination of C and R or the combination (C, R).

As suggested above, constitutive rules may be responsible for genuine conflicts when combined with a regulative set. In order to model this feature, it should be possible to verify regulative sets that are consistent but whose combination with a constitutive set implies an inconsistent conditional norm. To avoid confusions let us then qualify normative sets as consistent or inconsistent and combinations as coherent or incoherent.

Consistency is defined with respect to a given context. We say that a normative set N is b-consistent (accordingly a combination (C, R) b-coherent) if and only if $(b, \bot) \notin out_i(N)$ (accordingly $(b, \bot) \notin out_i(C, R)$). If we have a set of obligations N and a set of explicit permissions P, then such normative sets are b-consistent if $(b, \bot) \notin perm(O, P)$.

To demand $\bot$-consistency would limit the possibility to give opposite commands to logically independent conditions, since $N = \{(a, x), (b, \neg x)\}$ would be rendered inconsistent. On the other hand, to demand $\top$-consistency would be inadequate, since normative sets with genuine conflicts such as $N = \{(a, x), (a, \neg x)\}$ would be rendered consistent. So we consider a normative set N consistent if it is b-consistent for every b such that $b \in Cl(a)$ and $a \in bodyN$. By its turn a combination (C, R) is coherent if it is b-coherent for every b such that $b \in Cl(a)$ and $a \in bodyC$.

Therefore, we may have a consistent set R but an incoherent combination (C, R), which would demand a contraction to restore coherence.

Consider again the example on the admissibility of access by police officer to the message exchange stored in a cell phone collected at a searched premise. We may represent it as $C = \{(sms, data), (data, property), (sms, comm)\}$, $O = \{(comm, \neg access)\}$, and $P = \{(property, access)\}$. We may simplify it if we consider that police officers have the *obligation* to access any property found in the premises, so we would have $R = \{(property, access), (comm, \neg access)\}$. The figure below illustrates this normative combination, where each dash linking two nodes is a pair and each node is a propositional sentence:

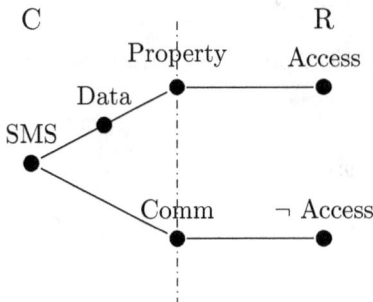

First we note that the regulative set is consistent. Incoherence, by its turn, would depend on the particular output operator employed. We have, for instance that in simple-minded output and basic output the combination is also coherent, since $(sms, \bot) \notin out_i(C, R)$, for $i \in \{1, 2\}$. Using reusable output or basic reusable output the combination is incoherent, considering that $(sms, \bot) \in out_i(C, R)$, for $i \in \{3, 4\}$. Considering the same example with permissions, if the output operator on the constitutive set is at least reusable, the regulative sets O and P are consistent but the combination $(C, O/P)$ is incoherent, since $(sms, \bot) \in out(C, O/P)$.

It is not the purpose of this paper to define which output operator would be applied to the constitutive set and which would be applied to the regulative set. Considering that we shall define contraction operators on bases and not on closed sets, the particular logic involved is not relevant.

4 Kernel contraction of normative sets

Models of belief contraction and revision are built in order to satisfy the demand of minimal change to keep consistency of a theory. There are two basic strategies to reach this goal. The first, called *"partial meet contraction"* consists in the selection of the resulting contraction among the maximal consistent sets that do not derive the sentence to be deleted. The second, employed both by *"safe contraction"* and by *"kernel contraction"*, consists in "breaking" the minimal subsets of the theory or base that derive the sentence to be deleted. We shall follow the second strategy, calling those minimal subsets "arguments", which here are the base of normative implications from the set of rules. The construction proceeds basically by making minimal withdrawals on those arguments.

Definition 4. (*argument*) $X \subseteq L \times L$ is an *argument* for (a, x) based on a normative set N if, and only if:
(i) $X \subseteq N$
(ii) $(a, x) \in out_i(X)$
(iii) if $X' \subset X$ then $(a, x) \notin out_i(X')$.

$Args_N(a, x)$ is the set of arguments for (a, x) based on N.

Definition 5. An incision σ is a choice-like function on $Args_N(a, x)$ to $\wp(L \times L)$ such that:
(i) $\sigma(Args_N(a, x)) \subseteq \bigcup Args_N(a, x)$
(ii) $\sigma(Args_N(a, x)) \cap X \neq \varnothing$, for all $X \in Args_N(a, x)$

Definition 6. Let N be a normative set and (a, x) a conditional norm. Then, the *contraction* of N by (a, x) is defined as:

$$N -_\sigma (a, x) = N \backslash \sigma(Args_N(a, x))$$

The *contraction* of a normative set N by a conditional norm (a, x) may also be defined by postulates on a contraction function, as follows.

Definition 7. The contraction of a normative set N by a conditional norm (a, x) is a function $N- : L \times L \longrightarrow \wp(L \times L)$ satisfying:

1. *success*: if $(a, x) \notin out_i(\varnothing)$, then $(a, x) \notin out_i(N - (a, x))$

2. *inclusion*: $N - (a, x) \subseteq N$

3. *core-retainment*: if $(b, y) \in N \backslash N - (a, x)$, then there is $N' \subset N$ such that $(a, x) \notin out_i(N')$, but $(a, x) \in out_i(N' \cup \{(b, y)\})$

4. *uniformity*: if for all $N' \subseteq N$, $(a, x) \in out_i(N')$ if, and only if $(b, y) \in out_i(N')$, then $N - (a, x) = N - (b, y)$.

Theorem 1. $N -_\sigma (a, x) = N - (a, x)$

Proof. Since out_i is a *Tarskian consequence operator*, the proof is straightforward by substituting the set of sentences to be contracted for a normative set and the sentence to be deleted for a conditional norm in the representation theorem for kernel contraction in Hansson [8, p. 172]. □

5 Contraction of combined normative sets

Now we are in place to construct and characterize contraction operators on combined normative sets. The first operator, called *constitutive contraction*, contracts only the constitutive set. The second operator, called *regulative contraction*, contracts the regulative set, while the *combined contraction* operator may contract both in order to delete a norm from the combination of the constitutive and regulative sets.

Definition 8. (*constitutive contraction*) The *constitutive contraction* of a combination (C, R) by a conditional norm (a, x) is a function $C-_R : L \times L \longrightarrow \wp(L \times L)$, satisfying:

1. *success*: if $(a,x) \notin out_i(\varnothing, R)$ then $(a,x) \notin out_i(C -_R (a,x), R)$

2. *inclusion*: $C -_R (a,x) \subseteq C$

3. *core-retainment*: if $(b,y) \in C \backslash C -_R (a,x)$, then there is $C' \subset C$ such that $(a,x) \notin out_i(C', R), but (a,x) \in out_i(C' \cup \{(b,y)\}, R)$

4. *uniformity*: if for all $C' \subseteq C$ it is the case that $(a,x) \in out_i(C', R)$ if, and only if, $(b,y) \in out_i(C', R)$ then $C -_R (b,y) = C -_R (a,x)$

Definition 9. (*regulative contraction*) The regulative contraction of a combination C, R by a conditional norm (a,x) is a function $R-_C : L \times L \longrightarrow \wp(L \times L)$, satisfying:

1. *success*: if $(a,x) \notin out_i(C, \varnothing)$ then $(a,x) \notin out_i(C, R -_C (a,x))$

2. *inclusion*: $R -_C (a,x) \subseteq R$

3. *core-retainment*: if $(b,y) \in R \backslash R -_C (a,x)$, then there is $R' \subset R$ such that $(a,x) \notin out_i(C, R')$, but $(a,x) \in out_i(C, R' \cup \{(b,y)\})$

4. *uniformity*: if for all $R' \subseteq R$, $(a,x) \in (C, R')$ if, and only if, $(b,y) \in out_i(C, R')$ then $R -_C (a,x) = R -_C (b,y)$.

We use the following conventions for the definition of the combined contraction of normative sets:
(i) if $(C, R) - (a,x) = (C^-, R^-)$, then $(C, R) \backslash (C, R) - (a,x) = (C \backslash C^-, R \backslash R^-)$
(ii) $\bigcup(C, R) = \bigcup\{C, R\}$

Definition 10. (*combined contraction*) The *combined contraction* of the combination (C, R) by a conditional norm (a,x) is a function $(C, R)- : L \times L \longrightarrow \wp(L \times L) \times \wp(L \times L)$, satisfying:

1. *success*: if $(a,x) \notin out_i(\varnothing)$, then $(a,x) \notin out_i((C, R) - (a,x))$

2. *inclusion*: if $(C, R) - (a,x) = (C^-, R^-)$, then $C^- \subseteq C$ and $R^- \subseteq R$

3. *core-retainment*: if $(b,y) \in \bigcup(C, R) \backslash (C, R) - (a,x)$, then there is $C' \subseteq C$ and $R' \subseteq R$ such that $(a,x) \notin out(C', R')$, but $(a,x) \in out_i(C' \cup \{(b,y)\}, R')$ or $(a,x) \in out_i(C', R' \cup \{(b,y)\})$

4. *uniformity*: if for all $C' \subseteq C$ and $R' \subseteq R, (a,x) \in out_i(C', R')$, if, and only if, $(b,y) \in out_i(C', R')$, then $(C, R) - (a,x) = (C, R) - (b,y)$

Now we define a general construction for kernel contraction of combined normative sets, from which we may specify constitutive, regulative and combined contraction operators.

Definition 11. (*combined argument*) A combination (X, Y) is a *combined argument* for (a, x) based on the combination (C, R) of a constitutive set C and a regulative set R if, and only if:
(i) $X \subseteq C$
(ii) $Y \subseteq R$
(iii) $(a, x) \in out_i(X, Y)$
(iv) if $X' \subset X$, then $(a, x) \notin out_i(X', Y)$
(v) if $Y' \subset Y$, then $(a, x) \notin out_i(X, Y')$.

We denote by $Args_{(C,R)}(a, x)$ the set of combined arguments for (a, x) based on (C, R).

Definition 12. An *incision* is a choice-like function on $Args_{(C,R)}(a, x)$ to $\wp(L \times L)$ such that:

(i) if $Args_{(C,R)}(a, x) = \{(X_i, Y_i) : i \in I\}$ then $\sigma(Args_{(C,R)}(a, x)) \subseteq \bigcup_{i \in I}(X_i \cup Y_i)$
(ii) $\sigma(Args_{(C,R)}(a, x)) \cap (X_i \cup Y_i) \neq \emptyset$ for every $(X_i, Y_i) \in Args_{(C,R)}(a, x)$

Definition 13. An incision on $Args_{(C,R)}(a, x)$ is *constitutive* if, and only if, $\sigma(Args_{(C,R)}(a, x)) \cap R = \emptyset$

Definition 14. An incision on $Args_{(C,R)}(a, x)$ is *regulative* if, and only if, $\sigma(Args_{(C,R)}(a, x)) \cap C = \emptyset$

Definition 15. (*contraction*). Let (C, R) be a combination of normative sets and (a, x) a conditional norm. Then, the *contraction* of (C, R) by (a, x) based on the incision σ, is defined as $(C, R) -_\sigma (a, x) = (C^-, R^-)$, where $C^- = C \backslash \sigma(Args_{(C,R)}(a, x))$ and $R^- = R \backslash \sigma(Args_{(C,R)}(a, x))$.

Theorem 2. A contraction of (C, R) by (a, x) based on a constitutive incision σ is a constitutive contraction, that is, $(C, R) -_\sigma (a, x) = (C -_R (a, x), R)$. Moreover, given a constitutive contraction, there is a constitutive incision σ such that $(C, R) -_\sigma (a, x) = (C -_R (a, x), R)$.

Proof. We note that since σ is a constitutive incision function, we have $R^- = R$. So we must show for the first part of the theorem that (i) $-_\sigma$ satisfy the $C -_R (a, x)$ postulates. For the second part we must show that an incision $\sigma(Args_{C,R}(a, x)) = C \backslash C -_R (a, x)$ is a constitutive incision function.
For the first part, *inclusion* is immediate by construction. For *success*, assume that $(a, x) \in out_i((C, R) -_\sigma (a, x))$. By finiteness, a combined argument (A, X) for (a, x) based on $(C, R -_\sigma (a, x)) = (C^-, R)$ would also be a combined argument for (a, x) based (C, R), where $A \subseteq C^- \subseteq C$. So, given that $(a, x) \notin out_i(\emptyset, R)$, there would be (b, y) such that both $(b, y) \in C^-$ and, provided that σ is a constitutive incision function, $(b, y) \in \sigma(Args_{(C,R)}(a, x))$. But that is impossible, since $C^- = C \backslash \sigma(Args_{C,R}(a, x))$.

For *core-retainment*, suppose there is $(b,y) \in C \backslash C^-$, for $(C,R) -_\sigma (a,x) = (C^-, R^-)$. Since σ is a constitutive incision function, $(b,y) \in \sigma(Args_{(C,R)}(a,x))$ and there is a combined argument (A, X) for (a,x) based on (C,R), where $(b,y) \in A$. Hence we have that $(a,x) \notin out_i(A\backslash(b,y), R)$ and $(a,x) \in out_i(A\backslash(b,y) \cup \{(b,y)\}, R)$. Finally, for *uniformity* if all subsets of C imply (a,x) iff imply (b,y), then $\sigma(Args_{C,R}(a,x)) = \sigma(Args_{C,R}(b,y))$ and provided that σ is a constitutive incision, $(C \backslash \sigma(Args_{(C,R)}(a,x)), R) = (C \backslash \sigma(Args_{(C,R)}(b,y)), R)$.

For the second part, it follows from *uniformity* of the constitutive contraction that $\sigma(Args_{C,R}(a,x)) = C \backslash C -_R (a,x)$ is a function. To show that $\sigma \subseteq \bigcup_{i \in I}(X_i, Y_i)$ of arguments for (a,x) based on (C,R), assume that $(b,y) \in C \backslash C -_R (a,x)$. Then, by *core-retainment*, there is $C' \subseteq C$ such that $(a,x) \in out(C' \cup \{(b,y)\}, R)$ but $(a,x) \notin out(C', R)$. So, there is a combined argument (X, Y) for (a,x) based on (C,R) with $(b,y) \in X$. Now, for the last clause of an incision, assume an arbitrary argument (X, Y) for (a,x) based on (C,R). Then $(a,x) \in out(X, R)$. By *success*, we know that $(a,x) \notin (C -_R (a,x), R)$, so $X \not\subseteq C -_R (a,x)$. Given that $X \subseteq C$, there is $(b,y) \in C \backslash C -_R (a,x)$. By construction, since $R^- = R$, it follows that σ is a constitutive incision and, therefore, by *inclusion*, $C -_R (a,x) = C \backslash \sigma$. Hence $(C,R) -_\sigma (a,x) = (C \backslash \sigma, R) = (C -_R (a,x), R)$. □

Theorem 3. A contraction of (C,R) by (a,x) based on a regulative incision σ is a regulative contraction, that is, $(C,R) -_\sigma (a,x) = (C, R -_C (a,x))$. Moreover, given a regulative contraction, there is a regulative incision σ such that $(C,R) -_\sigma (a,x) = (C, R -_C (a,x))$.

Proof. The proof is analogous to the proof of Theorem 2. □

Theorem 4. $(C,R) -_\sigma (a,x) = (C,R) - (a,x)$

Proof. The proof that the construction satisfies the postulates *success*, *inclusion* and *uniformity* is similar to the proof of the corresponding postulates in Theorem 2. So does the postulates to construction part. So we concentrate on the clauses of the construction that warrant minimal change, which is reflected by the postulate of *core-retainement*. From the construction to the postulate, assume that $(b,y) \in \bigcup(C,R) \backslash (C,R) -_\sigma (a,x)$. Then, $(b,y) \in \sigma(Args_{(C,R)}(a,x))$ and there is $(A, X) \in Args_{(C,R)}(a,x)$. Since $\sigma \cap (A \cup X) \neq \emptyset$, then by clauses (iv) and (v) of Definition 12, either $(b,y) \in A$ or $(b,y) \in X$. So $A\backslash\{(b,y)\}$ or $X\backslash\{(b,y)\}$ is the set to which adding (b,y) would give $(a,x) \in (A\backslash\{(b,y)\} \cup \{(b,y)\}, X)$ or $(a,x) \in (A, X\backslash\{(b,y)\} \cup \{(b,y)\})$. From the postulate to construction, defining $\sigma(Args_{(C,R)}(a,x)) = C \backslash (C,R) - (a,x)$, core-retainment is used to show that $\sigma \subseteq \bigcup_{i \in I}(X_i, Y_i)$ of arguments for (a,x) based on (C,R). Assume that $(b,y) \in C \backslash (C,R) - (a,x)$. Then, by *core-retainement*, there is $C' \subseteq C$ and $R' \subseteq R$ such that $(a,x) \notin (C', R')$ but $(a,x) \in (C' \cup \{(b,y)\}, R')$ or $(a,x) \in (C', R' \cup \{(b,y)\})$. In both cases, we conclude that there is $(X, Y) \in Args_{(C,R)}(a,x)$ such that $(b,y) \in (X \cup Y)$. □

6 Constitutive, regulative or combined contraction?

The contraction operators here discussed do not involve constraints for the choice of the incision. Therefore, there is no preference regarding a regulative over a constitutive or over a combined contraction.

This feature may be illustrated by the running example, where the constitutive set is $C = \{(sms, data), (data, property), (sms, comm)\}$, the regulative set is $R = \{(comm, \neg access), (property, access)\}$ and the i/o-logic is at least reusable. In this case, a contraction $(C, R) -_\sigma (sms, \bot)$ would provide five alternative outcomes, each of which determined by an unitary incision function. It is easy to check that the alternative incisions are either $\sigma = \{(sms, data)\}$, or $\sigma = \{(sms, comm)\}$, or $\sigma = \{(data, property)\}$, or $\sigma = \{(comm, \neg access)\}$, or $\sigma = \{(property, access)\}$. Hence the resulting contraction would be either constitutive or regulative.

However there may be contractions that would neither be regulative nor constitutive. Consider, for instance, a variation of the running example, where the constitutive set is $C = \{(sms, data), (data, property), (sms, comm)\}$ and the set of positive permissions is $P = \{(comm, sorder \to access), (property, sorder \to access)\}$. That is, it is permitted to have access to the content of the message exchange stored in the cell phone, provided that there is an specific order, either if the exchange is considered property or if its considered communication. Now suppose the legislator derogates the positive permission to access the content of message exchange or a interpreter consider that such permission is unconstitutional given that it violates the fundamental right to freedom of communication. In this case, a contraction $(C, P) - (sms, sorder \to access)$ has among its possible outcomes the following constitutive and regulative sets $C = \{(data, property), (sms, comm)\}$, $P = \{(property, sorder \to)\}$, as illustrated by the figure below:

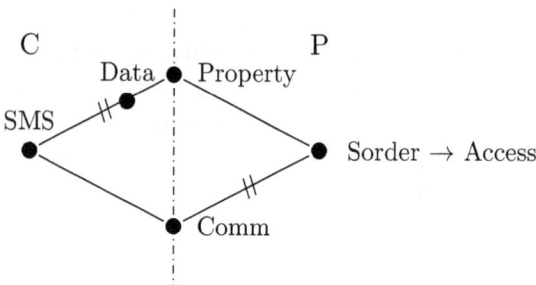

This would indeed be the most reasonable choice for the incision if the contraction is based on an argument favoring freedom of communication. A regulative contraction seems to be unreasonable, since to exclude also the explicit permission to access any property items found in the premises would make any search an seizure useless. On the other hand, a constitutive contraction would leave message exchange completely undefined by law, what would create difficulties with respect to the application of other rules regarding, for instance, data protection. That is, such constitutive contraction would have a significant impact on the amount of legal consequences with respect to this modality of communication. Lastly, the deletion of the constitutive rule to the effect that data counts as property would limit access to any document found in the premises.

Those are domain specific arguments to justify the choice of a particular incision. Nevertheless, the discussion provides a clue to an abstract constraint that might be explored. The discussion by Makinson and van der Torre [11] of constraints for i/o-logics suggests a distinction between rule maximization (*maxrule*: to maximize the preservation of rules in order to satisfy a constraint) and output maximization (*maxout*: to maximize the preservation of outputs in order to satisfy a constraint). The Mobius Strip example is a radical case and may be seen as a contraction. Consider $N = \{(\top, a), (a, b), (b, \neg a)\}$. The contraction $N - (\top, \bot)$ has two possible outcomes $N_1 = \{(\top, a)\}$ or $N_2 = \{(a, b), (b, \neg a)\}$. While N_1 satisfies *maxout* and fails *maxrule*, N_2 satisfies *maxrule* and fails *maxout*.

The qualitative discussion of the examples in this paper also involved arguments about the inconvenience of the loss of regulative rules *versus* the loss of outputs. In a combination of a constitutive and a regulative set, it is not difficult to find examples where a constitutive contraction would be inadequate in terms of *maxout*. See for instance, the example in the figure below, of a contraction $(C, R) - (a, x)$:

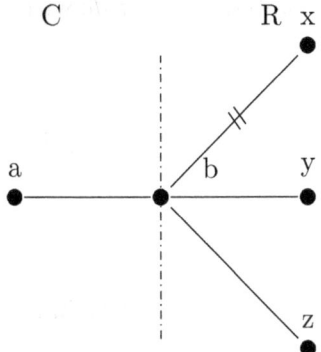

Indeed constitutive contractions tend to favor *maxrule* and sacrifice *maxout*, since intermediary concepts may be connected with different rules. The construction of the contraction operators for combined normative sets in this paper was based on rule maximization, but future investigations should try to find reasonable contraints to temper the demand for *maxrule* with the demand for *maxout*.

7 Conclusion and future research

Based on the framework of i/o-logics, we have characterized an operator of kernel contraction of normative sets and three operators of contraction of combined normative sets, called, respectively, constitutive contraction, regulative contraction and combined contraction.

The constitutive contraction operator only changes the constitutive set in order to avoid the implication of the conditional norm to be deleted. By its turn, the regulative contraction changes the regulative set. The combined contraction is a general operator that may change the regulative, the constitutive, or both.

We have suggested a definition of consistency of normative sets and coherence of combined normative sets which capture the intuition that constitutive rules may be responsible for creating genuine normative conflicts when combined with consistent regulative rules. This feature of regulative rules revives the problem of how to represent and handle such deontic conflicts: should we treat constitutive rules as defeasible, or should we treat them as undefeasible but subject to revision? The answer to such question is relevant for current work on the representation of legal interpretation as normative revision.

Approaches to the representation of legal interpretation as revision tended to model change of the normative system as changes of meaning ascription, considering constitutive rules as defeasible. The model proposed here makes room to represent normative interpretation as combined and coherent changes both on constitutive and regulative rules.

As interesting developments for future research it would be desirable:

1. to find constraints for the contraction operators combining *maxrule* with *maxout* goals;

2. to provide characterizations of contraction operators for closed normative sets and closed combined normative sets, whose properties would depend on the underlying i/o-logic adopted;

3. to develop revision operators for combined normative sets;

4. to connect changes of constitutive and regulative rules to underlying values that would provide constraints to the selection of incisions;

5. to develop other revision operators for combined constitutive, regulative and value sets such as "refinement", as proposed by [14].

References

[1] Alchourron,C. Gardenfors, P. and Makinson, D. On the logic of theory change: partial meet contraction and revision functions. *Journal of Symbolic Logic*, 50(2):510-530, 1985.

[2] Boella, G and Van der Torre, L. A logical Architecture of normative system, in. Goble, L and Meyer,J-J.C (eds): *DEON 2006*, pp-24-35, Springer, 2006.

[3] Boella, G. Pigozzi, G. and van der Torre, L. AGM contraction and Revision of Rules, *Journal of Logic, Language and Information*, 25: 273-297, 2016.

[4] Boella,G. Governatori, G., Rotolo, A. and van der Torre, L. Lex minus dixit quam voluit, lex magis dixit quam voluit: a formal study on legal compliance and interpretation, in Casanovas, P., Pagallo, U., Ajani,G and Sartor,G. (eds). *AI Aproaches to the complexity of legal systems*, Springer, 2010.

[5] Governatori, G. and Rotolo, A. A computational framework for institutional agency, *Artificial Intelligence and Law* 16: 25-52, 2008.

[6] Governatori, G. and Rotolo, A. Changing Legal Systems: Abrogation and Annulment. Part I: Revision of Defeasible Theories,in van der Mayden and van der Torre (eds.) *Deontic Logic in Computer Science*, Proceedings of DEON 2010, Springer, pp.3-19, 2008.

[7] Grossi, D. Meyer, J-J. and Dignun, F. The many faces of counts as: a formal analysis of constitutive rules. *Journal of applied logic* 6:192-217, 2008.

[8] Hansson, S.O. *A Texbook of Belief Dynamics: theory change and database updating*, Kluwer Academic Publishers, 1999.

[9] Lindahl, L. and Odelstad, J. The theory of joining systems, In Gabbay, D., Horty, J., Parent, X., van der Mayden, R., van der Torre, L. (eds.) *Handbook of Deontic Logic and Normative Systems*, pp. p. 545-634. College Publications, London (2013)

[10] Makinson, D and van der Torre, L. Input-output logics. *Journal of philosophical logic* 29: 383-408, 2000.

[11] Makinson, D. and van der Torre, L. Constraints for input-output logics. *Journal of Philosophical Logic* 30:155-185, 2001.

[12] Makinson, D. and van der Torre, L. Permissions from an input/output perspective. *Journal of philosophical logic* 32 (4): 391-416, 2003.

[13] Maranhão, J. Refinement: a tool to deal with inconsistencies, in *Proceedings of ICAIL'01*, p. 52-59, 2001.

[14] Maranhão, J. Defeasibility, Contributory Conditionals and Refinement of Legal Systems, in Beltran, J., Ratti, G.(eds) *The Logic of Legal Requirements: Essays on Defeasibility*, Oxford University Press,53-57, 2012,

[15] Maranhão, J. A logical architecture for dynamic legal interpretation, in *Proceedings of ICAIL'17*, London, United Kingdom, June 12-16, 2017.

[16] Parent, X. and van der Torre, L. "Sing and Dance!": input/output logics without weakening, in F.Cariani et.al. (eds), *Deontic Logic and Normative Systems*, Proceedingas of DEON 2014, Springer, pp.149-165, 2014.

[17] Searle, J. *Making the Social World*.Oxford University Press, 2010.

[18] Sun, X. and van der Torre, L. Combining Constitutive and Regulative Norms in Input/Output Logic, in Cariani, F. et. al (eds)*Deontic Logic and Normative Systems*, Proceedingas of DEON 2014, pp. 241-257, Springer, 2014.

Toward a Systematization of Logics for Monadic and Dyadic Agency & Ability (Preliminary Version)

Paul McNamara
University of New Hampshire, USA
paul.mcnamara@unh.edu

I prove a fundamental theorem for canonical models for logical frameworks for representing agency and ability, broadly in the tradition of Kanger, Pörn, Elgesem, etc. Special interest is given to dyadic agency (and in an extension, to monadic and dyadic ability), in both pure forms as well as mixed forms (e.g. monadic agency logics, dyadic agency logics, and monadic-dyadic agency logics). Employing the fundamental theorem, eighteen correspondence proofs are given, and these collectively are prolific in the myriad strong completeness proofs entailed. I use minimal models, but with an extra parameter to facilitate strong completeness proofs for some formulae where strong completeness would otherwise appear to stall. At DEON, I will present on the motivation, main results, and on applications and rationales for some of the principles, but in the paper, the focus is on the framework and key meta-theorems, with some philosophical remarks only in the footnotes.

I prove a fundamental theorem for canonical models for logical frameworks developed for representations of agency and ability broadly in the tradition of Kanger, Pörn, Elgesem, and others.[1] Special interest is given to *dyadic* agency (and in an extension of this paper, to monadic and dyadic ability), in both pure forms as well as mixed forms (e.g. monadic agency logics, dyadic agency logics and monadic-dyadic agency logics). Elgesem is the only author in this tradition who explored dyadic

---

This work owes a special debt to Lou Goble. Lou generously provided substantial comments that proved invaluable, as well as showing the author that a more complex employment of a parameter for propositions that this author had proposed was overkill. Some further particulars are mentioned below. Xavier Parent's comments led to my revisiting Elgesem's dissertation and realizing that I had overlooked the second less obvious place in his dissertation where he takes the subject of dyadic agency back up (Chapter 3, section II.2). I also benefited indirectly from work on counts-as conditionals by Jones and Parent (Jones, Andrew and Xavier Parent [7]), and from discussion of principle K's plausibility for agency with Andrew Jones, Mark Brown, and Marek Sergot. I am indebted to three anonymous referees, each of whom provided helpful comments, and one of whom went so far beyond the call in providing extensive and very helpful comments that I requested that he be asked to allow for identification by name, but s/he declined. Lastly, I am indebted to Sarah Blagdon for various helpful suggestions.

agency. In my judgment, this constitutes a gap in this tradition, especially in the area of systematic exploration of these logics.

In places where I list formulas that (to my knowledge) have not been considered in this tradition, or considered and rejected perhaps prematurely, I will remark on those in footnotes; otherwise the focus will be on proving the fundamental theorem and then proving a variety of correspondence theorems for formulas and frame constraints, thus providing the first systematic account of such logics. Because I will consider some formulas and systems that are not usually considered for agency or ability, or for completeness using minimal models at all (e.g. K, and system EK respectively), and for which completeness proofs appear to stall unless facilitated by some technical ploy (e.g. like that used within), the model structures employed here will be slightly non-standard minimal model structures.[2] There will be one extra parameter, P for propositions, and some frame constraints will be relativized to that parameter in relatively innocuous ways to facilitate direct strong completeness proofs. I will explain as the issue comes up, and indicate where I believe it is manifested in the proofs below, since the frames are slightly unusual, as are a few of the constraints, but the Appendix provides a quick preview.

---

[1]Kanger, Stig and Helle Kanger [11]; Kanger, Stig [10]; Pörn, Ingmar [14], [15], [16]; Elgesem, Dag [3], [4]; Jones, Andrew and Marek Sergot [8], [9]; Santos, Filipe and Jose Carmo [17]; Santos, Filipe, Andrew Jones and Jose Carmo [18]; Jones, Andrew and Xavier Parent [7]; Sergot, Marek [19],[20]. In Governatori, Guido and Antonino Rotolo [6], it is shown that an oft-cited basic monadic system for agency and ability is incomplete on Elgesem's semantics, and the authors then go on to prove completeness for that system extended with one axiom, both using Elgesem's semantics and that of standard minimal models. The work here has, as a consequence of the core theorems, provides completeness for a large class of such monadic agency logics, and goes on to do the same for logics for dyadic agency and dyadic-monadic agency, dyadic agency being a much-underexplored area generally. There has been no systematic metatheory for monadic much less dyadic agency logics using minimal models or minor variants thereof. Here I approach the subject systematically, closing gaps in the monadic work, and dyadic work and do this via a fundamental theorem for canonical models for a large class of such logics.

[2]The appendix contains a short note with such a stalled proof and then a "fixed" proof. This note or a similar version of it was shared with a number of logicians, including Brian Chellas, Eric Pacuit, Steven Kuhn, Ed Mares, Lou Goble, Xavier Parent, and various logicians at Bayreuth and Ghent, and the basic problem was outlined at Trends in Logic XVII in Lublin Sept 2017 [13]. There was no indication that the stalling of the apparent straightforward attempt at a correspondence proof for K, nor a solution (that would yield full completeness for say EK with standard semanticminimal model clause for K), was known. (Pacuit's online book draft *Neighborhood Semantics for Modal Logic*, retrieved March 21, 2017 from http://pacuit.org/modal/neighborhoods, did not mention the problem with K nor attempt completeness for EK, nor is it discussed in Arlo-Costa, Horacio and Eric Pacuit [1], although in both places completeness for the *normal* modal logic K (i.e. EMCN) is discussed). Chellas leaves a proof of EK's completeness as an exercise (Chellas, Brian F. [2]), apparently easy and in need of no hints, but in correspondence he was not sure what the solution was offhand. Segerberg passes over K in his classic text as uninteresting. There are various reasons why K does indeed have limited applications, but the ones offered against K for monadic agency are fallacious in my opinion, an opinion shared by at least two others who have worked in this area, Andrew Jones and Mark Brown. Furthermore, the problem with completeness proofs using standard minimal models appears to reiterate for other formulas, presumably indefinitely. The

I will cast things with an eye toward generalization to classical modal logics and especially to various conditional logics (broadly conceived) for the dyadic systems. For this reason, in Part I, the "Monadic and/or Dyadic Agency" logics (MDAs) are defined weakly, although the intended interpretation that is nonetheless primarily in focus is that of logics of monadic and dyadic *agency* operators. Also, there are three sub-types of MDA logics considered: (Pure) Monadic Agency logics (PMAs), (Pure) Dyadic Agency logics (PDAs), and Dyadic-Monadic Agency logics (DMAs) containing both a monadic operator and a dyadic one. Soundness for (and distinctness of) the logics is left out for now, and so determination theorems as well strictly, but these should be comparatively straightforward.

In an extension, the results are adapted and expanded to include monadic and dyadic ability operators and then amalgamations with the agency logics of Part I converging on monadic-dyadic logics of agency and ability. Most of the ability addition is generated by minor changes in the main metatheorems, and minor links to the agency operators (e.g. what Jane Doe brings about she is able to bring about), the importance of which is stressed rather persuasively in Governatori, Guido and Antonino Rotolo [6]).

1 The PMA, PDA, DMA Logics

We will be concerned with logics for two agency operators in what follows:

$\mathbf{BA}\varphi$: Jan Doe brings it about that φ

$\mathbf{BA}'_\psi \varphi$: Jan Doe brings it about that φ by bringing it about that ψ[3]

1.1 Monadic Agency

The monadic agency logic framework is as follows.

Formulas of Pure Monadic Agency (PMA) Logics:

(1) All Propositional Variables (PV), $P_1, \ldots, P_n, \ldots$ are PMA formulas[4]

(2) If φ, ψ are PMA formulas, so are: (a) $\neg\varphi$, $(\varphi \lor \psi)$, $(\varphi \land \psi)$, $(\varphi \to \psi)$, $(\varphi \leftrightarrow \psi)$ and (b) $\mathbf{BA}\varphi$.

Appendix illustrate the problem of stalled completeness for system EK using standard minimal models and a preview of the approach to mending things back together used here.

[3] Or *by bringing it about that ψ, Jane Doe brings it about that φ.*

[4] For each of the systems discussed, the set PV has as its elements the propositional variables of the language, $P_1, \ldots, P_n, \ldots$.

For generality, I define a weaker class of logics than those apt for agency:

A Normal PMA Logic, L: L is a set of PMA formulas such that

(1) All tautologous formulas are in L

(2) L is closed under MP and closed under $\text{RE}_{\mathbf{BA}}$, if $\vdash \varphi \leftrightarrow \psi$, then $\vdash \mathbf{BA}\varphi \leftrightarrow \mathbf{BA}\psi$

(3) L is non-trivial (does not contain all formulas)

We will consider as well the following further candidate axiom schemata for PMA logics, the first two of which are standard in most presentations of agency operators:

$\text{T}_{\mathbf{BA}}$: $\vdash \mathbf{BA}\varphi \to \varphi$ [**BA** is a "success" or alethic operator]
$\text{NO}_{\mathbf{BA}}$: $\vdash \neg \mathbf{BA}\top$ [Nothing logically necessary is agential]
$\text{C}_{\mathbf{BA}}$: $\vdash (\mathbf{BA}\varphi \wedge \mathbf{BA}\psi) \to \mathbf{BA}(\varphi \wedge \psi)$ [Agential Conjunction]
$\text{K}_{\mathbf{BA}}$: $\vdash \mathbf{BA}(\varphi \to \psi) \to (\mathbf{BA}\varphi \to \mathbf{BA}\psi)$ [Agential K principle]
$\text{CS}_{\mathbf{BA}}$: $\vdash \mathbf{BA}(\varphi \wedge \psi) \to (\neg \mathbf{BA}\varphi \to \mathbf{BA}\psi)$ [Agential Conjunctive Syllogism][5]

1.2 Dyadic Agency

We now turn to the dyadic systems. We also characterize the normal Pure Dyadic Agency logics (PDAs) very leanly with an eye to alternative interpretations of agency (e.g. where there is a time lag between the "antecedent" (means) and "consequent" (end)), but also with an eye to non-agential interpretations of the operator (e.g. as conditionals or propositionally-relativized modal operators), which may not involve some of the theses that are quite plausible for the instantaneous agency interpretation typical of the Kanger, Pörn, Elgesem tradition we have primarily in mind.

Formulas of Pure Dyadic Agency (PDA) Logics:

(1) $P_1, \ldots, P_n, \ldots$ are PDA formulas.

---

[5]$\text{CS}_{\mathbf{BA}}$ says that if I bring about the conjunction of two propositions, but not one of the conjuncts, then I bring about the other conjunct, or equivalently, if I bring about the conjunction, then I bring about at least one of the conjuncts (i.e. $\mathbf{BA}(\varphi \& \psi) \to (\mathbf{BA}\varphi \wedge \mathbf{BA}\psi)$). It would seem that if I do bring about a conjunction, but not (say) its first conjunct, then that is because that conjunct is rendered true independently of my agency. But then the only way the truth of the conjunction could result from my agency is if the truth of the other conjunct results from it. Conversely, if I neither bring about p nor bring about q, then it would seem that I can't bring about both p and q. At the least, this formula seems worth exploring. CS is not considered (or validated) in Jones and Sergot [8], Santos, Carmo and Jones [18], Santos and Carmo [17], Elgesem [3] or Elgesem [4]. I endorsed this principle in McNamara, Paul [12]. Note: should it be objected that one brings about a conjunction only if one brings about both conjuncts, that is, $\mathbf{BA}(\varphi \& \psi) \to (\mathbf{BA}\varphi \& \mathbf{BA}\psi)$, then CS is trivially true, but this is not the usual way of analyzing agential operators for conjunctions. Mark Brown reminded me of this possible objection in correspondence.

(2) If φ, ψ are PDA formulas, so are: $(a)\ \neg\varphi, \ldots (\varphi \leftrightarrow \psi)$ and $(b)\ \mathbf{BA}'_\varphi\psi$;

A Normal PDA Logic, L: L is a set of PDA formulas such that:

(0) L is non-trivial (does not contain all formulas)
(1) All tautologous PDA formulas are in L
(2) L is closed under MP
(3) L is closed under $\text{RE}^r_{\mathbf{BA}'}$ and $\text{RE}^l_{\mathbf{BA}'}$:

$\text{RE}^r_{\mathbf{BA}'}$: If $\vdash \varphi \leftrightarrow \psi$ then $\vdash \mathbf{BA}'_\chi\varphi \leftrightarrow \mathbf{BA}_\chi\psi$
$\text{RE}^l_{\mathbf{BA}'}$: If $\vdash \varphi \leftrightarrow \psi$ then $\vdash \mathbf{BA}'_\varphi\chi \leftrightarrow \mathbf{BA}'_\psi\chi$

We will consider the following candidate additional axiom schemata for PDA logics:

$\text{T}_{\mathbf{BA}'}$: $\vdash \mathbf{BA}'_\psi\varphi \to (\varphi \wedge \psi)$ [$\mathbf{BA}'$ is a Success/Alethic Modal Operator]
$\text{NO}_{\mathbf{BA}}$: $\vdash \neg(\mathbf{BA}'_\varphi\top \vee \mathbf{BA}'_\top\varphi)$ [Logical Truths are excluded from agency]
$\text{IR}_{\mathbf{BA}'}$: $\vdash \neg\mathbf{BA}'_\varphi\varphi$ [Irreflexivity]
$\text{CC}_{\mathbf{BA}'}$: $\vdash (\mathbf{BA}'_\varphi\psi \wedge \mathbf{BA}'_\varphi\chi) \to \mathbf{BA}'_\varphi(\psi \wedge \chi)$ [Conjunction of concurrent "Consequents"][6]
$\text{DC}_{\mathbf{BA}'}$: $\vdash (\mathbf{BA}'_\varphi\psi \wedge \mathbf{BA}'_{\varphi'}\psi') \to \mathbf{BA}'_{(\varphi \wedge \varphi')}(\psi \wedge \psi')$ [Double Conjunction][7]

---

[6]I regret that I have not been able to explore interactions between some of these schemata much here. For example, as one anonymous reviewer pointed out, irreflexivity coupled with conjunction of concurrent consequents entails asymmetry of dyadic agency: $\neg(\mathbf{BA}'_\varphi\psi\ \&\ \mathbf{BA}'_\psi\varphi)$.

[7]$\text{CC}_{\mathbf{BA}'}$ is a special case of this more general principle. Just let φ instantiate φ' and then apply $\text{RE}^l_{\mathbf{BA}'}$ to get $\text{CC}_{\mathbf{BA}'}$.

$DA_{BA'}$:	$\vdash (BA'_\psi \varphi \wedge BA'_\chi \varphi) \to BA'_{(\psi \vee \chi)} \varphi$	[Disjunction of concurrent "Antecedents"][8]
$DD_{BA'}$:	$\vdash (BA'_\varphi \psi \wedge BA'_{\varphi'} \psi') \to BA'_{(\varphi \vee \varphi')}(\psi \vee \psi')$	[Double Disjunction][9]
$TR_{BA'}$:	$\vdash (BA'_\varphi \psi \wedge BA'_\psi \chi) \to BA'_\varphi \chi$	[Transitivity]
$CT_{BA'}$:	$\vdash (BA'_\varphi \psi \wedge BA'_{(\varphi \wedge \psi)} \chi) \to BA'_\varphi \chi$	[Cumulative Transitivity]
$CK_{BA'}$:	$\vdash BA'_\chi (\varphi \to \psi) \to (BA'_\chi \varphi \to BA'_\chi \psi)$	[Conditional agential K principle]
$DCS_{BA'}$:	$\vdash BA'_\chi (\varphi \wedge \psi) \to (\neg BA'_\chi \varphi \to BA'_\chi \psi)$	[Dyadic Conjunctive Syllogism]
$SS_{BA'}$:	$\vdash BA'_\chi BA'_\varphi \psi \to BA'_{(\chi \wedge \varphi)} \psi$	[Stage-Setting Agency Principle][10]

1.3 Dyadic and Monadic Agency

Formulas of Dyadic-Monadic Agency (DMA) Logics:

(1) $P_1, \ldots, P_n, \ldots$ are DMA formulas

(2) If φ, ψ are DMA formulas, so are: (a) $\neg \varphi, \ldots (\varphi \leftrightarrow \psi)$, and (b) $BA\varphi$, $BA'_\varphi \psi$.

For the logics with both monadic and dyadic agency operators, one new schema will be central, one that says that if I bring about φ by bringing about ψ, then I bring about each:

S: $\vdash BA'_\psi \varphi \to (BA\varphi \wedge BA\psi)$ [Separation]

---

[8] This principle is unusual, but I think over-determination by the same agent is possible (e.g. I vote by raising each hand, ψ, χ); then the antecedent can be true where ψ and χ in the antecedent are not identical (not logically equivalent), and then the consequent plausibly holds—either of ψ and χ is sufficient; on the other hand, if such over-determination is not possible, then the principle is true because the only case where the antecedent can be true is where ψ and χ are identical (up to logical equivalence), and so the consequent would then hold trivially by $RE^1_{BA'}$. In either event, it serves to raise the single agent analog of the often discussed multi-agent over-determination (e.g. the killing of Caesar).

[9] $DA_{BA'}$ is a special case of this more general principle. Just let ψ instantiate ψ' and then apply $RE^r_{BA'}$ to get $DA_{BA'}$.

[10] This principle is unusual, but suppose as a chair of my department, following much discussion that appears to have ended, I request a show of hands by saying "All those in favor of the motion—raise your hand?" (χ), and I do this while raising my own hand (φ). I then, by requesting this vote, bring it about that by raising my hand, I bring it about that (ψ) I vote. This in turn implies that I bring it about that I vote by both saying what I said (χ) and raising my hand (φ). Neither of these acts along would suffice for my voting: calling for a show of hands is what *sets the stage* for the possibility that raising my hand can constitute voting in favor. As with DA, perhaps the important thing is to raise the issue, in this case, of stage-setting agency. With an ability operator, and dual agents, we can express my bringing it about by calling for a show of hands that you are able to vote by raising your hand. These sorts of agency cases are particularly important for changing normative positions by our actions in moral and institutional settings. Commitment for example can be analyzed along these lines with the addition of a personal (but not agential), obligation operator, **OB** (see McNamara, Paul [12] on such an operator): φ commits Jane Doe to ψ might be rendered as $BA'\varphi \, OB\psi$, so that for example, Jane's bringing it about that she promises to meet you would ordinarily bring it about that she is obligated to bring it about that she does meet you. We can of course inquire into the status of this commitment itself.

For the combined system, for generality, we define the normal DMA logics as:

A Normal DMA, L: L is a set of DMA formulas such that:

(0) L is non-trivial (does not contain all DMA formulas)

(1) L contains all tautologous DMA formulas

(2) L is closed under MP and $\text{RE}_{\mathbf{BA}}$ (see normal PMA logic above)

(3) L is closed under $\text{RE}^r_{\mathbf{BA'}}$ and $\text{RE}^l_{\mathbf{BA'}}$ (see normal PDA logic above)

The base logics for the three logic types are as follows.

Base Pure Monadic Agency Logic (PMA):
SL: All Tautologies
MP: If $\vdash \varphi$ and $\vdash \varphi \to \psi$ then $\vdash \psi$
$\text{RE}_{\mathbf{BA}}$: If $\vdash \varphi \leftrightarrow \psi$ then $\vdash \mathbf{BA}\varphi \leftrightarrow \mathbf{BA}\psi$

Base Pure Dyadic Agency Logic (PDA):
SL: All Tautologies
MP: If $\vdash \varphi$ and $\vdash \varphi \to \psi$ then $\vdash \psi$
$\text{RE}^r_{\mathbf{BA}}$: If $\vdash \varphi \leftrightarrow \psi$ then $\vdash \mathbf{BA}'_\chi \varphi \leftrightarrow \mathbf{BA}'_\chi \psi$ [Rule of Right Replacement for $\mathbf{BA}'$]
$\text{RE}^l_{\mathbf{BA}}$: If $\vdash \varphi \leftrightarrow \psi$ then $\vdash \mathbf{BA}'_\varphi \chi \leftrightarrow \mathbf{BA}'_\psi \chi$ [Rule of Left Replacement for $\mathbf{BA}'$]

Base Dyadic-Monadic Agency System (DMA):
SL: All Tautologies
MP: If $\vdash \varphi$ and $\vdash \varphi \to \psi$ then $\vdash \psi$
$\text{RE}_{\mathbf{BA}}$: If $\vdash \varphi \leftrightarrow \psi$ then $\vdash \mathbf{BA}\varphi \leftrightarrow \mathbf{BA}\psi$
$\text{RE}^r_{\mathbf{BA}}$: If $\vdash \varphi \leftrightarrow \psi$ then $\vdash \mathbf{BA}'_\chi \varphi \leftrightarrow \mathbf{BA}'_\chi \psi$
$\text{RE}^l_{\mathbf{BA}}$: If $\vdash \varphi \leftrightarrow \psi$ then $\vdash \mathbf{BA}'_\varphi \chi \leftrightarrow \mathbf{BA}'_\psi \chi$

Let me note here that, ignoring generality, on the intended interpretation we might expect the following basic formulas to hold in all of what we might call the "*Preferred* Normal DMA Logics":

$\text{T}_{\mathbf{BA}}$: $\vdash \mathbf{BA}\varphi \to \varphi$
$\text{NO}_{\mathbf{BA}}$: $\vdash \neg \mathbf{BA}\top$
$\text{IR}_{\mathbf{BA'}}$: $\vdash \neg \mathbf{BA}'_\varphi \varphi$ [Irreflexivity][11]
S: $\vdash \mathbf{BA}'_\psi \varphi \to (\mathbf{BA}\varphi \wedge \mathbf{BA}\psi)$ [Separation]

All instances of $\text{T}_{\mathbf{BA'}}$ and $\text{NO}_{\mathbf{BA'}}$ are theorems of all Preferred Normal DMA logics:

$T_{BA'}$: $\vdash BA'_\psi \varphi \to (\varphi \wedge \psi)$
Proof: Assume $BA'_\psi \varphi$. By S, $BA\varphi \wedge BA\psi$. So by T_{BA}, $\varphi \wedge \psi$.

$NO_{BA'}$: $\vdash \neg(BA'_\varphi \top \vee BA'_\top \varphi)$
Proof: Assume $BA'_\varphi \top \vee BA'_\top \varphi$. So by S, $BA\top$, contrary to (monadic) NO_{BA}.

2 Semantics for PMA, PDA, DMA Logics

We first define the frames for the logics

A Normal PMA Frame, $F = <W, P, f^1>$
(1) W is non-empty   [Worlds]
(2) $P \subseteq \text{Pow}(W)$ [The Propositions][12]
(3) $f^1: W \to \text{Pow}(P)$ [Maps worlds to sets of propositions]

A Normal PDA Frame, $F = <W, P, f^2>$:
(1) W is non-empty   [Worlds]
(2) $P \subseteq \text{Pow}(W)$ [The Propositions]
(3) $f^2: W \times P \to \text{Pow}(P)$ [Maps world-proposition pairs to sets of propositions]

A Normal DMA Frame, $F = <W, P, f^1, f^2>$:
(1) W is non-empty   [Worlds]
(2) $P \subseteq \text{Pow}(W)$ [The Propositions]
(3) $f^1: W \to \text{Pow}(P)$ [Maps worlds to sets of propositions]
(4) $f^2: W \times P \to \text{Pow}(P)$ [Maps world-proposition pairs to sets of propositions]

Additional normal PMA and DMA frames will be considered with one or more of these clauses for the monadic operator:

(t)   If $X \in f^1(w)$, then $w \in X$
(no) $W \notin f^1(w)$
(c)   If $X \in f^1(w)$ and $Y \in f^1(w)$, then $X \cap Y \in f^1(w)$
(k')   If $X \in P, Y \in P$, $-X \cup Y \in f^1(w)$ and $X \in f^1(w)$, then $Y \in f^1(w)$
(cs)   If $X \in P, Y \in P$, then if $X \cap Y \in f^1(w)$ then $X \in f^1(w)$ or $Y \in f^1(w)$.[13]

---

[11] I am well aware that certain efficiencies might be achieved by dropping irreflexivity (e.g. by then defining monadic agency via dyadic agency ($BA\varphi$ as $BA'_\varphi \varphi$), and although this might be worth exploring technically, I do not think irreflexivity can be maintained given the intended notion of dyadic agency, of bringing about φ by bringing about ψ. This notion, like that of "because", "in virtue of the fact that" and others, is intrinsically irreflexive. Put another way, if we allow irreflexivity, we then must read $BA'_\varphi \psi$ along some such lines as *Jane Doe brings it about that ψ by bringing it about that φ or Jane brings it about that ψ and Jane brings it about that φ*.

[12] Or designated propositions: P can be any subset of $\text{Pow}(W)$. This will be used to facilitate some correspondence proofs.

Additional normal PDA and DMA frames will be considered with one or more of these clauses for the dyadic operator:

- (**t′**)   If $Y \in f^2(w, X)$, then $w \in Y \cap X$
- (**no′**) $\neg \exists X(W \in f^2(w, X))$ and $f^2(w, W) = \emptyset$
- (**ir**) $X \notin f^2(w, X)$
- (**cc**)   If $Y \in f^2(w, X)$ and $Z \in f^2(w, X)$, then $Y \cap Z \in f^2(w, X)$
- (**dc**)   If $Y \in f^2(w, X)$ and $Y' \in f^2(w, X')$, then $Y \cap Y' \in f^2(w, X \cap X')$.
- (**da**)   If $X \in f^2(w, Y)$ and $X \in f^2(w, Z)$, then $X \in f^2(w, Y \cup Z)$
- (**dd**)   If $Y \in f^2(w, X)$ and $Y' \in f^2(w, X')$, then $Y \cup Y' \in f^2(w, X \cup X')$.
- (**tr**)   If $Y \in f^2(w, X)$ and $Z \in f^2(w, Y)$, then $Z \in f^2(w, X)$
- (**ct**)   If $Y \in f^2(w, X)$ and $Z \in f^2(w, X \cap Y)$, then $Z \in f^2(w, X)$
- (**ck′**)   If $X \in P$, $Y \in P$, $-X \cup Y \in f^2(w, Z)$, and $X \in f^2(w, Z)$, then $Y \in f^2(w, Z)$.
- (**dcs**)   If $X \in P$, $Y \in P$ and $X \cap Y \in f^2(w, Z)$ then $X \in f^2(w, Z)$ or $Y \in f^1(w, Z)$.
- (**ss**)   If $X \in P$, $Y \in P, Z \in f^2(w, U)$ and $Z = \{w' : Y \in f^2(w', X)\}$, then $Y \in f^2(w, U \cap X)$.

including this key clause linking f^1 and f^2,

- (**s**) f^1 and f^2 such that: If $Y \in f^2(w, X)$, then $X \in f^1(w)$ & $Y \in f^1(w)$. [Dyadic-Monadic Bridge]

However, for DMA frames with clauses **t**, **no**, **ir**, and **s**, clauses **t′** and **no′** are derivable, just as there sentential analogs, $T_{BA'}$ and $NO_{BA'}$, are derivable in the DMA logics containing T_{BA}, NO_{BA}, $IR_{BA'}$, and S.

A PMA Model, $M =< F, V >$, where F is an PMA frame, $<W, P, f^1>$: $V : PV \to P$ [V Maps Propositional Variables to elements of P (propositions)]

Similarly, for a âĂIJPDA ModelâĂİ and for a âĂIJDMA ModelâĂİ.

Truth on a PMA Model, $M =<< W, P, f^1 >, V >$:

---

[13]Note the relativization to P in the antecedent for these two constraints. I will use such underlining to remind the reader of those places where it appears that such relativization is essential to proving the correspondence theorems for the associated formulas in the canonical models to be defined, and thus to getting completeness. More on this below.

[PV] If $\varphi \in PV$, $M, w \models \varphi$ iff $w \in V(\varphi)$
[⊤] $M, w \models \top$, for each $w \in w$
[⊥] $M, w \not\models \bot$, for each $w \in w$
[¬] $M, w \models \neg\varphi$ iff $M, w \not\models \varphi$
...
[→] $M, w \models (\psi \to \chi)$ iff $M, w \not\models \psi$ or $M, w \models \chi$
[BA] $M, w \models \mathbf{BA}\psi$ iff $[\![\psi]\!]^M \in f^1(w)$, where $[\![\psi]\!]^M$ is $\{w : M, w \models \psi\}$.

Truth on an PDA Model, $M = \langle W, P, f^2 \rangle, V \rangle$:
Same but replace [BA] above with: [BA'] If $\varphi = \mathbf{BA}'_\psi \chi$, $M, w \models \varphi$ iff $[\![\chi]\!]^M \in f^2(w, [\![\psi]\!]^M)$

Truth on an DMA Model, $M = \langle W, P, f^1, f^2 \rangle, V \rangle$:
Same as PMA with the clause above for [BA'] added.

3 The Fundamental Theorems for Canonical Models

We now define the canonical models for the MDA logics (i.e the PMA, PDA, and DMA logics), and prove a fundamental theorem for such models, and then proceed to prove various theorems linking logics containing all instances of the preceding formulas schemata we listed as theorems and their canonical models.[14] In a familiar way, these will entail a large array of completeness results.

Let Σ^L be the set of maximal consistent sets (MCSs) of formulas for MDA logic, L. $|\varphi|L : \{w \in \Sigma^L : \varphi \in w\}$.

For any given MDA logic, L, from here on, the superscript "L" will be left as understood for L's canonical model and its components.

A *Canonical Model*, $M = \langle W, P, f^1, f^2, v \rangle$ for any DMA logic, L, is defined as follows:

(a) $W = \Sigma$

(b) $P = \{X : \exists \varphi (|\varphi| = X)\}$

(c) $f^1(w) = \{X : \exists \varphi (|\varphi| = X\ \&\ \mathbf{BA}\varphi \in w)\}$
  [So $X \in f^1(w)$ iff there is a formula, $\mathbf{BA}\varphi$, in w such that X is the set of the MCSs containing φ.][15]

(d) $f^2(w, X) = \{Y : \exists \psi (|\psi| = Y\ \&\ \exists \varphi (|\varphi| = X\ \&\ \mathbf{BA}'_\varphi \psi \in w))\}$
  [So $Y \in f^2(w, X)$ iff there is a formula $\mathbf{BA}'_\varphi \psi$ in w such that Y is the set of MCSs containing ψ and X is the set of MCSs containing φ.][16]

(e) $v(Pn) = |Pn|$.

Canonical Models for PMA logics: drop f^2 and d.
Canonical Models for PDA logics, drop f^1 and c.

We note these *Basic Properties* (BP) *of MCSs* apply to W in our canonical model:

$$\varphi \text{ iff } \forall w \in W : \varphi \in w;$$
$$|\neg \varphi| = \Sigma - |\varphi|;$$
$$|\varphi \wedge \psi| = |\varphi| \cap |\psi|;$$
$$|\varphi \vee \psi| = |\varphi| \cup |\psi|;$$
$$\vdash \varphi \leftrightarrow \psi \text{ iff } |\varphi| = |\psi|.$$

The following two-part lemma will come in handy.

Lemma 1
(a) For any φ and w: $\mathbf{BA}\varphi \in w$ iff $|\varphi| \in f^1(w)$.
Proof:
$\Rightarrow$: Suppose $\mathbf{BA}\varphi \in w$. Then since $|\varphi| = |\varphi|$, we have $\exists \varphi(|\varphi| = |\varphi| \,\&\, \mathbf{BA}\varphi \in w)$, which suffices by clause c for $|\varphi| \in f^1(w)$.
$\Leftarrow$: Assume $|\varphi| \in f^1(w)$. So by clause c, $\exists \varphi'(|\varphi'| = |\varphi| \,\&\, \mathbf{BA}\varphi' \in w)$. So fixing φ', we have $|\varphi'| = |\varphi| \,\&\, \mathbf{BA}\varphi' \in w$. So by BP, $\vdash \varphi \leftrightarrow \varphi'$, and then by RE$_{\mathbf{BA}}$, $(\mathbf{BA}\varphi \leftrightarrow \mathbf{BA}\varphi') \in w$, and so $\mathbf{BA}\varphi \in w$.

---

[14] We will often talk of a logic as containing a theorem (e.g. "containing theorem T$_{\mathbf{BA}}$'') where this will be understood as shorthand for a logic containing all instances of the schemata in question.
[15] Given RE$_{\mathbf{BA}}$, it will turn out that $|\varphi| \in f^1(w)$ iff $\mathbf{BA}\varphi \in w$. See Lemma 1a.
[16] Given RE$^\mathrm{r}_{\mathbf{BA}}{}'$ and RE$^\mathrm{l}_{\mathbf{BA}}{}'$, it will turn out that $|\psi| \in f^2(w, |\varphi|)$ iff $\mathbf{BA}'_\varphi \psi \in w$. See Lemma 1b.

(b) For any φ, ψ, and w: $\mathbf{BA}'_\varphi \psi \in w$ iff $|\psi| \in f^2(w, |\varphi|)$.
Proof:
$\Rightarrow$: Suppose $\mathbf{BA}'_\varphi \psi \in w$. Then since $|\psi| = |\psi|$ and $|\varphi| = |\varphi|$, we have $\exists \psi(|\psi| = |\psi|$ & $\exists \varphi(|\varphi| = |\varphi|$ & $\mathbf{BA}'_\varphi \psi \in w))$, which suffices by clause d for $|\psi| \in f^2(w, |\varphi|)$.
$\Leftarrow$: Assume $|\psi| \in f^2(w, |\varphi|)$. So by clause d, $\exists \psi'(|\psi'| = |\psi|$ & $\exists \varphi'(|\varphi'| = |\varphi|$ & $\mathbf{BA}'_{\varphi'} \psi' \in w))$. So fixing ψ' and φ', we have $|\psi'| = |\psi|$ & $|\varphi'| = |\varphi|$ & $\mathbf{BA}'_{\varphi'} \psi' \in w$. So by BP, $\vdash \varphi \leftrightarrow \varphi'$ and $\vdash \psi \leftrightarrow \psi'$, and hence by $\mathrm{RE^r_{BA'}}$ and $\mathrm{RE^l_{BA'}}$, we get $(\mathbf{BA}'_{\varphi'} \psi' \leftrightarrow \mathbf{BA}'_\varphi \psi) \in w$, and so $\mathbf{BA}'_\varphi \psi \in w$.

We now need to show that the canonical model of any MDA logic is an MDA model. In the proof below, it is also straightforward to separate the components for the three types of logics. For example, for PMA logics, only clauses $1-3$ are relevant, for DMA logics only clauses 1, 2, and 4 are relevant.

For any MDA logic, L, its canonical model, M, is an MDA model.
Proof:
(1) $W \neq \varnothing$. Since by definition, no MDA logic contains all MDA formulas, some formula φ will be a non-theorem, so there will be a maximal consistent extension of $\neg \varphi$, contained in W.
(2) $P \subseteq \mathrm{Pow}(W)$, for by definition, for each $X \in P$, there is a formula φ such that $|\varphi| = X$, and by design of W, $|\varphi| \subseteq W$.
(3) $f^1 : W \to \mathrm{Pow}(P)$, for by definition of f^1, its domain is W, and for any such w, $f^1(w)$ is $\{X : \exists \varphi(|\varphi| = X$ & $\mathbf{BA}\varphi \in w\}$, but by definition of P, it must contain all such Xs.
(4) $f^2 : W \times P \to \mathrm{Pow}(P)$, for by definition of f^2, its domain must be $W \times P$, for its image for any pair (w, X) in its domain is $\{Y : \exists \psi(|\psi| = Y)$ & $\exists \varphi(|\varphi| = X$ & $\mathbf{BA}'_\varphi \psi \in w)\}$, and by definition of P, it must contain all such values of X and Y.

We can now easily prove the fundamental theorem:

Fundamental Theorem for the Canonical Models for MDA Logics:

$$M, w \models \varphi \text{ iff } \varphi \in w, \text{ that is, } [\![\varphi]\!]^M = |\varphi|.$$

Proof: Assume the theorem is to be proved in the usual way by induction on the complexity of the formulas, and that it is already proved for formulas whose main connective is one of our truth-functional operators. (The base case holds by stipulation of clause e of the definition of a canonical model for any MDA logic.) We show that it holds for the remaining possible formula types, $\mathbf{BA}\psi$ and $\mathbf{BA}'_\chi \psi$.
(A) Suppose $\varphi = \mathbf{BA}\psi$, for some ψ. So by IH, for every $w, M, w \models \psi$ iff $\psi \in w$, thus $[\![\psi]\!]^M = |\psi|$. By the semantic clause for $\mathbf{BA}\psi, M, w \models \mathbf{BA}\psi$ iff $[\![\psi]\!]^M \in f^1(w)$. So $M, w \models \mathbf{BA}\psi$ iff $|\psi| \in f^1(w)$. But by lemma 1a, $|\psi| \in f^1(w)$ iff $\mathbf{BA}\psi \in w$. Thus

$M, w \models \mathbf{BA}\psi$ iff $\mathbf{BA}\psi \in w$.

(B) Suppose $\varphi = \mathbf{BA}'_\chi \psi$, for some χ, ψ. So by IH, $[\![\chi]\!]^M = |\chi|$ and $[\![\psi]\!]^M = |\psi|$. By the semantic clause for $\mathbf{BA}'_\chi \psi$, $M, w \models \mathbf{BA}'_\chi \psi$ iff $[\![\psi]\!]^M \in f^2(w, [\![\chi]\!]^M)$. So $M, w \models \mathbf{BA}'_\chi \psi$ iff $|\psi| \in f^2(w, |\chi|)$. But by lemma 1b, $\mathbf{BA}'_\chi \psi \in w$ iff $|\psi| \in f^2(w, |\chi|)$. So $M, w \models \mathbf{BA}'_\chi \psi$ iff $\mathbf{BA}'_\chi \psi \in w$.

Thus the theorem holds generally, and so the theorems of any MDA logic, L, are exactly those valid in L's canonical model, M^L, with A pertaining to the PMA logics, and B pertaining to the PDA logics and both A and B pertaining to the DMA logics.

With this in mind, we focus on correspondences between key axioms and clauses that govern the two modal operators to show various logics complete with respect to their associated semantics.

4 Correspondence Results for Completeness

For theorems where the proof depends on the propositions involved being relativized to para-meter P in the structures, we will label the theorem number with a superscript "P" (e.g. see T4P).

T1. Any canonical model for an MDA logic with $T_{\mathbf{BA}}$ satisfies the constraint **t**: If $X \in f^1(w)$, then $w \in X$. Suppose $X \in f^1(w)$. So $\exists \varphi(|\varphi| = X \ \& \ \mathbf{BA}\varphi \in w)$. Fixing φ, it follows that $|\varphi| = X$ and $\mathbf{BA}\varphi \in w$. But since $\vdash \mathbf{BA}\varphi \to \varphi$, $\mathbf{BA}\varphi \to \varphi \in w$, and thus $\varphi \in w$. Thus, $w \in |\varphi|$, that is $w \in X$.

T2. Any canonical model for an MDA logic with $NO_{\mathbf{BA}}$ satisfies the constraint **no** : $W \notin f^1(w)$. Assume that $W \in f^1(w)$. So $\exists \varphi(|\varphi| = W$ and $\mathbf{BA}\varphi \in w)$. Fixing φ, we have $|\varphi| = W$ and $\mathbf{BA}\varphi \in w$. But $|\varphi| = W$ iff $\vdash \varphi \leftrightarrow \top$, so by $RE_{\mathbf{BA}}$, $\vdash \mathbf{BA}\varphi \leftrightarrow \mathbf{BA}\top$, and so $\mathbf{BA}\top \in w$. Yet $\vdash \neg \mathbf{BA}\top$, so $\neg \mathbf{BA}\top \in w$, contrary to assumption that w is consistent.

T3. Any canonical model for an MDA logic with $C_{\mathbf{BA}}$ satisfies the constraint **c**: If $X \in f^1(w)$ and $Y \in f^1(w)$, then $X \cap Y \in f^1(w)$. Suppose $X \in f^1(w)$ and $Y \in f^1(w)$. So by the reasoning above in T1, we have that there exists a φ and a ψ such that: $|\varphi| = X$ and $|\psi| = Y$ and $\mathbf{BA}\varphi \in w$ and $\mathbf{BA}\psi \in w$. But $\vdash (\mathbf{BA}\varphi \land \mathbf{BA} \psi) \to \mathbf{BA}(\varphi \land \psi)$, so $\mathbf{BA}(\varphi \land \psi) \in w$. But $|\varphi \land \psi| = |\varphi \land \psi|$, so $\exists Z (Z = |\varphi \land \psi| \ \& \ \mathbf{BA}(\varphi \land \psi) \in w)$, and so $|\varphi \land \psi| \in f^1(w)$. But $|\varphi \land \psi| = |\varphi| \cap |\psi| = X \cap Y$, so $X \cap Y \in f^1(w)$.

The next two correspondence proofs utilize the parameter, P, and the next footnote explains its utility.

T4^P. Any canonical model for an MDA logic with $\mathbf{K_{BA}}$ satisfies the constraint $\mathbf{k}'$, that if $X \in P$ and $Y \in P$, and $-X \cup Y \in f^1(w)$ and $X \in f^1(w)$, then $Y \in f^1(w)$. Assume (1) $X \in P$ and $Y \in P$, and (2) $X \cup Y \in f^1(w)$ and $X \in f^1(w)$. From 2 by clause c of M, $\exists \varphi(|\varphi| = -X \cup Y \ \& \ \mathbf{BA}\varphi \in w)$ and $\exists \psi(|\psi| = X \ \& \ \mathbf{BA}\psi \in w)$. Fixing φ and ψ, we get (3) $|\varphi| = -X \cup Y \ \& \ \mathbf{BA}\varphi \in w$ and $(|\psi| = X \ \& \ \mathbf{BA}\psi \in w)$. From assumption 1 by clause b of M, $\exists \chi(|\chi| = X)$ and $\exists \chi'(|\chi'| = Y)$.[17] Fixing χ and χ', we have $|\chi| = X$ and $|\chi'| = Y$, and then by BP, also $|\neg\chi| = -X$. Substituting in assumption 3, we get $|\varphi| = |\neg\chi| \cup |\chi'| \ \& \ \mathbf{BA}\varphi \in w$ and $|\psi| = |\chi| \ \& \ \mathbf{BA}\psi \in w$. By BP again, $|\neg\chi| \cup |\chi'| = |\neg\chi \vee \chi'|$. Substituting again, we get $|\varphi| = |\neg\chi \vee \chi'|$. By BP, we then get $\vdash \varphi \leftrightarrow (\neg\chi \vee \chi')$ and $\vdash \psi \leftrightarrow \chi$. So by $\mathrm{RE}_{\mathbf{BA}'}$, we have $\vdash \mathbf{BA}\varphi \leftrightarrow \mathbf{BA}(\neg\chi \vee \chi')$ and $\vdash \mathbf{BA}\psi \leftrightarrow \mathbf{BA}\chi$. So by BP again and assumption 3 above, it follows that $\mathbf{BA}(\neg\chi \vee \chi') \in w$ and $\mathbf{BA}\chi \in w$; and from the former, $\mathbf{BA}(\chi \to \chi')$ follows by $\mathrm{RE}_{\mathbf{BA}}$. But then since we are assuming $\mathbf{K_{BA}}$ holds for this logic, by BP, $\mathbf{BA}(\chi \to \chi') \to (\mathbf{BA}\chi \to \mathbf{BA}\chi') \in w$, and hence $\mathbf{BA}\chi' \in w$. So by lemma 1a, it follows that $|\chi'| \in f^1(w)$, that is $Y \in f^1(w)$.

T5^P. Any canonical model for an MDA logic with $\mathbf{CS_{BA}}$ satisfies the constraint **cs**: If $X, Y \in P$ and $X \cap Y \in f^1(w)$, then $X \in f^1(w)$ or $Y \in f^1(w)$. Suppose for reductio that (1) $X, Y \in P$, (2) $X \cap Y \in f^1(w)$, and (3) $X \notin f^1(w)$ and (4) $Y \notin f^1(w)$. So from 2 we have $\exists \varphi(|\varphi| = X \cap Y \ \& \ \mathbf{BA}\varphi \in w)$, and then fixing φ, we get (2$'$) $|\varphi| = X \cap Y \ \& \ \mathbf{BA}\varphi \in w$. Given 3 and 4 it follows that $\neg\exists \psi(|\psi| = X \ \& \ \mathbf{BA}\psi \in w)$ and $\neg\exists \chi(|\chi| = X \ \& \ \mathbf{BA}\chi \in w)$, that is (3$'$) $\forall \psi(\text{if} |\psi| = X$, then $\mathbf{BA}\psi \notin w)$ and (4$'$) $\forall \chi(\text{if} |\chi| = X$ then $\mathbf{BA}\chi \notin w)$. But given 1, it follows from clause b of M that $\exists \psi(|\psi| = X)$ and $\exists \chi(|\chi| = Y)$; instantiating, we have $|\psi| = X$ and $|\chi| = Y$. Then from 3$'$ and 4$'$, we get $\mathbf{BA}\psi \notin w$ and $\mathbf{BA}\chi \notin w$. Given 2 and the identifications above for X, Y, we have $|\psi| \cap |\chi| \in f^1(w)$, and by BP, we get $|\psi \wedge \chi| \in f^1(w)$. Then from lemma 1a, it follows that $\mathbf{BA}(\psi \wedge \chi) \in w$. So we have $\mathbf{BA}(\psi \wedge \chi) \in w$, $\mathbf{BA}\psi \notin w$, and $\mathbf{BA}\chi \notin w$, but $\mathbf{CS_{BA}}$ is also in w, so w turns out to be inconsistent.

We next turn to correspondences involving candidate axioms for pure dyadic systems.

T6. Any canonical model for any MDA logic with $\mathbf{T_{BA}'}$ satisfies constraint $\mathbf{t}'$: If $Y \in f^2(w, X)$, then $w \in Y \cap X$. Suppose $Y \in f^2(w, X)$. Then $\exists \psi(Y = |\psi|) \ \& \ \exists \varphi(|\varphi| = $

---

[17]This is where the relativization to P in the frames comes in handy. Without it, although we have $|\varphi| = -X \cup Y$ and $\psi = X$, we have no guarantee there is any formula, ψ' such that $|\psi'| = Y$, so that in turn we can get $|\varphi| = -X \cup Y = |\neg\psi| \cup |\psi'| = |\psi \vee \psi'|$. With $X, Y \in P$, this is assured and so strong completeness results. See the Appendix of Goble, Lou [5] for an encounter with a similar problem, and an alternative strategy that generates weak completeness using the standard semantic constraint for K. The use of P here allows for strong completeness proofs, but only with the constraint involving the relativization to P.

X & $\mathbf{BA}'_\varphi \psi \in w$). Fixing ψ, φ, we have: $Y = |\psi|$ & $|\varphi| = X$ & $\mathbf{BA}'_\varphi \psi \in w$. But $\mathbf{T_{BA'}}$ is a thesis, so $\mathbf{BA}'_\varphi \psi \to (\varphi \wedge \psi) \in w$, and so $(\varphi \wedge \psi) \in w$. So $w \in |\varphi \wedge \psi|$, that is $w \in |\varphi| \cap |\psi|$, and then the consequent of $\mathbf{t}'$ follows: $w \in Y \cap X$.

T7. Any canonical model for any MDA logic with $\mathrm{NO_{BA'}}$ satisfies the constraint **no'**: $\neg \exists X$ such that $W \in f^2(w, X)$ and $f^2(w, W) = \varnothing$. For suppose instead that (a) $\exists X[W \in f^2(w, X)]$ or (b) $f^2(w, W) \neq \varnothing$. Fixing X in case a, we have $W = f^2(w, X)$. But since $|\top| = W$, we have $|\top| \in f^2(w, X)$. So by definition of the canonical models (and since $|\top| = |\top|$), we have $\exists \varphi(|\varphi| = X$ & $\mathbf{BA}'_\varphi \top \in w)$. Fixing φ, we have $|\varphi| = X$ & $\mathbf{BA}'_\varphi \top \in w$. But by schema $\mathrm{NO_{BA'}}$, we have $\neg(\mathbf{BA}'_\varphi \top \vee \mathbf{BA}'_\top \varphi) \in w$, and thus $\neg \mathbf{BA}'_\varphi \top \in w$, so w is not consistent. In case b, $X \in f^2(w, W)$, for some X. But since $|\top| = W$, $X \in f^2(w, |\top|)$. So by definition of the canonical models, $\exists \varphi(|\varphi| = X$ & $\mathbf{BA}'_\top \varphi \in w)$. Fixing φ, we have $|\varphi| = X$ & $\mathbf{BA}'_\top \varphi \in w$. But given $\mathrm{NO_{BA'}}$, $\neg(\mathbf{BA}'_\varphi \top \vee \mathbf{BA}'_\top \varphi) \in w$, and thus $\neg \mathbf{BA}'_\top \varphi \in w$, so w is not consistent.

T8. Any canonical model for any MDA logic with $\mathrm{IR_{BA'}}$ satisfies the constraint **ir**: $X \notin f^2(w, X)$. Suppose $X \in f^2(w, X)$, for any X. Then $\exists \psi(|\psi| = X)$ & $\exists \varphi(|\varphi| = X$ & $\mathbf{BA}'_\varphi \psi \in w)$. Fix ψ and φ. So $|\psi| = X$ & $|\varphi| = X$ & $\mathbf{BA}'_\varphi \psi \in w$. But since $|\psi| = |\varphi|$, $\vdash \psi \leftrightarrow \varphi$, and so by $\mathrm{RE^r_{BA'}}$, we have $\mathbf{BA}'_\varphi \varphi \in w$. But given $\mathrm{IR_{BA'}}$, $\neg \mathbf{BA}'_\varphi \varphi \in w$ too, so w is not consistent.

T9. Any canonical model for any MDA logic with $\mathrm{CC_{BA'}}$ satisfies constraint **cc**: If $Y \in f^2(w, X)$ and $Z \in f^2(w, X)$, then $Y \cap Z \in f^2(w, X)$. Suppose $Y \in f^2(w, X)$ and $Z \in f^2(w, X)$. Then $\exists \psi(Y = |\psi|)$ & $\exists \varphi(|\varphi| = X$ & $\mathbf{BA}'_\varphi \psi \in w)$ and $\exists \psi'(Z = |\psi'|)$ & $\exists \varphi'(|\varphi'| = X$ & $\mathbf{BA}'_{\varphi'} \psi' \in w)$. Fixing ψ, φ, ψ', and φ', we have: $Y = |\psi|$ & $|\varphi| = X$ & $\mathbf{BA}'_\varphi \psi \in w$ and $Z = |\psi'|$ & $|\varphi'| = X$ & $\mathbf{BA}'_{\varphi'} \psi' \in w$. Since $|\varphi| = X$ & $|\varphi'| = X$, $|\varphi| = |\varphi'|$, and thus $\vdash \varphi \leftrightarrow \varphi'$. Then given $\mathrm{RE^l_{BA'}}$, $\vdash \mathbf{BA}'_{\varphi'} \psi' \leftrightarrow \mathbf{BA}'_\varphi \psi'$ and thus $\mathbf{BA}'_{\varphi'} \psi' \leftrightarrow \mathbf{BA}'_\varphi \psi' \in w$. So we have $X = |\varphi|$ & $Y = |\psi|$ & $Z = |\psi'|$ & $\mathbf{BA}'_\varphi \psi \in w$ & $\mathbf{BA}'_\varphi \psi' \in w$. But given $\mathrm{CC_{BA}}$', $(\mathbf{BA}'_\varphi \psi$ & $\mathbf{BA}'_\varphi \psi') \to \mathbf{BA}'_\varphi(\psi \wedge \psi') \in w$, so $\mathbf{BA}'_\varphi(\psi \wedge \psi') \in w$. By lemma 1b, $\mathbf{BA}'_\varphi(\psi \wedge \psi') \in w$ iff $|\psi \wedge \psi'| \in f^2(w, |\varphi|)$. But $|\psi \wedge \psi'| = |\psi| \cap |\psi'|$, that is, $|\psi \wedge \psi'| = Y \cap Z$, and we already have $|\varphi| = X$. So the consequent of **cc** follows: $Y \cap Z \in f^2(w, X)$.

The following theorem generalizes the preceding one.

T10. Any canonical model for any MDA logic with $\mathrm{DC_{BA'}}$ satisfies constraint **dc**: If $Y \in f^2(w, X)$ and $Y' \in f^2(w, X')$, then $Y \cap Y' \in f^2(w, X \cap X')$. Assume $Y \in f^2(w, X)$ and $Y' \in f^2(w, X')$. Then $\exists \psi(Y = |\psi|)$ & $\exists \varphi(|\varphi| = X$ & $\mathbf{BA}'_\varphi \psi \in w)$ and $\exists \psi'(Y' = |\psi'|)$ & $\exists \varphi'(|\varphi'| = X'$ & $\mathbf{BA}'_{\varphi'} \psi' \in w)$. Fixing ψ, φ, ψ', and φ', we have: $Y = |\psi|$ & $|\varphi| = X$ & $\mathbf{BA}'_\varphi \psi \in w$ and $Y' = |\psi'|$ & $|\varphi'| = X'$ & $\mathbf{BA}'_{\varphi'} \psi' \in w$. So $Y \cap Y' = |\psi| \cap |\psi'|$, and so by BP, $Y \cap Y' = |\psi \wedge \psi'|$. Similarly,

$X \cap X' = |\varphi| \cap |\varphi'| = |\varphi \wedge \varphi'|$. But since $\mathbf{BA}'_\varphi \psi \in w$ and $\mathbf{BA}'_{\varphi'} \psi' \in w$ and all instances of $\mathrm{DC}_{\mathbf{BA}'}$ are in w, it follows that $\mathbf{BA}'_{(\varphi \wedge \varphi')}(\psi \wedge \psi') \in w$. So by lemma 1b, $|\psi \wedge \psi'| \in f^2(w, |\varphi \wedge \varphi'|)$, and then from the identities above, $Y \cap Y' \in f^2(w, X \cap X')$ follows.

T11. Any canonical model for any MDA logic with $\mathrm{DA}_{\mathbf{BA}'}$ satisfies constraint **da**: If $X \in f^2(w, Y)$ and $X \in f^2(w, Z)$, then $X \in f^2(w, Y \cup Z)$. Suppose $X \in f^2(w, Y)$ and $X \in f^2(w, Z)$. Then $\exists \psi(X = |\psi|) \,\&\, \exists \varphi(|\varphi| = Y \,\&\, \mathbf{BA}'_\varphi \psi \in w)$ and $\exists \psi'(X = |\psi'|) \,\&\, \exists \varphi'(|\varphi'| = Z \,\&\, \mathbf{BA}'_{\varphi'} \psi' \in w)$. Fixing ψ, φ, ψ', and φ', we have: $X = |\psi| \,\&\, |\varphi| = Y \,\&\, \mathbf{BA}'_\varphi \psi \in w$ and $X = |\psi'| \,\&\, |\varphi'| = Z \,\&\, \mathbf{BA}'_{\varphi'} \psi' \in w$. Since $|\psi| = X \,\&\, |\psi'| = X$, $|\psi| = |\psi'|$, and thus $\vdash \psi \leftrightarrow \psi'$. Then given $\mathrm{RE}^{\mathrm{r}}_{\mathbf{BA}'}$, $\vdash \mathbf{BA}'_{\varphi'} \psi \leftrightarrow \mathbf{BA}'_{\varphi'} \psi'$ and thus $\mathbf{BA}'_{\varphi'} \psi \leftrightarrow \mathbf{BA}'_{\varphi'} \psi' \in w$. So we have $X = |\psi| \,\&\, Y = |\varphi| \,\&\, Z = |\varphi'| \,\&\, \mathbf{BA}'_\varphi \psi \in w \,\&\, \mathbf{BA}'_{\varphi'} \psi \in w$. But given $\mathrm{DA}_{\mathbf{BA}'}$, $(\mathbf{BA}'_\varphi \psi \wedge \mathbf{BA}'_{\varphi'} \psi) \to \mathbf{BA}'_{(\varphi \vee \varphi')} \psi \in w$, so $\mathbf{BA}'_{(\varphi \vee \varphi')} \psi \in w$. By lemma 1b, $\mathbf{BA}'_{(\varphi \vee \varphi')} \psi \in w$ iff $|\psi| \in f^2(w, |\varphi \vee \varphi'|)$. But $|\varphi \vee \varphi'| = |\varphi| \cup |\varphi'|$, that is, $|\varphi \vee \varphi'| = Y \cup Z$, and we already have $|\psi| = X$. So the consequent of **da** follows: $X \in f^2(w, Y \cup Z)$.

The following theorem also generalizes the preceding one.

T12. Any canonical model for any MDA logic with schema $\mathrm{DD}_{\mathbf{BA}'}$ satisfies constraint **dd**: If $Y \in f^2(w, X)$ and $Y' \in f^2(w, X')$, then $Y \cup Y' \in f^2(w, X \cup X')$. Assume $Y \in f^2(w, X)$ and $Y' \in f^2(w, X')$. Then $\exists \psi(Y = |\psi|) \,\&\, \exists \varphi(|\varphi| = X \,\&\, \mathbf{BA}'_\varphi \psi \in w)$ and $\exists \psi'(Y' = |\psi'|) \,\&\, \exists \varphi'(|\varphi'| = X' \,\&\, \mathbf{BA}'_{\varphi'} \psi' \in w)$. Fixing ψ, φ, ψ', and φ', we have: $Y = |\psi| \,\&\, |\varphi| = X \,\&\, \mathbf{BA}'_\varphi \psi \in w$ and $Y' = |\psi'| \,\&\, |\varphi'| = X' \,\&\, \mathbf{BA}'_{\varphi'} \psi' \in w$. So $Y \cup Y' = |\psi| \cup |\psi'|$, and so by BP, $Y \cup Y' = |\psi \vee \psi'|$. Similarly, $X \cup X' = |\varphi| \cup |\varphi'| = |\varphi \vee \varphi'|$. But since $\mathbf{BA}'_\varphi \psi \in w$ and $\mathbf{BA}'_{\varphi'} \psi' \in w$ and using $\mathrm{DD}_{\mathbf{BA}'}$, it follows that $\mathbf{BA}'_{(\varphi \vee \varphi')}(\psi \vee \psi') \in w$. So by lemma 1b, $|\psi \vee \psi'| \in f^2(w, |\varphi \vee \varphi'|)$, and then from the identities above, $Y \cup Y' \in f^2(w, X \cup X')$ follows.

T13. Any canonical model for any MDA logic with $\mathrm{TR}_{\mathbf{BA}'}$ satisfies constraint **tr**: If $Y \in f^2(w, X)$ and $Z \in f^2(w, Y)$, then $Z \in f^2(w, X)$. Suppose $Y \in f^2(w, X)$ and $Z \in f^2(w, Y)$. Then $\exists \psi(Y = |\psi|) \,\&\, \exists \varphi(|\varphi| = X \,\&\, \mathbf{BA}'_\varphi \psi \in w)$ and $\exists \psi'(Z = |\psi'|) \,\&\, \exists \varphi'(|\varphi'| = Y \,\&\, \mathbf{BA}'_{\varphi'} \psi' \in w)$. Fixing ψ, φ, ψ', and φ', we have: $Y = |\psi| \,\&\, |\varphi| = X \,\&\, \mathbf{BA}'_\varphi \psi \in w$ and $Z = |\psi'| \,\&\, |\varphi'| = Y \,\&\, \mathbf{BA}'_{\varphi'} \psi' \in w$. Since $Y = |\psi| \,\&\, |\varphi'| = Y$, $|\psi| = |\varphi'|$, and thus $\vdash \psi \leftrightarrow \varphi'$. Then given $\mathrm{RE}^{\mathrm{l}}_{\mathbf{BA}'}$, $\vdash \mathbf{BA}'_{\varphi'} \psi' \leftrightarrow \mathbf{BA}'_\psi \psi'$ and thus $\mathbf{BA}'_\psi \psi' \in w$. So we have $X = |\varphi| \,\&\, Y = |\psi| \,\&\, Z = |\psi'| \,\&\, \mathbf{BA}'_\varphi \psi \in w \,\&\, \mathbf{BA}'_\psi \psi' \in w$. But given $\mathrm{TR}_{\mathbf{BA}'}$, $(\mathbf{BA}'_\varphi \psi \wedge \mathbf{BA}'_\psi \psi') \to \mathbf{BA}'_\varphi \psi' \in w$, so $\mathbf{BA}'_\varphi \psi' \in w$. By lemma 1b $\mathbf{BA}'_\varphi \psi' \in w$ iff $|\psi'| \in f^2(w, |\varphi|)$. But $|\psi'| = Z$ and $X = |\varphi|$, so the consequent of constraint **tr** follows: $Z \in f^2(w, X)$.

T14. Any canonical model for any MDA logic with $\mathrm{CT}_{\mathbf{BA}'}$ satisfies constraint **ct**:

If $Y \in f^2(w, X)$ and $Z \in f^2(w, X \cap Y)$, then $Z \in f^2(w, X)$. Suppose $Y \in f^2(w, X)$ and $Z \in f^2(w, X \cap Y)$, Then $\exists \psi(Y = |\psi|)$ & $\exists \varphi(|\varphi| = X$ & $\mathbf{BA}'_\varphi \psi \in w)$ and $\exists \psi'(Z = |\psi'|)$ & $\exists \varphi'(|\varphi'| = X \cap Y$ & $\mathbf{BA}'_{\varphi'} \psi' \in w)$. Fixing ψ, φ, ψ', and φ', we have: $Y = |\psi|$ & $|\varphi| = X$ & $\mathbf{BA}'_\varphi \psi \in w$ and $Z = |\psi'|$ & $|\varphi'| = X \cap Y$ & $\mathbf{BA}'_{\varphi'} \psi' \in w$. Since $Y = |\psi|$ & $|\varphi| = X$ & $|\varphi'| = X \cap Y$, we have $|\varphi'| = |\varphi| \cap |\psi|$, and thus $|\varphi'| = |\varphi \wedge \psi|$. So $\vdash \varphi' \leftrightarrow (\varphi \wedge \psi)$, and then by $\text{RE}^l_{\mathbf{BA}'}$, $\mathbf{BA}'_{\varphi'} \psi' \leftrightarrow \mathbf{BA}'_{(\varphi \wedge \psi)} \psi' \in w$. Hence we have $X = |\varphi|$ & $Y = |\psi|$ & $Z = |\varphi \wedge \psi|$ & $\mathbf{BA}'_\varphi \psi \in w$ & $\mathbf{BA}'_{(\varphi \wedge \psi)} \psi' \in w$. But given $\text{CT}_{\mathbf{BA}'}$, $(\mathbf{BA}'_\varphi \psi \wedge \mathbf{BA}'_{(\varphi \wedge \psi)} \psi') \to \mathbf{BA}'_\varphi \psi' \in w$, so $\mathbf{BA}'_\varphi \psi' \in w$. By lemma 1b, $\mathbf{BA}'_\varphi \psi' \in w$ iff $|\psi'| \in f^2(w, |\varphi|)$. But $|\psi'| = Z$ and $|\varphi| = X$, so the consequent of constraint **ct** follows: $Z \in f^2(w, X)$.

T15^P. Any canonical model for an MDA logic with $\text{CK}_{\mathbf{BA}'}$ satisfies constraint **ck'**: If $X \in P$, $Y \in P$, $-X \cup Y \in f^2(w, Z)$, and $X \in f^2(w, Z)$, then $Y \in f^2(w, Z)$. Assume (1) $X \in P$ and $Y \in P$, and (2) $-X \cup Y \in f^2(w, Z)$ and $X \in f^2(w, Z)$. So from 2 by clause d of M, $\exists \psi(|\psi| = -X \cup Y)$ & $\exists \varphi(|\varphi| = Z$ & $\mathbf{BA}'_\varphi \psi \in w)$ and $\exists \psi'(|\psi'| = X)$ & $\exists \varphi'(|\varphi'| = Z$ & $\mathbf{BA}'_{\varphi'} \psi' \in w)$. Fixing φ, ψ, φ', and ψ', we get (3) $|\psi| = -X \cup Y$ & $|\varphi| = Z$ & $\mathbf{BA}'_\varphi \psi \in w$ and (4) $|\psi'| = X$ & $|\varphi'| = Z$ & $\mathbf{BA}'_{\varphi'} \psi' \in w$. From 1 by, $\exists \chi(|\chi| = X)$ and $\exists \chi'(|\chi'| = Y)$. Fixing χ and χ', we have $|\chi| = X$ and $|\chi'| = Y$, and then by BP, also $|\neg \chi| = -X$. Substituting in 3 and 4, we get (3') $|\psi| = |\neg \chi| \cup |\chi'|$ & $|\varphi| = Z$ & $\mathbf{BA}'_\varphi \psi \in w$ and (4') $|\psi'| = |\chi|$ & $|\varphi'| = Z$ & $\mathbf{BA}'_{\varphi'} \psi' \in w$. By BP again, $|\neg \chi| \cup |\chi'| = |\neg \chi \vee \chi'|$. Substituting again in 3', we get (3'') $|\psi| = |\neg \chi \vee \chi'|$ & $|\varphi| = Z$ & $\mathbf{BA}'_\varphi \psi \in w$. By BP from the identities in 3', 4' and 3'', we get $\vdash \psi \leftrightarrow (\neg \chi \vee \chi')$, $\vdash \psi' \leftrightarrow \chi$, and $\vdash \varphi \leftrightarrow \varphi'$. So by $\text{RE}^l_{\mathbf{BA}'}$ and $\text{RE}^r_{\mathbf{BA}'}$, we have $\vdash \mathbf{BA}'_\varphi \psi \leftrightarrow \mathbf{BA}'_\varphi (\neg \chi \vee \chi')$, $\vdash \mathbf{BA}'_{\varphi'} \psi' \leftrightarrow \mathbf{BA}'_\varphi \chi$. It follows that $\mathbf{BA}'_\varphi (\neg \chi \vee \chi') \in w$, and $\mathbf{BA}'_\varphi \chi \in w$. Then using $\text{CK}_{\mathbf{BA}'}$, $\mathbf{BA}'_\varphi \chi' \in w$. But from $\mathbf{BA}'_\varphi \chi' \in w$, by lemma 1b, it follows that $|\chi'| \in f^2(w, |\varphi|)$, that is, $Y \in f^2(w, Z)$.

T16^P. Any canonical model for an MDA logic with $\text{DCS}_{\mathbf{BA}'}$ satisfies constraint **dcs**: If $X, Y \in P$ and $X \cap Y \in f^2(w, Z)$ then $X \in f^2(w, Z)$ or $Y \in f^2(w, Z)$. Suppose for reductio that (1) $X, Y \in P$, (2) $X \cap Y \in f^2(w, Z)$, and (3) $X \notin f^2(w, Z)$ and (4) $Y \notin f^2(w, Z)$. From 2 we have $\exists \psi(|\psi| = X \cap Y)$ & $\exists \varphi(|\varphi| = Z$ & $\mathbf{BA}'_\varphi \psi \in w)$, and then fixing ψ, φ, we get (2') $|\psi| = X \cap Y$ & $|\varphi| = Z$ & $\mathbf{BA}'_\varphi \psi \in w$. From 3 and 4 it follows that $\neg \exists \psi(|\psi| = X)$ & $\exists \varphi(|\varphi| = Z)$ & $\mathbf{BA}'_\varphi \psi \in w$ and $\neg \exists \psi(|\psi| = Y)$ & $\exists \varphi(|\varphi| = Z)$ & $\mathbf{BA}'_\varphi \psi \in w$, that is, (3') $\forall \psi(\text{if } |\psi| = X)$ then $\forall \varphi(|\varphi| = Z)$ only if $\mathbf{BA}'_\varphi \psi \notin w$ and (4') $\forall \psi(\text{if } |\psi| = Y)$ then $\forall \varphi(|\varphi| = Z)$ only if $\mathbf{BA}'_\varphi \psi \notin w$. From assumption 1 we have $\exists \psi'(|\psi'| = X)$ and $\exists \psi''(|\psi''| = Y)$; instantiating, we have $|\psi'| = X$ and $|\psi''| = Y$. Then applying these to 3' and 4', we get (3'') $\forall \varphi(\text{if } |\varphi| = Z)$ then $\mathbf{BA}'_\varphi \psi' \notin w$ and (4'') $\forall \varphi(\text{if } |\varphi| = Z$ then $\mathbf{BA}'_\varphi \psi'' \notin w)$. So from $|\varphi| = Z$ in 2', we have $\mathbf{BA}'_\varphi \psi' \notin w$ $\mathbf{BA}'_\varphi \psi'' \notin w$. Given 2 and the identifications above for X, Y, Z, we have $|\psi'| \cap |\psi''| \in f^2(w, |\varphi|)$, and by BP, it then follows that $|\psi' \wedge \psi''| \in f^2(w, |\varphi|)$. Invoking lemma 1b, it follows that $\mathbf{BA}'_\varphi (\psi' \wedge \psi'') \in w$. But since $|\psi'| = X$ and $|\psi''| = Y$, and $|\psi| = X \cap Y$, it follows that $|\psi| = |\psi'| \cap |\psi''|$, and then from BP, we get $|\psi| = |\psi' \wedge \psi''|$. By BP again, $\vdash \psi \leftrightarrow (\psi' \wedge \psi'')$ follows, and then by $\text{RE}^r_{\mathbf{BA'}}$, $\vdash \mathbf{BA}'_\varphi \psi \leftrightarrow \mathbf{BA}'_\varphi (\psi' \wedge \psi'')$ follows, and then from 2' and BP, we get $\mathbf{BA}'_\varphi (\psi' \wedge \psi'') \in w$. So now we have $\mathbf{BA}'_\varphi (\psi' \wedge \psi'')$ & $\mathbf{BA}'_\varphi \psi' \notin w$ & $\mathbf{BA}'_\varphi \psi'' \notin w$, but by $\text{DCS}_{\mathbf{BA}}{'}$, w is rendered inconsistent.

The following candidate dyadic formula involves embedding of a dyadic operator within the scope of a dyadic operator. I call this formula dyadic "stage setting."

T17^P. Any canonical model for an MDA logic with $\text{SS}_{\mathbf{BA'}}$ satisfies the constraint ss: if $X \in P$, $Y \in P$,[18] then if $Z \in f^2(w, U)$ and $Z = \{w' : Y \in f^2(w', X)\}$, then $Y \in f^2(w, U \cap X)$. Assume (1) $X \in P$ and $Y \in P$, (2) $Z \in f^2(w, U)$, and (3) $Z = \{w' : Y \in f^2(w', X)\}$. Given 2 by clause d of M, $\exists \psi(|\psi| = Z)$ & $\exists \varphi(|\varphi| = U$ & $\mathbf{BA}'_\varphi \psi \in w)$. Fixing φ and ψ, we get $|\psi| = Z$ & $|\varphi| = U$ & $\mathbf{BA}'_\varphi \psi \in w$. From 1, there must be an χ and χ' with $|\chi| = X$ and $|\chi'| = Y$. Then given $|\psi| = Z$, $|\varphi| = U$, $|\chi| = X$ and $|\chi'| = Y$, from 2 and 3, we get: (2') $|\psi| \in f^2(w, |\varphi|)$ and (3') $|\psi| = \{w' : |\chi'| \in f^2(w', |\chi|)\}$. But then applying lemma 1b, we get these equivalents: (2') $\mathbf{BA}'_\varphi \psi \in w$ and (3'') $|\psi| = \{w' : \mathbf{BA}'_\chi \chi' \in w'\}$. Then since by definition, $|\mathbf{BA}'_\chi \chi'| = \{w' : \mathbf{BA}'_\chi \chi' \in w'\}$, and so $|\psi| = |\mathbf{BA}'_\chi \chi'|$. Hence by BP, $\vdash \psi \leftrightarrow \mathbf{BA}'_\chi \chi'$, and then by $\text{RE}^r_{\mathbf{BA'}}$, $\vdash \mathbf{BA}'_\varphi \psi \leftrightarrow \mathbf{BA}'_\varphi \mathbf{BA}'_\chi \chi$, and so $\mathbf{BA}'_\varphi \psi \in w$ iff $\mathbf{BA}'_\varphi \mathbf{BA}'_\chi \chi' \in w$. But since $\text{SS}_{\mathbf{BA'}}$ is a thesis, we then get $\mathbf{BA}'_{(\varphi \wedge \chi)} \chi' \in w$. So by lemma 1b, $|\chi'| \in f^2(w, |\varphi \wedge \chi|)$. But by BP, $|\varphi \wedge \chi| = |\varphi| \cap |\chi|$, and since $|\varphi| = U$, $|\chi| = X$ and $|\chi'| = Y$, we have $Y \in f^2(w, U \cap X)$.

The following is the key thesis for combined monadic-dyadic agency logics:

T18. Any canonical model for any DMA logic with S satisfies constraint s: If $Y \in f^2(w, X)$, then $X \in f^1(w)$ and $Y \in f^1(w)$. Suppose $Y \in f^2(w, X)$, then $\exists \psi (Y = |\psi|)$ & $\exists \varphi (|\varphi| = X$ & $\mathbf{BA}'_\varphi \psi \in w)$. Fixing ψ and φ, we have: $Y = |\psi|$ & $|\varphi| = X$ & $\mathbf{BA}'_\varphi \psi \in w$. But given S, $(\mathbf{BA}'_\varphi \psi \to (\mathbf{BA}\varphi \wedge \mathbf{BA}\psi)) \in w$, so $(\mathbf{BA}\varphi \wedge \mathbf{BA}\psi) \in w$, and thus $\mathbf{BA}\varphi \in w$ & $\mathbf{BA}\psi \in w$. By lemma 1a, $\mathbf{BA}\varphi \in w$ iff $|\varphi| \in f^1(w)$ and $\mathbf{BA}\psi \in w$ iff $|\psi| \in f^1(w)$. So we have the consequent of s: $X \in f^1(w)$ and $Y \in f^1(w)$.

---

[18] Goble notes in correspondence that T17^P can be straightforwardly demonstrated without invoking the relativization to P, provided that we are looking at a system that also includes $\text{T}_{\mathbf{BA}}{'}$. Given supposition (2), you get $Z = |\psi|$ and $U = |\varphi|$ and $\mathbf{BA}'_\varphi \psi \in w$. Given $\text{T}_{\mathbf{BA}}{'}$, you get $\psi \in w$, hence $w \in |\psi|$. By 3, since $Z = |\psi| = \{w : Y \in f^2(w, X)\}$, you get $w \in \{w : Y \in f^2(w, X)\}$. So $Y \in f^2(w, X)$. That provides the χ such that $X = |\chi|$ and the χ' such that $Y = |\chi'|$ and the rest continues as before, without invoking assumption 1.

5 Correspondences & Completeness for Systems Summarized

Below is a table summarizing the eighteen correspondence theorems:

Formula Schema (and label):   Constraint on Frames

T_{BA}:	$BA\varphi \to \varphi$	(t)		If $X \in f^1(w)$, then $w \in X$
NO_{BA}:	$\neg BA\top$	(no)		$W \notin f^1(w)$
C_{BA}:	$(BA\varphi \wedge BA\psi) \to BA(\varphi \wedge \psi)$	(c)		If $X \in f^1(w)$ and $Y \in f^1(w)$, then $X \cap Y \in f^1(w)$
K_{BA}:	$BA(\varphi \to \psi) \to (BA\varphi \to BA\psi)$	(k')		If $X \in P, Y \in P, -X \cup Y \in f^1(w)$ and $X \in f^1(w)$, then $Y \in f^1(w)$
CS_{BA}:	$BA(\varphi \wedge \psi) \to (\neg BA\varphi \to BA\psi)$	(cs)		If $X \in P, Y \in P$, then if $X \cap Y \in f^1(w)$ then $X \in f^1(w)$ or $Y \in f^1(w)$
$T_{BA'}$:	$BA'_\psi \varphi \to (\varphi \wedge \psi)$	(t')		If $Y \in f^2(w, X)$, then $w \in Y \cap X$
$NO_{BA'}$:	$\neg(BA'_\varphi \top \vee BA'_\top \varphi)$	(no')		$\neg \exists X(W \in f^2(w, X))$ and $f^2(w, W) = \varnothing$
$IR_{BA'}$:	$\neg BA'_\varphi \varphi$	(ir)		$X \notin f^2(w, X)$

Formula Schema (and label):   Constraint on Frames

$CC_{BA'}$:	$(BA'_\varphi \psi \wedge BA'_\varphi \chi) \to BA'_\varphi(\psi \wedge \chi)$	(cc)		If $Y \in f^2(w, X)$ and $Z \in f^2(w, X)$, then $Y \cap Z \in f^2(w, X)$
$DC_{BA'}$:	$(BA'_\varphi \psi \wedge BA'_{\varphi'} \psi') \to BA'_{(\varphi \wedge \varphi')}(\psi \wedge \psi')$	(dc)		If $Y \in f^2(w, X)$ and $Y' \in f^2(w, X')$, then $Y \cap Y' \in f^2(w, X \cap X')$
$DA_{BA'}$:	$(BA'_\psi \varphi \wedge BA'_\chi \varphi) \to BA'_{(\psi \vee \chi)} \varphi$	(da)		If $X \in f^2(w, Y)$ and $X \in f^2(w, Z)$, then $X \in f^2(w, Y \cup Z)$
$DD_{BA'}$:	$(BA'_\varphi \psi \wedge BA'_{\varphi'} \psi') \to BA'_{(\varphi \vee \varphi')}(\psi \vee \psi')$	(dd)		If $Y \in f^2(w, X)$ and $Y' \in f^2(w, X')$, then $Y \cup Y' \in f^2(w, X \cup X')$
$TR_{BA'}$:	$(BA'_\varphi \psi \wedge BA'_\psi \chi) \to BA'_\varphi \chi$	(tr)		If $Y \in f^2(w, X)$ and $Z \in f^2(w, Y)$, then $Z \in f^2(w, X)$
$CT_{BA'}$:	$(BA'_\varphi \psi \wedge BA'_{(\varphi \wedge \psi)} \chi) \to BA'_\varphi \chi$	(ct)		If $Y \in f^2(w, X)$ and $Z \in f^2(w, X \cap Y)$, then $Z \in f^2(w, X)$
$CK_{BA'}$:	$BA'_\chi(\varphi \to \psi) \to (BA'_\chi \varphi \to BA'_\chi \psi)$	(ck')		If $X \in P, Y \in P, -X \cup Y \in f^2(w, Z)$, and $X \in f^2(w, Z)$, then $Y \in f^2(w, Z)$.
$DCS_{BA'}$:	$BA'_\chi(\varphi \wedge \psi) \to (\neg BA'_\chi \varphi \to BA'_\chi \psi)$	(dcs)		If $X \in P, Y \in P$ and $X \cap Y \in f^2(w, Z)$ then $X \in f^2(w, Z)$ or $Y \in f^1(w, Z)$.
SS:	$BA'\chi BA'_\varphi \psi \to BA'_{(\chi \wedge \varphi)} \psi$	(ss)		If $X \in P, Y \in P, Z \in f^2(w, U)$ and $Z = \{w' : Y \in f^2(w', X)\}$, then $Y \in f^2(w, U \cap X)$
S:	$BA'_\psi \varphi \to (BA\varphi \wedge BA\psi)$	(s)		f^1 and f^2 such that: If $Y \in f^2(w, X)$, then $X \in f^1(w)$ & $Y \in f^1(w)$

6 Conclusion

With the fundamental theorem and the eighteen correspondence theorems provided above, we can establish completeness for very large number (thousands) of distinct monadic and dyadic and monadic-dyadic logics, using combinations of the schemata covered in the eighteen theorems.[19] This is a strength of the framework. The semantic framework is weak in allowing for a great deal of independence and thus serving well to provide a characterization of the domain of agency logics, at least to a first approximation. Other stronger and more robust semantic frameworks like those in the STIT tradition will in some ways be more philosophically attractive and informative because they provide a more specific model of agency, but at the same time, there is a cost too in greater specificity since it will rule out or interlink formulas in ways that are more substantive and contentious. It is thus useful to have a weaker more general framework as a backdrop or reference point for stronger frameworks, which are of course also worth exploring, but doing so is beyond the scope of the current paper. As it is, the current framework obviously needs to be expanded to represent monadic and dyadic *ability* and its interactions with monadic and dyadic agency. Other natural additional directions would be to add multi-agents and/or temporal operators, not to mention exploring other aspects of the metatheory for the framework (distinctness of logics, decidability). What we have here is first step toward systematizing a neighborhood semantic framework for monadic as well as dyadic agency logics.

7 Appendix on Stalled Proof

Natural but Failed Attempt at Showing Correspondence of K with Condition (k):

We show that in the canonical model M^L for EK, if $W^L - X \cup Y \in N^L(w)$ and $X \in N^L(w)$, then $Y \in N^L(w)$. For any w, suppose (a) $X \in N^L(w)$ and (b) $W^L - X \cup Y \in N^L(w)$. So by a, $X \in \{|\varphi| : \Box\varphi \in w\}$, and so $X = |\varphi|$, for some φ such that $\Box\varphi \in w$. Fix φ. So $W^L - X \cup Y = W^L - |\varphi| \cup Y = |\neg\varphi| \cup Y$; then by b, $|\neg\varphi| \cup Y \in \{|\psi| : \Box\psi \in w\}$, that is, $|\neg\varphi| \cup Y = |\psi|$, for some ψ such that $\Box\psi \in w$. Fix ψ...???

The rub: What can assure that if $|\neg\varphi| \cup Y = |\psi|$ then $\exists \chi$ such that $|\neg\varphi \vee \chi| = |\neg\varphi| \cup Y = |\psi|$? How do we know that Y is expressible by some χ?

A Fix:

Suppose instead the frames have an additional parameter, $P \subseteq \text{Pow}(W)$, and then in the canonical model we assure that P is the set of proof sets: $X \in P$ iff $\exists\varphi(|\varphi| = X)$.

$M^L = <W^L, P^L, N^L, V^L>$ is a canonical model for an E logic iff:

(a) W^L = the set of MCSs for L

---

[19] Although as noted in footnote 19 regarding theorem 17^P, not all combinations will involve non-redundant schemata (schemata not derivable from the remainder).

(b) $P^L = \{X : \exists \varphi(|\varphi| = X)\}$, where $X \subseteq W^L$

(c) $N^L(w) = \{X : \exists \varphi(|\varphi| = X \ \& \ \Box \varphi \in w)\}$, where $X \subseteq W^L$

(d) $V(Pn) = |Pn|$.

The Fundamental Theorem for Canonical Models goes through as before. If we then cast the frame condition for formulas K as follows:

(k′) If $X, Y \in P$, *then* if $W - X \cup Y \in N_{(w)}$ and $X \in N_{(w)}$, then $Y \in N_{(w)}$

The correspondence proof for EK goes through smoothly.

References

[1] H. Arlo-Costa and E. Pacuit. First-order classical modal logic: Applications in logics of knowledge and probability. In *TARK '05: 10th Conference on Theoretical Aspects of Rationality and Knowledge*, pages 262–278. National University of Singapore, 2007.

[2] B.F. Chellas. *Modal Logic: An Introduction*. Cambridge University Press, 1980.

[3] D. Elgesem. *Action Theory and Modal Logic*. PhD thesis, University of Oslo, 1993.

[4] D. Elgesem. The modal logic of agency. *Nordic Journal of Philosophical Logic*, 2:1–46, 1997.

[5] L. Goble. A proposal for dealing with deontic dilemmas. In A. Lomuscio and D. Nute, editors, *DEON 2004: 7th International Workshop on Deontic Logic in Computer Science*, volume 3065 of *LNAI*, pages 74–113, Madeira, Portugal, 2004. Springer.

[6] G. Governatori and A. Rotolo. On the axiomatisation of elgesem's logic of agency and ability. *Journal of Philosophical Logic*, 34:403–431, 2005.

[7] A. Jones and X. Parent. A convention-based approach to agent communication languages. *Group Decision and Negotiation*, 16:101–141, 2007.

[8] A. Jones and M. Sergot. On the characterisation of law and computer systems: The normative systems perspective. In J.-J.C. Meyer and R.J. Wieringa, editors, *Deontic Logic in Computer Science: Normative System Specification*, pages 275–307, Chichester, New York, 1993. John Wiley and Sons.

[9] A. Jones and M. Sergot. A formal characterization of institutionalized power. *Logic Journal of the IGPL*, 4:429–445, 1996.

[10] S. Kanger. Law and logic. *Theoria*, 38:105–132, 1972.

[11] S. Kanger and H. Kanger. Rights and parliamentarism. In G. Holmström-Hintikka, S. Lindström, and R. Sliwinski, editors, *Collected Papers of Stig Kanger with Essays on His Life and Work,II*, pages 120–145. Kluwer Academic Publishers, Dordrecht, 1966/2001. Originally published in 1972 in *Theoria*, 32: 85-115.

[12] P. McNamara. Agential obligation as non-agential personal obligation plus agency. *Jounal of Applied Logic*, 2:117–152, 2004.

[13] P. McNamara. Acting beyond the call and kindred notions: Some reflections on their representation. In *Trends in Logic, XVII*, Lublin, Poland, 2017. John Paul II Catholic University. Keynote Address.

[14] I. Pörn. *The logic of Power*. Blackwell, Oxford, 1970.

[15] I. Pörn. *Action Theory and Social Science: Some Formal Models*. D. Reidel, 1977.

[16] I. Pörn. On the nature of a social order. In J. E. Fenstad, T. Frolov, and R. Hilpinen, editors, *Logic, Methodology and Philosophy of Science*, volume VIII, pages 553–567. Elsevier Science, New York, 1989.

[17] F. Santos and J. Carmo. Indirect action, influence and responsibility. In M.A. Brown and J. Carmos, editors, *Deontic Logic, Agency and Normative Systems*, pages 194–215. Springer Verlag, New York, 1996.

[18] F. Santos, A.J.I. Jones, and J. Carmo. Responsibility for action in organizations: a formal model. In G. HolmstrÃűm-Hintikka and R. Tuomela, editors, *Contemporary Action Theory, Vol II (Social Action)*, pages 333–350. Kluwer Academic Publishers, Dordrecht, 1997.

[19] M. Sergot. Normative positions. In P. McNamara and H. Prakken, editors, *Norms, Logics, and Information Systems: New Studies in Deontic Logic and Computer Science*, pages 194–215. Amsterdam, 1999.

[20] M. Sergot. Normative positions. In D. Gabbay, J. Horty, X. Parent, R. van der Meyden, and L. van der Torre, editors, *Handbook of Deontic Logic and Normative Systems*, volume I, pages 353–406. College Publications, London, 2013.

I/O Logics with a Consistency Check

Xavier Parent
University of Luxembourg
`xavier.parent@uni.lu`

Leendert van der Torre
University of Luxembourg
`leon.vandertorre@uni.lu`

Norm-based semantics to deontic logic typically come in an unconstrained and constrained version, where the unconstrained version comes with a proof system, and the constraints handle phenomena such as dilemmas, contrary-to-duty reasoning, uncertainty and defeasibility. This is analogous to the use of rule-based languages in non-monotonic logic such as logic programming or default logic, but in contrast to the traditional modal framework. Traditionally, for example, specific modal deontic logics have been defined that make dilemmas inconsistent, as well as other modal deontic logics representing dilemmas in a consistent way. This issue was raised recently in the input/output logic framework, and weaker unconstrained logics have been defined handling phenomena like dilemmas and contrary-to-duty reasoning. In this paper we introduce a semantics and proof theory for a system with various desirable properties. We show that our new deontic logic satisfies a criterion posed several years ago by Broersen and van der Torre, allowing deontic detachment while preventing Prakken and Sergot's pragmatic oddities as well as Sergot's drowning problem.

1 Introduction

This paper deals with I/O logic, initially devised by Makinson and van der Torre [7]. It falls within the category of what Hansen [6] calls a "norm-based" deontic logic. The central question of a norm-based deontic logic is : what obligations can be detached from a set N of (explicitly given) rules or conditional norms in a given context? The approach is in this regard very different from the more traditional one, aiming at identifying a set of "logical laws" using a possible worlds semantics.

Thanks to three anonymous reviewers for valuable comments. This work is supported by the European Union's Horizon 2020 research and innovation programme under the Marie Curie grant agreement No: 690974 (Mining and Reasoning with Legal Texts, MIREL).

1.1 Aim of paper

We first explain the main purpose of this paper, and its contribution to the literature on I/O logic. An overview of the I/O systems that have been studied so far in the literature is shown in Table 1 along with the system studied in this paper. A pair (a, x) denotes a conditional obligation. (a, x) is read: "if a, then x is obligatory". The columns show the rules characterising a system. The symbol "+" indicates the presence of a rule, and the symbol "-" its absence. The right-most column gives the paper(s) where the system has been studied. Each system comes with a semantics, and a completeness theorem linking the semantics with the proof theory. The last system mentioned in the table is the strongest one. It collapses with the system of classical propositional logic: (a, x) is derivable from N if and only if $m(N) \vdash a \to x$, where $\vdash$ is the deducibility relation used in classical propositional logic and $m(N)$ is the set of materializations of N, viz $\{b \to y : (b, y) \in N\}$. We list the rules in the order in which they appear in the table, starting from the left-most column.

- EQ (equivalence of the output): from (a, x), $x \vdash y$ and $y \vdash x$, infer (a, y)
- SI (strengthening of the input): from (a, x) and $b \vdash a$, infer (b, x) [1]
- WO (weakening of the output): from (a, x) and $x \vdash y$, infer (a, y)
- R-AND (restricted AND): from (a, x), (a, y) and $a \wedge x \wedge y$ consistent, infer $(a, x \wedge y)$
- R-AND': from (a, x), (a, y), $a \wedge x$ consistent and $a \wedge y$ consistent, infer $(a, x \wedge y)$
- AND: from (a, x) and (a, y), infer $(a, x \wedge y)$
- OR: from (a, x) and (b, x), infer $(a \vee b, x)$
- R-ACT (restricted ACT'): from (a, x), $(a \wedge x, y)$ and $a \wedge x \wedge y$ consistent, infer $(a, x \wedge y)$
- ACT (aggregative CT): from (a, x), $(a \wedge x, y)$, infer $(a, x \wedge y)$
- MCT (mediated CT): from (a, x'), $(a \wedge x, y)$, $x' \vdash x$, infer (a, y)
- CT (cumulative transitivity): from (a, x) and $(a \wedge x, y)$, infer $(a, x \wedge y)$
- ID (identity): (a, a)

In this paper, we want to investigate the effects of adding R-ACT to {SI,EQ}. (Given SI, R-ACT entails R-AND.) Our aim is to define an I/O operation validating the triplet of rules {SI, R-ACT, EQ}, and to establish a completeness theorem showing the equivalence between the semantics and the proof theory. The I/O operation is called reusable, because the output can be recycled as input (under suitable conditions). The system characterised by {SI, R-ACT, EQ} is weaker than the one characterised by {SI, ACT, EQ}, but stronger than the one characterised by {SI, R-AND, EQ}. In the presence of SI, R-ACT implies R-AND, but not conversely.

---

[1] Given SI, the analog of EQ holds for the input.

EQ	SI	WO	R-AND	R-AND'	AND	OR	R-ACT	ACT	MCT	CT	ID	References
+	+	-	+	-	-	-	-	-	-	-	-	[14]
+	+	-	+	-	-	-	+	-	-	-	-	this paper
+	+	-	+	+	-	-	-	-	-	-	-	[14]
+	+	-	+	+	+	-	-	-	-	-	-	[18, 17, 12]
+	+	-	+	+	+	-	+	+	-	-	-	[12]
+	+	-	+	+	+	-	+	+	+	+	-	[18, 17]
+	+	-	+	+	+	+	-	-	-	-	-	[12]
+	+	+	+	+	+	-	-	-	-	-	-	[7]
+	+	+	+	+	+	-	-	-	-	-	+	[7]
+	+	+	+	+	+	-	+	+	+	+	-	[7, 11]
+	+	+	+	+	+	-	+	+	+	+	+	[7]
+	+	+	+	+	+	+	-	-	-	-	-	[7, 11, 19]
+	+	+	+	+	+	+	-	-	-	-	+	[7]
+	+	+	+	+	+	+	+	+	+	+	-	[7]
+	+	+	+	+	+	+	+	+	+	+	+	[7]

Table 1: Overview of I/O systems

1.2 Broersen and van der Torre's open problem

In order to motivate this work, we explain how the pair of rules {SI, R-ACT} handles a problem pointed out by Broersen and van der Torre in their survey of open problems in deontic logic [2]. For each problem, they discuss traditional and new research questions. The one we will focus on is related to the topic of contrary-to-duty reasoning. After having introduced the traditional problems surrounding this topic, and identified a few new questions, they make the following observation:

> "The pragmatic oddity is the derivation of the conjunction 'you should keep your promise and apologize for not keeping it' from 'you should keep your promise', 'if you do not keep your promise you should apologize' and 'you do not keep your promise' [15]. Note that the sentences of this problem have the same structure as those of the Chisholm scenario. The drowning problem [also called by Parent and van der Torre [13] the violation detection problem][2] is that many solutions of the pragmatic oddity cancel the obligation in case of violation, such that for violations $\neg p \wedge \bigcirc p$, the violated obligation $\bigcirc p$ is no longer derivable." [2, p. 64]

They go on to ask the following question:

---

[2]The name "drowning problem" was suggested orally by M. Sergot to the first author of the present paper, in relation with the non-monotonic approaches to contrary-to-duty reasoning.

"New question 11: how to prevent the pragmatic oddity without creating the drowning problem?"

Example 1 shows how the proposed system handles this problem. To keep things simple, we make our point using R-AND, which is derivable from the pair of rules {SI,R-ACT}.

Example 1 (Broken promise). *Let k and a stand for keeping one's promise and for apologizing, respectively. Consider the following derivation, in which a blocked derivation step is represented by a dashed line.*

$$SI \frac{(\top, k)}{\underbrace{(\neg k, k)}_{\text{-----}} \quad (\neg k, a)}_{(\neg k, k \wedge a)} \text{ R-AND}$$

On the one hand, the drowning problem (or the violation detection problem) does not occur, because SI allows us to move from $(\top, k)$ to $(\neg k, k)$. Constrained I/O logic [8] blocks such a move: k is not consistent with $\neg k$. On the other hand, the pragmatic oddity is avoided, because R-AND cannot be applied to get $(\neg k, k \wedge a)$: $k \wedge \neg k \wedge a$ is not consistent.

Three remarks are in order:

- The rule R-AND' ("from (a, x), (a, y), $a \wedge x$ consistent and $a \wedge y$ consistent, infer $(a, x \wedge y)$") characterizing the second of the two I/O systems discussed in Parent and van der Torre [14] also blocks the pragmatic oddity.

- This treatment of the pragmatic oddity is very much in the spirit of Prakken and Sergot [16]'s own treatment. They (rightly, in our view) stress that primary and CTD obligations are of a different kind. Our proposal is not to allow them to aggregate using the AND rule, because of this difference in nature. This point was already made by Parent and van der Torre [14].

- R-AND is enough to tackle Broersen and van der Torre's problem. But there is an independent reason for using the stronger rule R-ACT. It is that a system with SI and R-AND only does not allow to chain norms together, and does not support any form of transitivity. This motivates the attempt made in this paper to extend the account described by Parent and van der Torre [14] so it can handle iterations of successive detachments. Since (given SI) ACT implies AND, the combination {ACT, R-AND} must be ruled out. Prima facie, there are other combinations that might be worthwhile studying, like {CT, R-AND}. However, it is still unknown what the corresponding I/O operations would be like. This is why they are not considered in this paper.

The present paper is technical. Our main interest is in formal systems. The proof of completeness we give is not a straightforward adaptation of proofs given elsewhere. As always the devil is in the details. There are two challenging complicating factors that make the proof non-trivial, and worth reporting. First, when calculating the output, one looks at what is "triggered" not by $Cn(A)$, the set of consequences of input set A, but by some $B \subseteq Cn(A)$. This is needed to resolve the violation detection (or drowning) problem: SI is supported. Second, the proposed I/O operations have a built-in "consistency check". They are thus close in spirit to the constrained I/O operations developed by Makinson and van der Torre [8]. The objective is the same: to filter excess output using consistency checks. There are similarities between the two frameworks, but also important differences. First, all the constrained I/O operations face the violation detection problem. Second, in contrast to the constrained I/O operations, the I/O operations studied in this paper have a proof theory. Their built-in consistency check (used to filter excess output) translates into a consistency proviso restraining the application of a rule. This explains the running title of this paper, "Consistent reusability".

This paper follows a straightforward structure. Section 2 gives the required background. Section 3 presents the system, and shows soundness and completeness.

2 Background

In this section we recall some basic definitions and a result from Parent and van der Torre [14], which will be used in the paper.

A normative code is a set N of pairs (a, x), where a and x are two formulae of classical propositional logic. Each pair represents a conditional obligation. a is called the body, and x is called the head. Given $M \subseteq N$, $h(M)$ denotes the set of all the heads of the pairs in M, and $b(M)$ denotes the set of all the bodies of the pairs in M. We use the standard notation $(\top, x)$ for the unconditional obligation of x, where $\top$ stands for a tautology like $a \vee \neg a$. $\mathcal{L}$ is the set of all formulae of classical propositional logic. Given an input $A \subseteq \mathcal{L}$, and a normative system N, $N(A)$ denotes the image of N under A, i.e., $N(A) = \{x : (a, x) \in N \text{ for some } a \in A\}$. $Cn(A)$ denotes the set $\{x : A \vdash x\}$, where $\vdash$ is the deducibility relation used in classical propositional logic. The notation $x \dashv\vdash y$ is short for $x \vdash y$ and $y \vdash x$. We use PL as an abbreviation for (classical) propositional logic.

In input/output logic, the main semantical construct has the form: $x \in O(N, A)$. This is read as follows: given input A (state of affairs), x (obligation) is in the output under norms N. The proof-theory is given in terms of inference rules manipulating pairs of Boolean formulas instead of formulas.

Definition 1 reformulates one of the two new I/O operations put forth by Parent and van der Torre [14]–both formulations are equivalent. The I/O operation is written O_1. The definition says the following. Given input A, x is outputted if the following condition holds: x is logically equivalent to the conjunction of all the heads of all the pairs in a non-empty and finite $M \subseteq N$, whose bodies are all in $Cn(A)$, and which are "collectively" consistent with x. Formally:

Definition 1 (Semantics, single-step detachment). $x \in O_1(N, A)$ iff there is some finite $M \subseteq N$ and a set $B \subseteq Cn(A)$ such that $M \neq \emptyset$, $B = b(M)$, $x \dashv\vdash \wedge h(M)$ and $\{x\} \cup B$ is consistent.

As usual, $O_1(N) = \{(A, x) : x \in O_1(N, A)\}$.

In Parent and van der Torre [14], the account has been applied to a number of benchmark examples from literature. The proposed account has been devised to handle simultaneously the two main categories of benchmark problems discussed in deontic logic, the group of those pertaining to contrary-to-duty reasoning [16, 3, 5], and the group of those dealing with (unresolved) conflicts between obligations [4]. These two categories of problems are usually considered separately one from the other. We believe it is a virtue of the present framework that it covers them both.

We turn to the proof-theory. A derivation of a pair (a, x) from N, given a set R of rules, is understood to be a tree with (a, x) at the root, each non-leaf node related to its immediate parents by the inverse of a rule in R, and each leaf node an element of N.

Definition 2 (Proof system). $(a,x) \in D_1(N)$ if and only if (a,x) is derivable from N using the rules {SI, EQ, R-AND}

$$SI \frac{(a,x) \quad b \vdash a}{(b,x)}$$

$$EQ \frac{(a,x) \quad x \dashv\vdash y}{(a,y)}$$

$$R\text{-}AND \frac{(a,x) \quad (a,y)}{(a, x \wedge y)} \; a \wedge x \wedge y \text{ is consistent}$$

Furthermore it is required that all the leaves of the derivation of (a,x) have (in the terminology of Makinson and van der Torre [8]) a consistent "fulfilment". That is, for all the leaves (b,y), $b \wedge y$ must be consistent.

When A is a set of formulas, $(A,x) \in D_1(N)$ means that $(a,x) \in D_1(N)$, for some conjunction a of elements of A. Furthermore, $D_1(N,A) = \{x : (A,x) \in D_1(N)\}$.

The requirement that the leaves of a derivation have a consistent fulfilment implies that D_1 fails inclusion, that is $N \subseteq D_1(N,A)$ does not necessarily hold. Put $N = \{(x, \neg x)\}$; we have $(x, \neg x) \notin D_1(N)$. This is in line with the semantics, which yields $(x, \neg x) \notin O_1(N)$.

The following applies.

Theorem 1 (Soundness and completeness). $D_1(N,A) = O_1(N,A)$.

Proof. The proof is a re-run of the one given in Parent and van der Torre [14], suitably adapted to take into account the changes made to the definition of the I/O operation. □

3 Recycling the output as input

The system described in the previous section has an important limitation: it does not allow the output to be recycled as input. In other words, it cannot handle iteration of successive detachments. In this section, we show how to remove this limitation. We describe a semantics and a proof system, and we establish the soundness and completeness of the second with respect to the first.

We start with the semantics. The I/O operation is written O_3. We use the same subscript as Makinson and van der Torre [7]–our O_3 echoes their reusable I/O operation out_3.

Definition 3 (Semantics, iterated detachments). $x \in O_3(N,A)$ iff there is a finite $M \subseteq N$ and a set $B \subseteq Cn(A)$ such that $M(B) \neq \emptyset$, $x \dashv\vdash \wedge h(M)$, and

i) $\forall B'(B \subseteq B' = Cn(B') \supseteq M(B') \Rightarrow b(M) \subseteq B')$

ii) $\{x\} \cup B$ is consistent

Observation 1. Let M, B and B' be such that $B \subseteq B' \supseteq M(B') \cup b(M)$. We have $h(M) = M(B')$.

Proof. The inclusion $M(B') \subseteq h(M)$ holds by definition. For the converse inclusion, let $y \in h(M)$. We have $(a, y) \in M$ for some $a \in b(M)$. Since $b(M) \subseteq B'$, $a \in B'$, and thus $y \in M(B')$ as required. □

As before, $O_3(N) = \{(A, x) : x \in O_3(N, A)\}$.

Observation 2. If $x \in O_3(N, A)$ and $A \cup \{x\}$ is consistent, then there is a finite $M \subseteq N$ such that $M(Cn(A)) \neq \emptyset$, $x \dashv\vdash \wedge h(M)$ and

$$\forall B'(Cn(A) \subseteq B' = Cn(B') \supseteq M(B') \Rightarrow b(M) \subseteq B') \quad (1)$$

Proof. Let $x \in O_3(N, A)$ and $A \cup \{x\}$ be consistent. By definition 3, there is a finite $M \subseteq N$ and a set $B \subseteq Cn(A)$ such that $M(B) \neq \emptyset$, $x \dashv\vdash \wedge h(M)$, and

i) $\forall B'(B \subseteq B' = Cn(B') \supseteq M(B') \Rightarrow b(M) \subseteq B')$

We have $M(Cn(A)) \neq \emptyset$, since $M(B) \neq \emptyset$. (1) follows from i) and $B \subseteq Cn(A)$. □

We now present the proof theory.

Definition 4 (Proof system). $(a, x) \in D_3(N)$ if and only if (a, x) is derivable from N using the rules $\{SI, EQ, R\text{-}ACT\}$:

$$R\text{-}ACT \frac{(a, x) \quad (a \wedge x, y) \quad a \wedge x \wedge y \text{ is consistent}}{(a, x \wedge y)}$$

As before it is required that all the leaves of the derivation of (a, x) have a consistent "fulfilment". That is, for all the leaves (b, y), $b \wedge y$ must be consistent.

As before, when A is a set of formulas, $(A, x) \in D_3(N)$ means that $(a, x) \in D_3(N)$ for some conjunction a of elements of A. Furthermore, $D_3(N, A) = \{x : (A, x) \in D_3(N)\}$.

The requirement that the leaves of a derivation have a consistent fulfilment implies that D_3 fails inclusion, that is $N \subseteq D_3(N)$ does not necessarily hold. Put $N = \{(x, \neg x)\}$; we have $(x, \neg x) \notin D_3(N)$. This is in line with the semantics, which yields $(x, \neg x) \notin O_3(N)$.

Observation 3. O_3 (for an input formula a) verifies the rules of D_3.

Proof. SI and EQ are straightforward. We show R-ACT. Assume

$$x \in O_3(N, a) \tag{HYP 1}$$
$$y \in O_3(N, a \wedge x) \tag{HYP 2}$$
$$a \wedge x \wedge y \text{ consistent} \tag{HYP 3}$$

From HYP 3, $a \wedge x$ is consistent. By observation 2, HYP 1 implies:

$$\exists M_1 \subseteq N \text{ such that } M_1(Cn(a)) \neq \emptyset \text{ and } x \Vdash \wedge h(M_1) \text{ and} \\ \forall B'(Cn(a) \subseteq B' = Cn(B') \supseteq M_1(B') \Rightarrow b(M_1) \subseteq B') \tag{2}$$

Similarly, by observation 2, HYP 2 and HYP 3 imply:

$$\exists M_2 \subseteq N \text{ such that } M_2(Cn(a,x)) \neq \emptyset \text{ and } y \Vdash \wedge h(M_2) \text{ and} \\ \forall B'(Cn(a,x) \subseteq B' = Cn(B') \supseteq M_2(B') \Rightarrow b(M_2) \subseteq B') \tag{3}$$

Put $M_3 = M_1 \cup M_2$. We have $M_3(Cn(a)) \neq \emptyset$. Also, $x \wedge y \Vdash \wedge h(M_3)$.

Let B' be such that $Cn(a) \subseteq B' = Cn(B') \supseteq M_3(B')$. We have $B' \supseteq M_1(B')$. By (2), $b(M_1) \subseteq B'$. By observation 1, $x \Vdash \wedge M_1(B')$. But $B' = Cn(B') \supseteq M_1(B')$. So $x \in B'$. Also $a \in B'$. Hence $a \wedge x \in B'$. So $Cn(a,x) \subseteq B'$. On the other hand, $B' \supseteq M_2(B')$. By (3), $b(M_2) \subseteq B'$, so that $b(M_3) \subseteq B'$. Last, $\{x \wedge y\} \cup Cn(a)$ is consistent, by HYP 3. Hence, $x \wedge y \in O_3(N, a)$ as required. □

Theorem 2 (Soundness). $D_3(N, A) \subseteq O_3(N, A)$.

Proof. The proof follows the usual format in I/O logic, using theorem 3. The requirement that all the leaves of the derivation of (a, x) must have consistent fulfilment is needed to handle the case where (a, x) is in N. Details are omitted. □

The remainder of the paper is devoted to the proof of completeness.

Lemma 1. *If $x \in D_3(M, A)$, then $h(M) \vdash x$.*

Proof. By induction on the length of the derivation of (A, x). Details are omitted. □

Theorem 3 (Completeness). $O_3(N, A) \subseteq D_3(N, A)$.

Proof. Assume $x \in O_3(N, A)$, viz. $(A, x) \in O_3(N)$. There is a finite $M \subseteq N$ and a set $B \subseteq Cn(A)$ such that $M(B) \neq \emptyset$, $x \Vdash \wedge h(M)$ and

i) $\forall B'(B \subseteq B' = Cn(B') \supseteq M(B') \Rightarrow b(M) \subseteq B')$

ii) $B \cup \{x\}$ is consistent

Define $B^\dagger = Cn(B \cup D_3(M, B))$.

Lemma 2. $M(B^\dagger) \subseteq B^\dagger$.

Proof of lemma 2. Let $y \in M(B^\dagger)$. Hence $(c, y) \in M$ and $c \in B^\dagger$. So $B \cup D_3(M, B) \vdash c$. Thus $b, y_1, ..., y_n \vdash c$, where b is a conjunction of elements in B, and $y_1, ..., y_n \in D_3(M, B)$. For all $i \leq n$, $y_i \in D_3(M, b_i)$, where b_i is a conjunction of elements in B. For the sake of conciseness, we define $\flat$ as a shorthand of $b \wedge (\wedge_{i=1}^n b_i)$. By PL, $\wedge_{i=1}^n y_i \vdash \flat \to c$, and thus $\wedge_{i=1}^n y_i \dashv\vdash \wedge_{i=1}^n y_i \wedge (\flat \to c)$.

Now, for two sub-lemmas.

Lemma 2.1. $\flat \wedge (\wedge_{i=1}^n y_i) \wedge (\flat \to c) \wedge y$ *is consistent.*

Proof of lemma 2.1. Proof by contradiction:

$\flat \wedge (\wedge_{i=1}^n y_i) \wedge (\flat \to c) \wedge y \vdash \bot$	assumption
$\flat \wedge (\wedge_{i=1}^n y_i) \wedge y \vdash \bot$	since $\wedge_{i=1}^n y_i \vdash \flat \to c$
$B \cup \{y_1, ..., y_n, y\} \vdash \bot$	since $B \vdash \flat$
$B \cup h(M) \cup \{y\} \vdash \bot$	by lemma 1
$B \cup h(M) \vdash \bot$	since $y \in h(M)$
$B \cup \{x\} \vdash \bot$	since $h(M) \dashv\vdash x$
= contradiction	

□

Lemma 2.2. $c \wedge y$ *is consistent.*

Proof of lemma 2.2. Proof by contradiction:

$c \wedge y \vdash \bot$	assumption
$\flat \wedge (\flat \to c) \wedge y \vdash \bot$	since $\flat \wedge (\flat \to c) \vdash c$
= contradiction	

□

The argument for lemma 2 continues thus. Now, we have

$$\text{EQ} \frac{\dfrac{(b_1, y_1)}{(\flat, y_1)} \text{SI} \quad ... \quad \dfrac{(b_n, y_n)}{(\flat, y_n)} \text{SI}}{\dfrac{(\flat, \wedge_{i=1}^n y_i)}{(\flat, \wedge_{i=1}^n y_i \wedge (\flat \to c))}} \text{R-AND, lemma 2.1}$$

Each (b_i, y_i) is the root of a derivation from leaves which (by definition) satisfy the requirement that they have a consistent fulfilment. Furthermore, the pair (c, y) has a consistent fulfilment, lemma 2.2. Thus,

$$\frac{(\flat, \wedge_{i=1}^n y_i \wedge (\flat \to c)) \quad \dfrac{(c, y)}{(\flat \wedge (\wedge_{i=1}^n y_i) \wedge (\flat \to c), y)} \text{ SI}}{(\flat, \wedge_{i=1}^n y_i \wedge (\flat \to c) \wedge y)} \text{ R-ACT, lemma 2.1}$$

$\flat$ is a conjunction of formulas in B. This implies that

$$\wedge_{i=1}^n y_i \wedge (\flat \to c) \wedge y \in D_3(M, B)$$

and so $y \in B^\dagger$ as required.

This completes the proof of lemma 2. $\square$

Lemma 3. $\flat(M) \subseteq B^\dagger$.

Proof of lemma 3. This follows from the fact that $B^\dagger$ meets all the conditions mentioned in the antecedent of the implication i). $\square$

Lemma 4. $B \cup D_3(M, B)$, and hence also $B^\dagger$, is consistent.

Proof of lemma 4. We establish the claim for $B \cup D_3(M, B)$ by contradiction. In the following derivation, $b_1, \ldots b_n$ are elements of $D_3(M, B)$.

$$\begin{array}{ll} B \cup \{b_1, \ldots, b_n\} \vdash \bot & \text{assumption} \\ B \cup h(M) \vdash \bot & \text{by lemma 1} \\ B \cup \{x\} \vdash \bot & \text{since } h(M) \dashv\vdash x \\ = \text{contradiction} & \end{array}$$

The claim for $B^\dagger$ follows from that for $B \cup D_3(M, B)$. $\square$

Lemma 5. $h(M) \subseteq B^\dagger$.

Proof of lemma 5. This follows from observation 1, $h(M) = M(B^\dagger)$, combined with the above. $\square$

Lemma 6. $\flat(M) \cup h(M)$ is consistent.

Proof of lemma 6. By lemmas 3 and 5, $\flat(M) \cup h(M) \subseteq B^\dagger$. By lemma 4, $B^\dagger$ is consistent. It immediately follows that $\flat(M) \cup h(M)$ is consistent. $\square$

Lemma 7. $\flat(M) \cup \{x\}$ is consistent.

Proof of lemma 7. Immediate from lemma 6 and $x \dashv\vdash \wedge h(M)$. $\square$

With lemmas 3 and 7 in hand, one then gets:

$$x \in O_1(N, B \cup D_3(M, B)) \qquad \text{by definition 1}$$
$$x \in D_1(N, B \cup D_3(M, B)) \qquad \text{by theorem 1}$$
$$x \in D_3(N, B \cup D_3(M, B))$$

This means that $x \in D_3(N, \{b\} \cup \{a_1, ..., a_n\})$, where b is a conjunction of elements of B and, for each a_i, $a_i \in D_3(M, B)$. For each a_i, there is a conjunction b_i of elements in B such that $a_i \in D_3(M, b_i)$.

At this point, one last lemma is needed:

Lemma 8. $\wedge_{i=1}^n b_i \wedge (\wedge_{i=1}^n a_i) \wedge b \wedge x$ is consistent.

Proof of lemma 8. Proof by contradiction:

$$\{a_1, ..., a_n, b_1, ..., b_n, b, x\} \vdash \bot \qquad \text{assumption}$$
$$h(M) \cup \{b_1, ..., b_n, b, x\} \vdash \bot \qquad \text{by lemma 1}$$
$$h(M) \cup B \cup \{x\} \vdash \bot \qquad \text{since } b_1, ..., b_n, b \in Cn(B)$$
$$B \cup \{x\} \vdash \bot \qquad \text{since } h(M) \dashv\vdash x$$
$$= \text{contradiction}$$

□

The following is thus derivable from M:

$$\dfrac{\dfrac{(b_1, a_1)}{(\wedge_{i=1}^n b_i, a_1)} \text{ SI} \quad \cdots \cdots \quad \dfrac{(b_n, a_n)}{(\wedge_{i=1}^n b_i, a_n)} \text{ SI}}{\dfrac{(\wedge_{i=1}^n b_i, \wedge_{i=1}^n a_i)}{(\wedge_{i=1}^n b_i \wedge b, \wedge_{i=1}^n a_i)} \text{ SI}} \text{ R-AND, lemma 8}$$

The following is also derivable from N:

$$\dfrac{(b \wedge (\wedge_{i=1}^n a_i), x)}{(\wedge_{i=1}^n b_i \wedge b \wedge (\wedge_{i=1}^n a_i), x)} \text{ SI}$$

By Lemma 1, for each a_i, $h(M) \vdash a_i$, and thus $h(M) \vdash \wedge_{i=1}^n a_i$. Hence, $x \vdash \wedge_{i=1}^n a_i$, and so $x \dashv\vdash x \wedge (\wedge_{i=1}^n a_i)$. Furthermore, $a^\star \vdash \wedge_{i=1}^n b_i \wedge b$, where $a^\star$ is a conjunction of elements of A. The following may, then, be derived.

$$\text{R-ACT, lemma 8} \dfrac{(\wedge_{i=1}^n b_i \wedge b, \wedge_{i=1}^n a_i) \quad (\wedge_{i=1}^n b_i \wedge b \wedge (\wedge_{i=1}^n a_i), x)}{\text{EQ} \dfrac{(\wedge_{i=1}^n b_i \wedge b, \wedge_{i=1}^n a_i \wedge x)}{\text{SI} \dfrac{(\wedge_{i=1}^n b_i \wedge b, x)}{(a^\star, x)}}}$$

Since $a^\star$ is a conjunction of elements of A, the pair (A, x) is derivable. □

This completes the proof of the main result of this paper, theorem 3.

4 Topics for future research

We end this paper with a number of topics for future research.

- Other restricted forms of chaining can be considered, like

$$\text{R-AT } \frac{(a,x) \quad (x,y) \quad a \wedge x \text{ (resp. } x \wedge y) \text{ is consistent}}{(a, x \wedge y)}$$

  R-AT is short for "Restricted aggregative transitivity". Is there an I/O operation validating this rule?

- In constrained I/O logic, there is the idea of a constraint set C filtering excess output. In the present paper, only the body and the head of a rule is treated as a constraint. Could one generalize the I/O logics with a consistency check in such a way that one can also work with an independent "constraint set"? Parent [10] and Dustin [20] use this technique to model defeasible reasoning. They take the traditional I/O logics as a starting point. What happens if the system described in the present paper is taken as a starting point? Would it yield new insights into our understanding of defeasible reasoning?

- What about the I/O logics for positive (static and dynamic) permission described by Makinson and van der Torre [9]? The various meta-results they establish in their paper (like the axiomatisation of positive static permission with the subverse rules) hold if a traditional I/O logic for obligation is used. Do similar meta-results can be obtained, using the system described in the present paper?

- Benzmüller and Parent [1] report some first results regarding the question of how to "implement" I/O logic using the so-called Shallow Semantic Embedding approach developed by Benzmüller. Their focus is on the traditional I/O logics. Can a similar embedding be obtained for the system described in the present paper?

References

[1] C. Benzmüller and X. Parent. I/O logic in HOL — first steps. Technical report, CoRR, 2018. https://arxiv.org/abs/1803.09681.

[2] J. Broersen and L. van der Torre. Ten problems of deontic logic and normative reasoning in computer science. In N. Bezhanishvili and V. Goranko, editors, *Lectures on Logic and Computation*, volume 7388 of *Lecture Notes in Computer Science*, pages 55–88. Springer, Berlin, Heidelberg, 2012.

[3] J. Carmo and A. J. I. Jones. Deontic logic and contrary-to-duties. In D. M. Gabbay and F. Guenthner, editors, *Handbook of Philosophical Logic: Volume 8*, pages 265–343. Springer Netherlands, Dordrecht, 2002.

[4] L. Goble. Prima facie norms, normative conflicts, and dilemmas. In D. Gabbay, J. Horty, X. Parent, R. van der Meyden, and L. van der Torre, editors, *Handbook of Deontic Logic and Normative Systems*, pages 241–352. College Publications, London. UK, 2013.

[5] G. Governatori and A. Rotolo. Logic of violations: A Gentzen system for reasoning with contrary-to-duty obligations. *Australasian Journal of Logic*, 3, 2005.

[6] J. Hansen. Reasoning about permission and obligation. In S. O. Hansson, editor, *David Makinson on Classical Methods for Non-Classical Problems*, pages 287–333. Springer, 2014.

[7] D. Makinson and L. van der Torre. Input/output logics. *Journal of Philosophical Logic*, 29(4):383–408, 2000.

[8] D. Makinson and L. van der Torre. Constraints for input/output logics. *Journal of Philosophical Logic*, 30(2):155–185, 2001.

[9] D. Makinson and L. van der Torre. Permission from an input/output perspective. *Journal of Philosophical Logic*, 32(4):391–416, 2003.

[10] X. Parent. Moral particularism in the light of deontic logic. *Artif. Intell. Law*, 19(2-3):75–98, 2011.

[11] X. Parent, D. Gabbay, and L. van der Torre. Intuitionistic basis for input/output logic. In Sven Ove Hansson, editor, *David Makinson on Classical Methods for Non-Classical Problems*, pages 263–286. Springer Netherlands, Dordrecht, 2014.

[12] X. Parent and L. van der Torre. "Sing and dance!". In F. Cariani, D. Grossi, J. Meheus, and X. Parent, editors, *Deontic Logic and Normative Systems*, pages 149–165, Cham, 2014. Springer International Publishing.

[13] X. Parent and L. van der Torre. Detachment in normative systems: examples, inference patterns, properties. In G. Pigozzi and L. van der Torre (guest editors), editors, *IfCoLog Journal of Logics and their Applications, Volume 4 (9)*, volume 4, pages 2295–3038. 2017.

[14] X. Parent and L. van der Torre. The pragmatic oddity in a norm-based semantics. In G. Governatori, editor, *Proceedings of the 16th Edition of the International Conference on Artificial Intelligence and Law*, ICAIL '17, pages 169–178, New York, NY, USA, 2017. ACM.

[15] H. Prakken and M. Sergot. Contrary-to-duty obligations. *Studia Logica*, 57(1):91–115, 1996.

[16] H. Prakken and M. Sergot. Dyadic deontic logic and contrary-to-duty obligations. In D. Nute, editor, *Defeasible Deontic Logic*, pages 223–262. Springer Netherlands, Dordrecht, 1997.

[17] A. Stolpe. Normative consequence: The problem of keeping it whilst giving it up. In R. van der Meyden and L. van der Torre, editors, *Deontic Logic in Computer Science, 9th International Conference, DEON 2008*, volume 5076 of *Lecture Notes in Computer Science*, pages 174–188. Springer, 2008.

[18] A. Stolpe. *Norms and Norm-System Dynamics*. PhD thesis, Department of Philosophy, University of Bergen, Norway, 2008.

[19] A. Stolpe. A concept approach to input/output logic. *J. of Applied Logic*, 13(3):239–258, 2015.

[20] D. Tucker. Variable priorities and exclusionary reasons in input/output logic. To appear in *Journal of Philosophical Logic*.

A LOGIC FOR REASONING ABOUT GROUP NORMS

DANIELE PORELLO
Free University of Bozen-Bolzano
`daniele.porello@unibz.it`

We present a number of modal logics to reason about group norms. As a preliminary step, we discuss the ontological status of the group to which the norms are applied, by adapting the classification made by Christian List of collective attitudes into aggregated, common, and corporate attitudes. Accordingly, we shall introduce modality to capture aggregated, common, and corporate group norms. We investigate then the principles for reasoning about those types of modalities. Finally, we discuss the relationship between group norms and types of collective responsibility.

Keywords. Group norms, Group agency, Non-normal modal logics, Collective Attitudes, Deontic logic, Logics of agency, Collective responsibility.

1 Introduction

Logics for representing and reasoning about norms are very important in knowledge representation and multiagent systems as they allow for compactly express desirable properties of the agents' behaviour as well as the fine principles of interaction among agents. In this paper, we propose a logic to reason about group norms, that is, norms that apply to collectivities of individuals. In order to conceptually understand the nature of the group to which the norms are ascribed, we use the distinction made by Christian List among collective propositional attitudes [10]. We view norms as propositional attitudes, endorsing the tradition in analytic philosophy that legitimates the representation of beliefs, desires, and also norms, by means of a modal logic constructed on top of a classical propositional logic. The modalities are in fact intended to capture the mode of relationship between an agent and a propositional content, allowing us to represent the attitude of the agent with respect to the proposition. In case of deontic attitudes, the modality express the normative force holding between an agent and a state of affairs represented by a proposition.

A collective propositional attitude is, generally speaking, a propositional attitude that is ascribed to a collective entity. A map of the most salient notions of collective attitudes was proposed in [10], by distinguishing between three kinds of collective attitudes: *aggregate*, *common*, and *corporate* attitudes. Corporate attitudes presuppose that the collectivity to which the attitudes are ascribed is an *agent* in its own right, an agent who is ontologically distinguished from the mere individuals that compose the collectivity. Common attitudes are ascribed to collectivities by requiring that every member of the group share the same attitude. Common attitudes have been presupposed for instance by the debate on joint action and collective intentionality [23, 10, 12]. In this view, possible divergences among the attitudes of the members of the group are excluded. For instance, under this reading, the sentence "PC members are supposed to return the review by the deadline" is true only if every individual who is a PC member is actually committed to meet the constraint.

By contrast, aggregative attitudes do not presuppose that every member of the group share the same attitude. In this case, a propositional attitude can be ascribed to the collectivity by solving the possible disagreement by means of a voting procedure such as the majority rule. For instance, a sentence like "the parliament decided to reduce taxation" does not require that every member of the parliament actually endorses the proposal, rather, it means that a suitable winning coalition of members of the parliament votes for the proposal.

We may view the three types of collective attitudes as generating three kinds of groups that differ in the relationship between the group and its members. We shall use this distinction in order to approach a taxonomy of group norms. We shall follow in particular the analysis of group norms provided by [2, 1] where group norms are classified according to a number of parameters, such as the addressee of the norm, those that are responsible for the commitment to the norm, and those who are subject to the norm.

In this paper, we shall assume that norms constrain the *actions* of individuals or groups. For that reason, as a preliminary investigation, we shall study the principle of agency of individuals and groups by discussing the logic of agency that we may assume for those types of agents. The notion of action that we endorse in this paper is very general and abstract, as we do not want to commit to any specific ontological view of actions. For this reason, we shall place our analysis within the tradition of the bringing-it-about modality [5, 8], which is in fact a logic of *agency*, in the sense that it does not specify what an action is, rather it models the effects of acting.

We introduce three logics to model the actions of groups defined in a common, aggregative, and corporate way. By means of this logics, we shall discuss the logical relationship between the actions of the group and the actions of its members. Then, we will approach a deontic logic for modelling group norms by making explicit how the collective responsibility may or may not transfer to the individuals that are members of the group.

On the technical side, the contribution of this paper is the following. We introduce three logics to discuss group actions that reflect the common, the aggregate, and the corporate view. To model those logics, we shall use non-normal modal logics defined by means of neighbourhood semantics, see [4] and [14] for an introduction. A number of specific principles for the modalities that express common, aggregated, and corporate actions is proposed, by specifying both an Hilbert system to reason about those modalities and a semantic framework based on neighbourhood semantics to ensure soundness and completeness. Although the presentation of this paper is rather informal, we shall present the conditions on the neighbourhood functions that are required in order to prove soundness and completeness of the systems that we introduce. Finally, we discuss the principles of the deontic modalities that relate collective and individual responsibility, we study the Hilbert system for them, and we present the relevant semantic conditions.

The remainder of this paper is organised as follows. In Section 2, we remind the basics of non-normal or minimal modal logics and we present the logics to treat individual actions and obligations. Section 3 is dedicated to present the logics of group agency, by distinguishing common, aggregated, and corporate actions. Section 4 approaches the treatment of group norms by introducing a number of modalities for collective obligation. Section 5 concludes and indicates future work.

2 Minimal modal logics

We propose a number of logics to reason about actions and obligations of individuals and collectives. We assume a (finite) set of agents $\mathbf{N}$, and we consider the power set of $\mathbf{N}$, $2^{\mathbf{N}}$, to talk about coalitions of agents. Thus, our labels that denote agents shall range over $2^{\mathbf{N}}$. To express individual propositional attitudes, we admit singleton coalitions; in that case, the meaning of a coalition C is $\{i\}$.[1]

We shall use minimal (or non-normal) modalities in order to ensure a number of basic principles to reason about agency of groups and individuals.

The semantics of the modalities that we are going to introduce is defined by means of a *neighbourhood semantics* [4]. Let W be a set of possible wolds (or states), a neighbourhood function is a mapping $N : W \to \mathcal{P}(\mathcal{P}(W))$ that associates a world m with a set of sets of worlds (see [4, 14]). The intuitive meaning of the neighbourhood function is that it associates to each world a set of propositions that are those propositions designated to hold at w. In this setting, a neighbourhood function associates to a world w the propositions that express the available actions or the salient norms at w.

The language of propositional logic is defined as follows. Let $Prop$ be a set of propositional atoms,

$$\mathcal{L} ::= p \in Prop \mid \neg \varphi \mid \varphi \wedge \varphi$$

A valuation v is a function that associates a possible world and propositional atom to a truth-value in the set $\{t, f\}$, that is $v : W \times Prop \to \{t, f\}$.

We define the extension of a formula φ in a model by $||\varphi|| = \{w \mid w \models \varphi\}$. The semantic definition of the modalities that we shall encounter follows the following pattern, where $\Box$ be one of the modality that we will introduce.

$$w \models \Box\varphi \text{ iff } ||\varphi|| \in N(w) \tag{1}$$

In non-normal modal logics, soundness and completeness are basically achieved by means of selecting the suitable conditions on the neighbourhood functions, see for instance [4]. In the following sections, we shall spell out the relevant conditions to achieve soundness

---

[1]This move is quite similar to the approach in [22] to discuss coalitional ability.

and completeness of the proposed logics, however, for reasons of space, we shall leave a detailed proof of those results for future work.

2.1 Individual Actions and Obligations

The logic to reason about actions that we use here is based on the minimal logic of *bringing-it-about*, which was traditionally developed in [5, 8]. The principles of this logic aim to capture a very weak view of actions that, for instance, does not presuppose intentionality or explicit goals. We apply this minimal view to conceptualise collective actions of different types of groups. For instance, this weak view is adequate also for an aggregative perspective on collective actions, for which the collective is not assumed, in general, to have joint intentionality nor any shared goal, [10].

Four principles of agency are captured by the classical bringing-it-about logic [5], which are here summarised in Table 1. The first corresponds to the axiom T of modal logics: $\mathsf{E}_i \varphi \to \varphi$, it states that if an action is brought about, then the action affects the state of the world, i.e. the formula φ that represents the effects of the action holds. The second principle (E2) allows for combining actions. The third principle corresponds to the axiom E3. It amounts to assuming that agents cannot bring about tautologies. The motivation is that a tautology is always true, regardless what an agent does, so if acting is construed as something that affects the state of the world, tautologies are not apt to be the content of something that an agent actually does. The fourth item allows for viewing bringing it about as a modality, validating the rule of equivalents: if $\vdash \varphi \leftrightarrow \psi$ then $\vdash \mathsf{E}_i \varphi \leftrightarrow \mathsf{E}_i \psi$.

The language of the logic of bringing it about BIAT, $\mathcal{L}_{\mathsf{BIAT}}$ simply extends the language of propositional logic by adding a formula $\mathsf{E}_i \varphi$ for each individual $i \in \mathbf{N}$.

The Hilbert system for BIAT is obtained by adding the following axioms (Table 1) and the following rule to those of classical propositional logic.

- All the propositional tautologies

E1 $\mathsf{E}_i \varphi \to \varphi$

E2 $\mathsf{E}_i \varphi \wedge \mathsf{E}_i \psi \to \mathsf{E}_i(\varphi \wedge \psi)$

E3 $\neg \mathsf{E}_i \top$

Table 1: Axioms of BIAT

$$\frac{\vdash \varphi \leftrightarrow \psi}{\vdash \mathsf{E}_i \varphi \leftrightarrow \mathsf{E}_i \psi} \mathsf{E}_i(\mathrm{re})$$

The semantics of BIAT is obtained by adding a number of neighbourhood functions N_i^E, one for each agent $i \in \mathbf{N}$. Each neighbourhood function represents the actions available to each agent at a certain world. The semantics clause for action modalities is then the following one:

$$w \models \mathsf{E}_i\varphi \text{ iff } ||\varphi|| \in N_i^{\mathsf{E}}(w) \qquad (2)$$

To ensure soundness and completeness for this E, a number of conditions on the neighbourhood functions has to be ensured. For the details, we refer to [8].

2.2 Individual norms

We extend the language of classical propositional logic by adding a number of modalities for obligations O_i, for $i \in \mathbf{N}$. For the sake of simplification, we use the standard deontic logic to model individual obligations. The Hilbert system for OL extends the case of propositional logic by adding the axioms in Table 2 and by adding the following rule.

- All the propositional tautologies

O1 $\mathsf{O}_i(\varphi \to \psi) \to (\mathsf{O}_i\varphi \to \mathsf{O}_i\psi)$

O2 $\mathsf{O}_i\varphi \to \neg\mathsf{O}_i\neg\varphi$

Table 2: Axioms of OL

$$\frac{\vdash \varphi}{\vdash \mathsf{O}_i\varphi} \mathsf{O}_i(\text{nec})$$

Although standard deontic logic is a *normal* modal logic, we can present its semantics in terms of neighbourhood functions as well [4, 14]. Moreover, a condition on the neighbourhood function for validating O2 is required [2]

The semantics of OL can be obtained by adding a number of neighbourhood functions N_i^{O}, one for each agent $i \in \mathbf{N}$. In this case, the neighbourhood functions represent the norms that are salient for an agent at a certain world. The semantic definitions for deontic modalities are then the following one:

$$w \models \mathsf{O}_i\varphi \text{ iff } ||\varphi|| \in N_i^{\mathsf{O}}(w) \qquad (3)$$

---

[2]The use of non-normal modal logic to express deontic modalities was motivated in [4, 7, 18]. Moreover, non-normal deontic logics have been used to discuss institutional agency in [3] and to model weak permissions in [21]. We present here the semantic definitions in terms of neighbourhood semantics as it will be useful for simplifying the subsequent arguments.

3 Group actions

We introduce three modalities for capturing a number of features of group actions, that shall be related to group norms. Here we are going to distinguish between common, aggregate, and corporate group actions, and we are going to introduce three modalities and three logics that capture their minimal principles. In particular, we highlight the principles that relate the group action with the actions of the individuals that are part of the group.

3.1 Common group actions

Common group actions are intended as those actions for which every agent of the group is indeed performing a same type of action. The axioms that govern this modality are presented in Table 3.

- All the propositional tautologies

COM1 $[COM]_C \varphi \to \varphi$

COM2 $[COM]_C \varphi \wedge [COM]_C \psi \to [COM]_C (\varphi \wedge \psi)$

COM3 $[COM]_C \varphi \wedge [COM]_D \varphi \to [COM]_{C \cup D} \varphi$

COM4 $\neg [COM]_C \top$

COM5 $[COM]_C \varphi \to \bigwedge_{i \in C} \mathsf{E}_i \varphi$

COM6 $\bigwedge_{i \in C} \mathsf{E}_i \varphi \to [COM]_C \varphi$

Table 3: Axioms of [COM]

COM1 again reflects the effectivity of acting. COM2 and COM3 specify how to combine common actions. COM4, again, prevents tautologies to be brought about.

COM5 may in principle be questionable, as it forces the idea that the group in this case cannot do anything more that what its members jointly do. We assume it here, by endorsing a strict view of common group actions, which are in fact entirely reducible to the joint actions of the members of the group. COM6 is again questionable. For instance, suppose that every member of a parliament orders a pizza, would we infer that the parliament as a group is ordering a pizza? To account for this delicate aspects, we need to separate actions that are done by individuals *qua* members of the group. We leave this points for a future dedicated work.

The rule of equivalents for common actions is expressed in the following $[COM]_C(re)$ rule.

$$\frac{\vdash \varphi \leftrightarrow \psi}{\vdash [\text{COM}]t_C\varphi \leftrightarrow [\text{COM}]_C\psi} \ [\text{COM}]_C(\text{re})$$

The semantics of the $[\text{COM}]_C$ modalities is defined as follows. For each modality, we introduce a number of neighbourhood function N_C^{COM}, one for each coalition of agents.

$$w \models [\text{COM}]_C\varphi \text{ iff } ||\varphi|| \in N_C^{\text{COM}}(w) \qquad (4)$$

To semantically validate axioms from COM1 to COM4, the conditions are similar to those presented for the individual logics of action and for the extension to coalitions proposed by [22].

To ensure the validity of axioms COM5 and COM6, a new condition on the functions N_C^{COM} is required.

$$N_C^{\text{COM}}(w) = \bigcap_{i \in C} N_i^{\text{E}}(w) \text{ for every } w \in W \qquad (5)$$

By means of 5, we can show that axioms COM5 and COM6 are valid as follows. For instance, we show that, for every model and every $w \in W$, $w \models [\text{COM}]_C\varphi \rightarrow \bigwedge_{i \in C} \text{E}_i\varphi$ (which, in fact, provides the soundness of axiom COM5 and COM6). Assume $w \models [\text{COM}]_C\varphi$, then $||\varphi|| \in N_C^{\text{COM}}(w)$. Then, by condition 5, $||\varphi|| \in N_i^{\text{E}}(w)$, therefore $w \models \text{E}_i\varphi$.

3.2 Aggregated group actions

Aggregated actions are those that result from the outcome of an aggregation procedure, such as the majority rule, applied to the actions of the individuals. We write $[AGG]_C^f\varphi$ to express that φ is the action performed by the group C under the aggregation procedure f. An *aggregation function* is a function that maps N-tuples of 0s and 1s associated to formulas to the set $\{0, 1\}$, i.e., $f : \{1, 0\}^N \rightarrow \{1, 0\}$. That is, f maps patterns of individual acceptance or rejections of formulas to a collective acceptance or rejection of a formula. For instance, in the simple majority rule, we assume that maj returns 1 on a majority of 1s, and it returns 0 otherwise. As a simplification move, we suppose in this paper that N is odd.

$$maj(x_1, \ldots, x_N) = \begin{cases} 1 \text{ if } |\{x_i \mid x_i = 1\}| > N/2; \\ 0 \text{ otherwise} \end{cases}$$

By adjusting the acceptance threshold of $n/2$, we can define the class of uniform quota rules, where each q provides a distinct aggregation procedure.

$$quota_q(x_1, \ldots, x_N) = \begin{cases} 1 \text{ if } |\{x_i \mid x_i = 1\}| > q; \\ 0 \text{ otherwise} \end{cases}$$

One may discuss the properties of such aggregators, along the lines of the traditional arguments in social choice theory and judgment aggregation [9]. For instance, the previous aggregators are *anonymous*, namely any permutation of the individual values provides the same output value. By contrast, the following two classes of aggregators, oligarchies and dictatorships, depend on specific choices of the agents.

Oligarchies: let $\{i_1, \ldots, i_L\}$ be a set of indexes with $L \leq N$,
$olig(x_1, \ldots, x_N) = x$ if $olig(x_{i_1}, \ldots, x_{i_L}) = x$.

Dictatorships of j: $d_j(x_1, \ldots, x_N) = x_j$.

In the case of $olig$, the oligarchy of agents $x_{i_1}, \ldots, x_{i_L}$ decides the outcome; in the case of d_j, the sole agent j is decisive.

We extend the language of propositional logic by adding a number of modal operators $[\text{AGG}]_C^f$ that depend on the aggregator f.

$$\mathcal{L}_{[\text{AGG}]_F} ::= \varphi \in \mathcal{L} \mid [\text{AGG}]_C^F \varphi$$

With this definition, as a simplification move, we are excluding possible nesting of modalities (cf. for instance [15]). By means of the aggregation function f, we can provide the semantics of the aggregated action modality as follows. Firstly, we associate to each modality $[AGG]_C^f$ a neighbourhood function N_C^f. The semantic clause is then, as usual, the following one.

$$w \models [AGG]_C^f \varphi \text{ iff } ||\varphi|| \in N_C^f(w) \qquad (6)$$

Denote by $\chi_X(N_i^\mathsf{E}(w))$ the function that returns 1 if $X \in N_i^\mathsf{E}(w)$ and 0 otherwise. A winning coalition of agents wrt. an aggregator f is, informally, a set of agents that can determine the outcome. The neighbourhood functions that we consider for this logic have to satisfy the following constraint.

$$X \in N_C^f(w) \text{ iff } f(\chi_X(N_{i_1}(w)), \ldots, \chi_X(N_{i_l}(w))) = 1 \text{ for some } \{i_1, \ldots, i_l\} \subseteq C \qquad (7)$$

That is, a proposition is accepted by the group C under an aggregative view that depends on the procedure f if and only if there is a winning coalition wrt. f contained in C. In particular, in case f is the majority rule, a set of words X (i.e. roughly, a proposition) is in $N_C^f(w)$ iff X is in a majority of individual neighbourhoods $N_i(w)$, for $i \in C$.

The axiomatisation of aggregated group actions depends on the specific aggregation function that we select. For instance, in [15] an axiomatisation of the majority rule is provided. Note that it is well known from the social choice and judgment aggregation literature that the aggregation of general propositions may return inconsistent outcomes (e.g. discursive dilemmas [11].) Therefore, for preventing the logic from being inconsistent, we shall discuss which axioms of a logic of agency to drop. One solution is to drop an axiom of the form $[AGG]_C^f \varphi \to \varphi$, and permit that agents in some cases may collectively accept contradictory propositions, without making the logic inconsistent. This solution applies to any aggregation procedure that may return inconsistent outcomes and permit contradictory propositions only within the scope of the $[AGG]_C^f$ modalities. A second solution is to prevent inconsistent propositions to be collectively accepted, by excluding them also from the scope of the $[AGG]_C^f$ modalities. This shall depend on the specific aggregation procedure and on the conditions under which it may return inconsistent sets [6].[3]

Further principles of aggregated attitudes are left for a future dedicated work. For instance, a combination axiom such as $[AGG]_C^f \varphi \land [AGG]_D^f \psi \to [AGG]_{C \cup D}^f (\varphi \land \psi)$ requires a careful examination of the effect of combining the outcomes of coalitions C and D wrt. f, cf. [20].

We show at least that the rule of equivalents holds for this definition of modality. Hence, aggregated group actions are legitimate modal operators.

$$\frac{\vdash \varphi \leftrightarrow \psi}{\vdash [AGG]_C^f \varphi \leftrightarrow [AGG]_C^f \psi} \; [AGG]_C^f(re)$$

Suppose that $\vdash \varphi \leftrightarrow \psi$, we have to show that, for every f, $\|\varphi\| \in N_C^f(w)$ iff $\psi \in N_C^f(w)$. The assumption entails that $\|\varphi\| = \|\psi\|$.

We have the following chain of equivalences, which allows us to conclude:

$$\|\varphi\| \in N_C^f(w) \text{ iff } f(\chi_{\|\varphi\|}(N_{i_1}(w)), \ldots, \chi_{\|\varphi\|}(N_{i_l}(w))) = 1$$

iff

$$f(\chi_{\|\psi\|}(N_{i_1}(w)), \ldots, \chi_{\|\psi\|}(N_{i_l}(w))) = 1 \text{ iff } \|\psi\| \in N_C^f(w)$$

To relate aggregative group actions to individual actions, we discuss the following two alternative assumptions. Firstly, we may view only the winning coalition of agents that were actually supporting φ as involved in the collective action resulting in φ. Secondly, we may view the entire group of agents, namely also those that were not voting for φ, as collectively bringing it about that φ. For instance, in case of a parliament passing a bill, we may view only those that voted for the bill as bringing it about, or we may view the entire parliament as acting so that the bill has passed. This distinction is reflected by selecting one between the following two axioms.

AGG 1 $[AGG]_C^f \varphi \to \bigwedge_{i \in C} E_i \varphi$

AGG 2 $[AGG]_C^f \varphi \to \bigwedge_{i \in D} E_i \varphi$ where D is a winning coalition wrt f.

The conditions on the neighbourhood functions that are are required in order to make AGG 1 and AGG 2 valid are, respectively, the following two.

$$N_C^f(w) \subseteq N_i^E(w) \text{ for every } i \in C. \tag{8}$$

---

[3]To prevent inconsistency and develop an axiom system to reason about aggregated group actions, a third solution is to use fragments of weak relevant and linear logics, cf. [16, 17]. To design logical principles that only combine collectively accepted propositions and maintain consistency, see [18].

$$N_C^f(w) \subseteq N_i^{\mathsf{E}}(w) \text{ for every } i \in D, \text{ s.t. } D \text{ is a winning coalition wrt. } f. \qquad (9)$$

Note that we can view aggregated actions as modalities because we decided to define aggregation procedures by means of f. This forces a property of *systematicity* on the aggregation procedure [15]. Namely, the collective acceptance only depends on the patterns of individual acceptance. In particular, any two propositions that exhibit the same pattern of acceptance and rejections are equally accepted or rejected. This condition restricts the class of aggregation functions that we are considering (e.g. by rejecting non-independent or non-neutral aggregators, see for instance [13]), but it allows us to view aggregation functions as modalities.

3.3 Corporate group actions

A corporate view of group actions requires the commitment to the existence of a *single* agent a who is the bearer of all the collective actions [10], who is in principle ontologically distinguished by the group of agents that are members of the corporate agent [19]. The group agent may be viewed as the reification of the group as a whole, distinguished from any individual of the set of agents, or it may be a specific agent who acts as a representative of the group.[4] To model this view, we enrich the set of agents **N** with a sufficient number of labels for group agents $\{a_1, \ldots, a_l\}$ and we assume that for each coalition of agents $C \in 2^{\mathbf{N}}$, there is a single group agent a_C that depends on C.

The agency of the group agent a_C is then expected to satisfy the same principles of agency of a standard individual agent. The motivation for this assumption is that the individual principles of agency are those that allow us, in this setting, to view a modality as truly agentive.

The language of corporate action modalities is defined as follows.

$$\mathcal{L}_{[\mathsf{COR}]} ::= \varphi \in \mathcal{L} \mid [\mathsf{COR}]_C \varphi \text{ where } C \in 2^{\mathbf{N}}$$

To capture the principles of corporate agents, we propose axiom [COR 1], that means that the agency of the corporate agent reduces to the agency of one individual and that the agency of such an individual can be captured by the reasoning principles of the E modality. Corporate agents modalities are again assumed to satisfy the rule of equivalents.

$$\frac{\vdash \varphi \leftrightarrow \psi}{\vdash [\mathsf{COR}]_C \varphi \leftrightarrow [\mathsf{COR}]_C \psi} \; [\mathsf{COR}]_a (\text{re})$$

The semantics of corporate actions modalities is then defined as usual by introducing neighbourhood functions N_C^{COR} with the following semantic constraint.

$$w \models [\mathsf{COR}]_C \varphi \text{ iff } ||\varphi|| \in N_C^{\mathsf{COR}}(w) \qquad (10)$$

---

[4]Notice that a group agent is distinguished from a dictator in the sense of the dictatorial aggregation procedure of Section 3.2. For instance, the group agent may not be a member of the group C.

- All the propositional tautologies

COR 1 $[COR]_C\varphi \to E_{a_C}\varphi$ where $a_C \in \mathbf{N} \cup \{\mathbf{a_1}, \ldots, \mathbf{a_l}\}$ is a designated agent for coalition C.

Table 4: Axioms of $[COR]_C$

It is easy to see that corporate agents modalities satisfy again the rule of equivalents. To make axiom [COR 1] valid, we need to assume the following constraints on the neighbourhood functions.

$$N_C^{COR}(w) \subseteq N_{a_C}^{E}(w) \text{ for } a_C \in \mathbf{N} \text{ and for every } w \in W. \tag{11}$$

By means of condition (11), the rule of equivalents immediately follows.

Axiom [COR 1] is shown to be valid as follows. Suppose $w \models [COR]_C\varphi$, then $||\varphi|| \in N_C^{COR}(w)$, then by condition (11), $||\varphi|| \in N_{a_C}^{E}(w)$, so $w \models E_{a_C}\varphi$.

We assumed the inclusion, rather then the equality, in condition 11, because we do not want to rule out the case where a_C is a standard individual agent, who acts as a representative of C in certain situations, and only part of her or his actions counts as the representative of C.

4 Group norms

We introduce three obligation modalities that relate the normative force of the collective obligation to the relevant type of action associated to the specific type of group.

The taxonomy of group norms proposed by [2] separates the dimension of agency and the dimension of type of responsibility. Here, we discussed the dimension of agency by means of the logics for [COM], [AGG], and [COR], and we approach the normative force by introducing the following modal operators O_C^{IR}, O_C^{RR}, and O_C^{CR} to select whether the obligation induces *individual*, *representative*, or *collective* responsibility (IR, RR, or CR, respectively). We also admit the case where the collective obligation transfers to a winning coalition of agents that can be blamed to be responsible of the collective action, we label this situation by WR. For instance, in the case of the aggregated view of collective actions, one may view as responsible of the course of action only the (winning) coalition of agents that actually supported the proposal at issue and not the whole collectivity that takes part in the decision.

We start by presenting the general principles for transferring obligations from collective agents to individuals or subgroups and then we shall discuss the interaction between types of collective obligations and collective actions. We extend the language of our logic by adding formulas of the type $O_C^Y\varphi$, where $Y \in \{\text{IR, RR, CR, WR}\}$ and $C \in 2^{\mathbf{N}}$.

The axioms that capture in general how the collective obligation transfers to individuals or to subgroups are the following.

IR $\quad O_C^{IR}\varphi \to \bigwedge_{i\in C} O_i\varphi$

RR $\quad O_C^{RR}\varphi \to O_a\varphi$ where $a \in \mathbf{N} \cup \{a_1, \ldots, a_l\}$.

WR $\quad O_C^{WR}\varphi \to O_D\varphi$ where $D \subseteq C \subseteq \mathbf{N}$.

The notion of collective responsibility CR is not approached here by any specific axiom: It is rather defined by the lack of transferability to any individual, representative, or winning coalitions of agents.

To provide a semantics of this new modalities, we assume a number of neighbourhood functions N_C^Y, where $Y \in \{\text{IR}, \text{RR}, \text{CR}, \text{WR}\}$ and $C \in 2^{\mathbf{N}} \cup \{a_1, \ldots, a_l\}$. The truth condition of the new modal formulas is as well presented as follows:

$$w \models O_C^Y \varphi \text{ iff } ||\varphi|| \in N_C^Y(w) \text{ for all } w \in W \qquad (12)$$

The constraints on the neighbourhood function that are required for the validity of the relevant axioms are then the following.

$$N_C^{IR}(w) \subseteq \bigcap_{i \in C} N_i^O(w) \qquad (13)$$

$$N_C^{RR}(w) \subseteq N_a^O(w) \text{ for a designated agent } a \in \mathbf{N} \cup \{a_1, \ldots, a_l\} \qquad (14)$$

$$N_C^{WR}(w) \subseteq N_D^O(w) \text{ for a designated } D \subseteq \mathbf{N} \qquad (15)$$

It is easy to see that all this modalities satisfy the rule of equivalents:

$$\frac{\vdash \varphi \leftrightarrow \psi}{\vdash O_C^Y\varphi \leftrightarrow O_C^Y\psi} O_C^Y(\text{re})$$

We show for instance that condition (15) allows for establishing the validity of Axiom WR. Assume that $w \models O_C^{WR}\varphi$, then $||\varphi|| \in N_C^{WR}(w)$. By means of condition 15, we have that $||\varphi|| \in N_D^O(w)$, thus $w \models O_D\varphi$.

Whether the collective obligations, as such, shall also satisfy the principles of standard deontic logic or whether a weaker deontinc logic is needed is left for future work. Here we approached obligations based on individual responsibility (IR) and corporate responsibility (CR) by reducing them to individual obligations. The case of group obligation based on winning coalitions (WR), by contrast, requires understanding the principles of group obligations along the lines of [18].

4.1 Discussion

We conclude by discussing a number of examples. In principle, we can permit every combination of the modalities O_C^{IR}, O_C^{RR}, and O_C^{CR} with the types of group agency [COM], [AGG], and [COR], therefore reconstructing in this framework the full taxonomy of [2]. In fact, we shall see that a few cases are delicate.

Firstly, by means of IR and of the view of common group actions, we can infer that, if the group has an obligation towards φ, then in this case every agent has an obligation towards φ.

$$\vdash O_C^{IR}[\text{COM}]_C\varphi \rightarrow O_i E_i \varphi \tag{16}$$

We can show that (16) is valid as follows. Suppose $w \models O_C^{IR}[\text{COM}]_C\varphi$, then by IR, we conclude $w \models O_i[\text{COM}]_C\varphi$. By Axiom O1 and COM 5, we infer that $w \models O_i \bigwedge_{i \in C} E_i \varphi$ and, again by O1, we conclude $O_i E_i \varphi$.

Moreover, if we view the responsibility of the group action as ascribed to a representative of the agents, say a, we can infer the following principle.

$$\vdash O_C^{RR}[\text{COR}]_C\varphi \rightarrow O_a E_a \varphi \tag{17}$$

Assume $w \models O_C^{RR}[\text{COM}]_C\varphi$. By Axiom RR, we infer that $w \models O_a[\text{COR}]_C\varphi$. Then, by COR1 and O1 we conclude $O_a E_a \varphi$.

Consider now the following formula.

$$\vdash O_C^{RR}[\text{COM}]_C\varphi \rightarrow O_a E_a \varphi \text{ only if } a \in C \tag{18}$$

Can we ascribe a representative responsibility to an group action defined by means of a common action? Formula (18) is derivable in our framework, only if the representative agent is among those in C. This makes sense since a common action of the group C is supposed to refer to the actions of the individuals in C. Therefore, in principle, we may allow for a representative agent of the common action, although it has to be part of the group.

For aggregated group actions, again we may select whether the responsibility is at the individual, coalitional, or representative level. For instance, the following formula is derivable only in case we assume that every individual in C is actually bringing about φ, even if she or he is voting against φ (cf. axiom AGG 1).

$$\vdash O_C^{IR}[\text{AGG}]_C\varphi \rightarrow O_i E_i \varphi \tag{19}$$

By contrast, the following formula holds in case we assume axiom AGG 2.

$$\vdash O_C^{WR}[\text{AGG}]_C^f\varphi \rightarrow O_i E_i \varphi \; D \subseteq \mathbf{N} \text{ winning coalition for } f \text{ and } i \in D. \tag{20}$$

To establish (20), we reason as follows. From $w \models O_C^{WR}[\text{AGG}]_C^f\varphi$, then by WR, we have that $O_D[\text{AGG}]_C^f\varphi$, where D is a winning coalition wrt. f. By means of AGG2 and O1, we infer that $O_i E_i \varphi$, for $i \in D$. In this case, i.e. by assuming AGG 2 instead of AGG 1, formula (20) fails in case the agent i is not a member of the winning coalition D.

Aggregation procedure are in fact quite versatile, as they can also be viewed as abstract representation of the decision mechanisms of an organisation. For instance, an oligarchic aggregation can in principle represent decisions taken at the level of the board of directors of a company.

5 Conclusion

We have introduced three logics to reason about common, aggregative, and corporate actions, by relating the agency of the group to the agency of the individuals that are members of the group. We have then introduced a number of deontic principles that relate collective responsibility to individual responsibility and we have discussed a few combinations of type of group action and type of responsibility. We have informally introduced those systems, however the conditions on the neighbourhood functions that we have presented are those that are required in order to establish soundness and completeness of our systems. By means of the logics that we have introduced, we can provide a logical foundation, which is also grounded in the philosophical analysis of groups developed in [10], of the taxonomy of group norms provided in [2].

Future work shall present a detailed proof of the completeness results that we suggested in this paper. Moreover, due to the simplicity of the systems that we have introduced, we conjecture that they are all decidable, future work shall establish this fact. Finally, we are interested in studying in detail the principles that relate the group action and the individual action, in particular, by expanding the analysis of aggregated group actions and of corporate group actions, which constitute the delicate cases.

References

[1] H. Aldewereld, V. Dignum, and W. Vasconcelos. We ought to; they do; blame the management! In *Coordination, Organizations, Institutions, and Norms in Agent Systems IX*, pages 195–210. Springer, 2014.

[2] H. Aldewereld, V. Dignum, and W. W. Vasconcelos. Group norms for multi-agent organisations. *ACM Transactions on Autonomous and Adaptive Systems (TAAS)*, 11(2):15, 2016.

[3] J. Carmo and O. Pacheco. Deontic and action logics for organized collective agency, modeled through institutionalized agents and roles. *Fundamenta Informaticae*, 48(2-3):129–163, 2001.

[4] B. Chellas. *Modal Logic: An Introduction*. Cambridge University Press, 1980.

[5] D. Elgesem. The modal logic of agency. *Nordic J. Philos. Logic*, 2(2), 1997.

[6] U. Endriss, U. Grandi, and D. Porello. Complexity of judgment aggregation. *Journal of Artificial Intelligence Research*, 45:481–514, 2012.

[7] L. Goble. A proposal for dealing with deontic dilemmas. In *Deontic Logic in Computer Science*, pages 74–113. Springer, 2004.

[8] G. Governatori and A. Rotolo. On the Axiomatisation of Elgesem's Logic of Agency and Ability. *Journal of Philosophical Logic*, 34:403–431, 2005.

[9] C. List. The theory of judgment aggregation: An introductory review. *Synthese*, 187(1):179–207, 2012.

[10] C. List. Three kinds of collective attitudes. *Erkenntnis*, 79(9):1601–1622, 2014.

[11] C. List and P. Pettit. Aggregating sets of judgments: An impossibility result. *Economics and Philosophy*, 18(1):89–110, 2002.

[12] C. List and P. Pettit. *Group Agency. The possibility, design, and status of corporate agents.* Oxford University Press, 2011.

[13] C. List and C. Puppe. Judgment aggregation: A survey. In *Handbook of Rational and Social Choice*. Oxford University Press, 2009.

[14] E. Pacuit. *Neighborhood Semantics for Modal Logics.* Springer, 2017.

[15] M. Pauly. Axiomatizing collective judgment sets in a minimal logical language. *Synthese*, 158(2):233–250, 2007.

[16] D. Porello. A proof-theoretical view of collective rationality. In *IJCAI 2013, Proceedings of the 23rd International Joint Conference on Artificial Intelligence, Beijing, China, August 3-9, 2013*, 2013.

[17] D. Porello. Judgement aggregation in non-classical logics. *Journal of Applied Non-Classical Logics*, 27(1-2):106–139, 2017.

[18] D. Porello. Logics for modelling collective attitudes. *Fundam. Inform.*, 158(1-3):239–275, 2018.

[19] D. Porello, E. Bottazzi, and R. Ferrario. The ontology of group agency. In *Formal Ontology in Information Systems - Proceedings of the Eighth International Conference, FOIS 2014, September, 22-25, 2014, Rio de Janeiro, Brazil*, pages 183–196, 2014.

[20] D. Porello and N. Troquard. Non-normal modalities in variants of linear logic. *Journal of Applied Non-Classical Logics*, 25(3):229–255, 2015.

[21] O. Roy, A. J. Anglberger, and N. Gratzl. The logic of obligation as weakest permission. In *Deontic Logic in Computer Science*, pages 139–150. Springer, 2012.

[22] N. Troquard. Reasoning about coalitional agency and ability in the logics of "bringing-it-about". *Autonomous Agents and Multi-Agent Systems*, 28(3):381–407, 2014.

[23] R. Tuomela. *Social ontology: Collective intentionality and group agents.* Oxford University Press, 2013.

HOW TO TAKE HEROIN (IF AT ALL). HOLISTIC DETACHMENT IN DEONTIC LOGIC

FREDERIK VAN DE PUTTE
Ghent University
`frederik.vandeputte@ugent.be`

STEF FRIJTERS
Ghent University
`stef.frijters@ugent.be`

JOKE MEHEUS
Ghent University
`joke.meheus@ugent.be`

The aim of the present paper is to investigate the logic of *holistic detachment*, i.e. detachment that is triggered by *all and only* those circumstances that are fixed (unalterable, unavoidable). To this end, we present the (monotonic) modal logic **HD** that captures the distinction between mere facts and fixed circumstances, and features a non-normal "all and only"-operator. We give a sound and (strongly) complete axiomatization of **HD** and discuss its most salient properties. We show that **HD** is rather weak when applied to realistic scenarios and we argue against what we call the "enthymematic approach" to mitigate this weakness. In contrast, it is shown that **HD** can and should be strengthened non-monotonically, in order to capture deontic reasoning.

1 Introduction

Consider the following dialogue, inspired by [22]:

David: Caroline ought not to take heroin. However, if she does take heroin, then she ought to take a light dose and use a clean needle.
Lou: Caroline does take heroin.
David: Then she ought to use a clean needle and take a light dose.

We are indebted to Christian Straßer, John Horty, Eric Pacuit, and Federico Faroldi for valuable comments on previous drafts of the paper.

Lou: No. She ought not to care about needles or dosage – she simply ought not to take heroin in the first place!

David: But if she does not use a clean needle, she might get HIV-infected...

Arguably, Lou and David are talking past one another in this little dialogue. What David means is that, if Caroline's taking heroin is taken for granted, if this is a fixed feature of the situation she is in, then Caroline ought to use a clean needle and take a light dose. In contrast, Lou entertains the possibility of Caroline *not* taking heroin, and that possibility is clearly preferable to her taking a light dose of heroin and using a clean needle.

This point is not new. One of the insights brought up by the discussion on contrary-to-duties and the associated paradoxes is that, whether one can ratioinally detach a conditional obligation in the light of factual information, depends on what one takes to be "fixed circumstances" [13, 10, 9, 22, 7]. Not all facts have the same status: one may e.g. consider Caroline's current unemployment as a fixed fact, but not her drug abuse. Unfortunately, it turns out to be very hard to pin down when one should consider some truth φ merely contingent and when φ is a circumstance that can trigger detachment.[1] The obvious way out of this puzzling question – at least for the deontic logician – is to build this distinction into one's logic. Once there, detachment can be formalized by relying on the following principle:[2]

Restricted Detachment: one should take *only* fixed circumstances[3] into account, when detaching oughts.

An obvious next question is: *how much* of the fixed circumstances do we need to take into account when applying detachment? To continue the above example: suppose that Caroline has a severe heroin addiction. This means that if she takes only a small dose of heroin, in the absence of medical supervision, she will suffer from dangerous withdrawal effects. Let us agree on the following conditional, and let us moreover agree that the factual claims below it are fixed:

(C) If Caroline takes heroin, and if she is a heroin addict, then she ought to take a sufficiently large dose of heroin.

(F1) Caroline takes heroin.

---

[1]See [7, pp. 283-284] for an attempt to do so.

[2]Restricted detachment (RD) is a refinement of *factual* detachment (FD), i.e. the derivation of "actual" or "situational" obligations from conditional obligations and information about the facts at hand [23, p. 118]. Factual detachment is usually opposed to *deontic* detachment (DD), which concerns the derivation of actual obligations from conditional obligations and other actual obligations. It is well-known, at least since [1] that (FD) and (DD) cannot easily be combined. Whereas some argue that (RD) can be combined with (DD), we leave the issue of (DD) for another occasion.

[3]In the remainder, we simply use "fixed circumstances" as shorthand for "those circumstances that are taken to be fixed by the person who reasons about the situation in question".

(F2) Caroline is a heroin addict.

In this extension of our first scenario, it seems Caroline ought *not* to take a light dose of heroin; in fact, she ought to take a dose that is sufficiently large, in order to avoid serious withdrawal effects. This intuition can be explained by the following credo:[4]

> **Holistic Detachment:** one should take *all and only* the fixed circumstances into account, when detaching oughts.

Although there are various logics that capture the principle of restricted detachment in one way or another, only few have attempted to target its holistic counterpart and explicate it in exact, formal terms.[5] This is the aim of the present paper. We present the (monotonic) modal logic **HD** which captures the distinction between mere facts and fixed circumstances, and which validates a rule of detachment triggered by "all that is fixed" (Section 2). In Section 3, we show that this logic is rather weak when applied to realistic scenarios and we argue against what we call the "enthymematic approach" to mitigate this weakness. In contrast, we show that **HD** can and should be strengthened non-monotonically. Section 4 gives a brief survey of related work and Section 5 concludes the paper.

Before we proceed, two disclaimers are in order. First, unlike many other existing accounts, ours yields a conservative extension of Standard Deontic Logic. This implies that it inherits some paradoxes of the latter, but also all its inferential power. In the current paper, we will remain mostly silent on those paradoxes, since we consider their solution to be orthogonal to the issue of detachment. We do however believe that giving up on the inferential power of **SDL** should not be taken too lightly.

Second, in this paper we focus on conditional, defeasible oughts that concern one particular agent – often these claims can be interpreted as forms of "advice" to the person in question. So we focus on tentative claims of the type "if you do X, then you ought to do Y" (where the agent is the same in the antecedent and consequent, or neither involves any agency). This can be contrasted with legal claims such as "if you drive above the speed limit, and if you are not in circumstance $X_1, \ldots,$ or X_n, then you must be fined". In the latter case, the exceptions are usually made explicit in the normative system, and the consequent of the conditional concerns an action of the legislator or an agent-independent proposition, not an action of the one who

---

[4] In his discussion of ethical reasoning, Jonsen seems to argue in favour of a principle akin to our holistic detachment, where he writes that "[...] the ultimate view of the case and its appropriate resolution comes, not from a single principle, nor from a dominant theory, but from the converging impression made by all of the relevant facts and arguments [...]" [26, p. 245]. Simplifying Jonsen's view somewhat, one can take arguments to be deontic conditionals, and relevant facts to be the fixed circumstances.

[5] We consider approaches similar or related to ours in Section 4.

violates the norm in question. We will not discuss this distinction here in detail, but merely flag it to avoid any confusion.[6]

2 The monotonic logic HD

In this section we present a monotonic logic for holistic detachment. Even though its underlying intuitions seem straightforward, they give rise to a rich system with some surprising interaction principles (cf. Section 2.3).

2.1 Formal language

Fix a countable set $\mathcal{S} = \{p, q, r, \ldots\}$ of sentential variables. The set of wffs of **HD**, $\mathcal{W}$, is obtained by closing $\mathcal{S} \cup \{\top, \bot\}$ under the classical truth-functional connectives $\neg, \vee, \wedge, \rightarrow, \leftrightarrow$, the unary operators $\mathsf{U}, \vec{\Box}, \vec{\Box}, \mathsf{O}$ and the binary operator $\mathsf{O}(.|.)$. We treat $\bot$, $\neg$ and $\vee$ as primitive, $\top$ and the other classical connectives are defined in the usual way.

U is a global modality in the sense of [18]; it simplifies the axiomatization of the logic and will turn out highly useful in defining the non-monotonic extensions of **HD**.[7] $\vec{\Box}$ is a normal modality of the type **KT**, and is used to express the properties of the situation that are fixed – one may also call those properties unalterable or unavoidable, cf. [7]. $\mathsf{O}(.|.)$ allows us to express the conditional oughts that are used in our deliberation, in order to determine the obligations that apply to the case at hand – the latter are then formalized using O, which is a normal modality of the type **KD**. Following [22], we read O as an operator for "situation-specific obligation", or more briefly, "situational obligation".

$\vec{\Box}$ is an "all and only" modality in the sense of [28] and [25]. The formula $\vec{\Box}\varphi$ allows us to express that φ is *all* that is fixed, and plays a crucial role in the detachment rule of **HD** (see the axiom (DET) in Section 2.3).

One can express various types of violations in $\mathcal{W}$. $\mathsf{O}(\varphi|\top) \wedge \neg\varphi$ stands for "the general obligation that φ is violated"; $\mathsf{O}(\varphi|\top) \wedge \vec{\Box}\neg\varphi$ expresses that this violation is fixed. $\mathsf{O}\varphi \wedge \neg\varphi$ should be read as "the situational obligation that φ is violated". Finally, $\mathsf{O}(\varphi|\top) \wedge \mathsf{O}\neg\varphi$ expresses that one has the situational obligation to violate the general obligation that φ. Each of these expressions are contingent in our logic (for contingent φ). One may further refine the formal language and distinguish various levels of "fixedness" (essentially generalizing the picture drawn in [7]), but we leave that aside here.

One may also define a different monadic operator for obligation:

$$\mathsf{O}_a\varphi =_{df} \mathsf{O}\varphi \wedge \neg\vec{\Box}\varphi$$

---

[6] Due to space limitations we had to omit the Appendix with meta-proofs in this manuscript. They are included in the online version of this article, available at www.clps.ugent.be/.

[7] See our definition of Δ_2 on page 12.

The operator O_a speaks of situational obligations that are not vacuous, in the sense that $O_a\varphi$ is true iff φ is true in all acceptable worlds relative to the case at hand, but φ is not fixed. Such a φ may however still be more or less specific. As a result, O_a still suffers from some paradoxes akin to the Ross paradox in Standard Deontic Logic.[8] As noted in the introduction, we will largely ignore those paradoxes, and focus on O in most of what follows. We do however briefly return to O_a in Section 4.

2.2 Semantics

Definition 1. *An **HD**-model is a tuple $M = \langle W, R, f, V \rangle$ where*

(C1) $W \neq \emptyset$ is the domain of M
(C2) $R \subseteq W \times W$ is reflexive
(C3) $f : W \times \wp(W) \to \wp(W)$ is a function that satisfies the following conditions:

   (C3.1)   where $w \in W$ and $\emptyset \neq X \subseteq W$: $f(w, X) \neq \emptyset$
   (C3.2)   where $w \in W$ and $X \subseteq W$: $f(w, X) \subseteq X$

(C4) $V : \mathcal{S} \to \wp(W)$ is a valuation function

Definition 2. *Where $M = \langle W, R, f, V \rangle$ is an **HD**-model and $w \in W$,*

(SC0) $M, w \models \varphi$ iff $w \in V(\varphi)$ for all $\varphi \in \mathcal{S}$
(SC1) $M, w \not\models \bot$
(SC2) $M, w \models \neg\varphi$ iff $M, w \not\models \varphi$
(SC3) $M, w \models \varphi \vee \psi$ iff $M, w \models \varphi$ or $M, w \models \psi$
(SC4) $M, w \models \mathsf{U}\varphi$ iff for all $w' \in W$, $M, w' \models \varphi$
(SC5) $M, w \models \vec{\square}\varphi$ iff $R(w) \subseteq \|\varphi\|^M$
(SC6) $M, w \models \overleftrightarrow{\square}\varphi$ iff $R(w) = \|\varphi\|^M$
(SC7) $M, w \models \mathsf{O}(\varphi|\psi)$ iff $f(w, \|\psi\|^M) \subseteq \|\varphi\|^M$
(SC8) $M, w \models \mathsf{O}\varphi$ iff $f(w, R(w)) \subseteq \|\varphi\|^M$

where, $\|\varphi\|^M = \{w' \in W \mid M, w' \models \varphi\}$ and $R(w) = \{w' \in W \mid Rww'\}$.

Semantic consequence ($\Gamma \Vdash \varphi$) and validity ($\Vdash \varphi$) are defined in the standard way. The interesting (since non-standard) clauses are those for $\overleftrightarrow{\square}$ (which corresponds to our intuitive reading of "all that is fixed"), and the one for O, which refers to both R and f.

Intuitively, $R(w)$ corresponds to the set of worlds $w' \in W$ that are available, in view of those circumstances that are fixed at w. The requirement that R is reflexive

---

[8] For instance, we have the following variant of the Ross paradox: $O_a\varphi \wedge \neg\overleftrightarrow{\square}(\varphi \vee \psi) \vdash O_a(\varphi \vee \psi)$. So if (according to this reading) it is obligatory that one mails the letter, and if mailing the letter or burning it is not fixed, then it is obligatory that one mails or burns it.

is motivated by the idea that, from the perspective of the person who reasons about a situation, whatever is fixed also obtains in the current world.

The function f is used to interpret deontic conditionals. Intuitively, $f(w, X)$ is the set of worlds $w' \in X$ that would be acceptable from the viewpoint of w, if X would coincide with one's options at w. We require that, unless φ is impossible in M, $f(w, \|\varphi\|^M) \neq \emptyset$. In other words: conditional on a proposition that is possible, one cannot be obliged to do the impossible.

Semantic clause (SC8) shows that situational obligations are a function of conditional obligations and the fixed circumstances. Note that, since R is reflexive, $R(w)$ is guaranteed to be non-empty, and hence so is $f(w, R(w))$ by condition (C3.1). In view of (SC8) this guarantees that one gets a normal modal operator of type **KD**. As a result, **HD** is a conservative extension of Standard Deontic Logic.

Recall that, in our example from the introduction, David and Lou had different views on what counts as fixed circumstances for their deontic reasoning. Such differences correspond, in our semantics, to a difference concerning the set $R(w)$. At one extreme, everything that happens to be true in our current world is fixed, and hence $R(w) = \{w\}$. This will trivialize the concept of situational obligation, since $\varphi \to \mathsf{O}\varphi$ becomes valid under this condition. At the other extreme, one only considers those circumstances fixed that are logically unavoidable: $R(w) = W$. This implies that $\mathsf{O}\varphi$ becomes equivalent to $\mathsf{O}(\varphi|\top)$. Note that in general, $\mathsf{O}\varphi$ and $\mathsf{O}(\varphi|\top)$ are logically independent in **HD**. Whereas $\mathsf{O}\varphi$ expresses that φ is obligatory in view of the fixed circumstances, $\mathsf{O}(\varphi|\top)$ can be read as "absent further information, φ is obligatory".

We do not explicitly model the difference in view between David and Lou at the object level of our logic. Rather, we see this as a difference in the premises they endorse, or alternatively, as a difference in the models each of them considers. $\vec{\Box}$ and $\overset{\leftrightarrow}{\Box}$ should hence be interpreted here in a metaphysical, not in an epistemological or doxastic sense: they express what is true of the situation at hand, not what a given agent knows or believes to be true. Accordingly, O does not represent belief-based or knowledge-based obligation in the sense of [32, 12].

2.3 Axiomatization

A sound and strongly complete axiomatization of **HD** is obtained by closing a complete axiomatization for classical logic together with all instances of the axiom schemata in Table 1 under necessitation for U and modus ponens.[9] φ is an **HD**-theorem ($\vdash \varphi$) iff φ can be derived from the **HD** axioms and rules. φ is **HD**-derivable from Γ ($\Gamma \vdash \varphi$) iff there are $\psi_1, \ldots, \psi_n \in \Gamma$ such that $\vdash (\psi_1 \wedge \ldots \wedge \psi_n) \to \varphi$.[10]

(CG) follows from the fact that the function f operates on sets of worlds, rather than formulas. The axioms (CK) and (UC) together with necessitation for U imply that, given that one holds the antecedent φ fixed, one can read $\mathsf{O}(.|\varphi)$ as a normal

---

[9]Note that this entails necessitation for $\vec{\Box}$ and O as well, in view of axiom (UB), resp. (BO).
[10]Note that this syntactic consequence relation is by definition compact.

	S5 for U	(UB)	$\mathsf{U}\varphi \to \vec{\Box}\varphi$
	KT for $\vec{\Box}$	(UC)	$\mathsf{U}\varphi \to \mathsf{O}(\varphi\|\psi)$
	KD for O	(BO)	$\vec{\Box}\varphi \to \mathsf{O}\varphi$
(CG)	$\mathsf{U}(\varphi \leftrightarrow \psi) \to (\mathsf{O}(\tau\|\varphi) \to \mathsf{O}(\tau\|\psi))$	(AO1)	$\overleftrightarrow{\Box}\varphi \to \vec{\Box}\varphi$
(CK)	$(\mathsf{O}(\psi\|\varphi) \wedge \mathsf{O}(\psi \to \tau\|\varphi)) \to \mathsf{O}(\tau\|\varphi)$	(AO2)	$(\vec{\Box}\varphi \wedge \overleftrightarrow{\Box}\psi) \to \mathsf{U}(\psi \to \varphi)$
(CP)	$\neg\mathsf{U}\neg\varphi \to (\mathsf{O}(\psi\|\varphi) \to \neg\mathsf{O}(\neg\psi\|\varphi))$	(AO3)	$\mathsf{U}(\varphi \leftrightarrow \psi) \to (\overleftrightarrow{\Box}\varphi \leftrightarrow \overleftrightarrow{\Box}\psi)$
(CI)	$\mathsf{O}(\varphi\|\varphi)$	(DET)	$(\overleftrightarrow{\Box}\varphi \wedge \mathsf{O}(\psi\|\varphi)) \to \mathsf{O}\psi$
		(ATT)	$(\overleftrightarrow{\Box}\varphi \wedge \mathsf{O}\psi) \to \mathsf{O}(\psi\|\varphi)$

Table 1: Axiom schemata for **HD**.

modal operator. (CP) and (CI) correspond to conditions (C3.1), respectively (C3.2) in Definition 1.

(UB) and (UC) follow from the fact that U is a global modality, and that both $\vec{\Box}$ and $\mathsf{O}(.\|\varphi)$ (for fixed φ) are normal modalities. The bridging principle (BO) follows from the semantic clause for O: if a given alternative is acceptable, conditional on all that is fixed, then it must be one that is still available given all that is fixed; hence it must make all the fixed circumstances true.

(AO1) and (AO2) express interactions between the normal modal operator $\vec{\Box}$ and its "all and only"-counterpart. Together with Necessitation for U, (AO3) entails that $\overleftrightarrow{\Box}$ is a classical operator in the sense of [11]. Note that, using (AO1)-(AO3), one can derive the following theorem:

(AO4) $\quad (\overleftrightarrow{\Box}\varphi \wedge \overleftrightarrow{\Box}\psi) \to \mathsf{U}(\varphi \leftrightarrow \psi)$

(DET) corresponds to our notion of holistic detachment. It can be seen as an introduction rule for O and as an elimination rule for $\mathsf{O}(.|.)$. Interestingly, with the current semantics we also get an elimination rule for O that allows us to introduce new conditionals, viz. (ATT) (for "attachment"). This rule says that, if you are in a situation where ψ is an unconditional obligation, and if φ provides an adequate and complete description of the fixed circumstances in that situation, then the conditional $\mathsf{O}(\psi|\varphi)$ is true.

Before closing this section, let us briefly mention some possibilities for varying on the above semantics. First, one may consider weaker or stronger requirements on the accessibility relation R, in line with traditional distinctions in normal modal logics. In [7, p. 291] it is argued that fixed propositions need not be true. Technically, this option – i.e. to give up reflexivity and the associated T-schema for $\vec{\Box}$ – poses no problems; the completeness proof can be run just as before. Alternatively, one may consider more restricted classes of models, where e.g. R is required to be transitive and/or symmetric. For instance, the logic of all models where R is an equivalence relation is characterized by adding the **S5**-axioms for $\vec{\Box}$ to our axiomatization of

HD, together with the following two axiom schemata for $\overleftrightarrow{\Box}$:[11]

(S5$\overleftrightarrow{\Box}$-1) $\overleftrightarrow{\Box}\varphi \leftrightarrow \overleftrightarrow{\Box}\overleftrightarrow{\Box}\varphi$
(S5$\overleftrightarrow{\Box}$-2) $\neg\overleftrightarrow{\Box}\varphi \to \vec{\Box}\neg\overleftrightarrow{\Box}\varphi$

Another type of variation would be obtained by imposing additional requirements on the deontic function f. In particular, one may require that f is constant, in the sense that for all $w, w' \in W$ and all $X \subseteq W$, $f(w, X) = f(w', X)$. This condition makes it possible to capture (an abstract form of) reasoning from cases to conditional norms, as it validates all instances of the following schema:

$$\neg\vec{\Box}\neg(\overleftrightarrow{\Box}\varphi \wedge \mathsf{O}\psi) \to \mathsf{O}(\psi|\varphi)$$

An exploration of these and other frame conditions, and the axiomatization of the resulting logics is left for future work.

3 Strengthening HD

Although **HD** has interesting features and is very expressive (cf. supra), it is also inferentially weak in at least two respects. We first explain why, after which we consider various ways one can strengthen **HD** and thus allow for a more realistic formalization of reasoning with deontic conditionals.[12]

3.1 All that is fixed?

If we were to formalize David's view in the example from the introduction in **HD**, the following premises seem natural:

(P1) $\mathsf{O}(\neg p|\top)$ — "In general, Caroline ought not to take heroin."
(P2) $\mathsf{O}(q \wedge s|p)$ — "If Caroline takes heroin, then she ought to take a light dose and use a clean needle."
(P3) p — "Caroline takes heroin."
(P4) $\vec{\Box}p$ — "It is fixed that Caroline takes heroin."

Note that Lou agrees with David on (P1)-(P3), but rejects (P4). Now, does it follow from (P1)-(P4) that Caroline ought to take a light dose and use a clean needle? In other words, does $\mathsf{O}(q \wedge s)$ follow from these premises? Intuitively it might, but it does not in **HD**. The reason is simple: the premises leave it open that there are propositions other than (and independent of) p that are also fixed.

---

[11]We sketch the completeness proof for this variant in the Appendix to the full version of this paper, available at www.clps.ugent.be/.

[12]What we write below applies just as well to the stronger logics obtained by imposing one or more frame conditions like the ones we discussed at the end of Section 2.3. So this is really a problem of the approach in general, not of the specificities of **HD**.

Fully in line with the idea behind **HD**, one needs to know that p expresses *all* that is fixed, before one can apply detachment. But unfortunately, one can never derive $\boxdot p$ from (P1)-(P4). More generally:

Theorem 1. *Let $\Gamma \subseteq \mathcal{W}$ be an **HD**-consistent set such that $\boxdot$ occurs in no member of Γ. Then there is no φ such that $\Gamma \vdash \boxdot \varphi$.*

In view of this theorem, there is a logical gap between formulas that express fixed circumstances and formulas of the form $\boxdot \varphi$. So if we formalize David's reasoning, and if we want to arrive at the appropriate conclusion using **HD**, then we should add $\boxdot p$ as a premise. We will return to this point below, but first consider a different problem for **HD**.

3.2 Applying general norms to specific cases

Suppose that we add the following premise to (P1)-(P4):

(P5) $\boxdot(p \wedge r)$ — "Caroline takes heroin and has a child, and this is all that is fixed"

In this case, we do have information about all that is fixed. Still, we cannot detach that Caroline ought to use a clean needle and take a light dose of heroin. The reason is that the deontic conditional $\mathsf{O}(q \wedge s|p)$ only speaks about those situations in which all that is fixed coincides with p. One obvious way out would be to assume that conditional obligations are closed under strengthening of the antecedent (henceforth, SA): from $\mathsf{O}(\varphi|\psi)$, to infer $\mathsf{O}(\varphi|\psi \wedge \tau)$. Semantically, this corresponds to the condition: if $X \subseteq Y$, then $f(w, X) \subseteq f(w, Y)$.

The problem with this move is that it implies a very strong reading of the deontic conditional: what advice can one ever give that is not overruled in certain very specific circumstances? Consider our second example from the introduction: there we have a clear exception to the general rule concerning heroin, which is made explicit only after the general rule was stated. This exception does *not* generate a conflict at the level of the eventual advice one will give: it simply *blocks* the application of the more general rule. If (SA) is built into the logic, then either one must rule out the possibility of such posterior exceptions – and hence, have all exception clauses built into one's deontic conditionals from the start –, or one should treat exceptions as merely "other considerations" that are on equal footing with the specific variant of the general rule that can be derived by (SA). Moreover, if all exception clauses are explicitly stated as part of the general rule, then one should also assume that all the negations of those clauses are fixed circumstances, in order to solve the problem noted in Section 3.1.

3.3 Tacit premises?

Each of the above problems can easily be tackled if we just add certain premises to our formalization of the examples in question. For the problem of specificity (Section

3.2), this means one would add the following premise:

(P6) $O(q|p \wedge r)$ — "If Caroline takes heroin and has a child, then she ought to take a light dose."

In other words, the argument from (P1)-(P5) to Oq is treated as an *enthymeme*: an argument that draws on a tacit premise – i.c. (P6) – that is endorsed by anyone who reasons about the example.

The enthymematic approach (as we shall call it) to deontic reasoning is not new. For instance, in his work on conflict-tolerant deontic logics, Goble developed logics which allow for restricted forms of aggregation (from $O\varphi, O\psi$ to infer $O(\varphi \wedge \psi)$) and restricted forms of inheritance (from $O\varphi$ and $\varphi \vdash \psi$, to infer $O\psi$).[13] Goble's restrictions are of the type "it is possible that τ" or "it is permitted that τ". In order to make natural examples of deontic reasoning work, one then has to treat such possibility or permissibility claims as tacit premises. In a similar vain, Carmo and Jones [7] need to add the premises $\neg\vec{\Box}\varphi$ and $\neg\vec{\Box}\neg\varphi$ in order to get the inference from $\vec{\Box}\psi, O(\varphi|\psi)$ to $O\varphi$ off the ground.

In itself, the enthymematic approach should not be rejected: it is a fact of life that we do not always make all our premises explicit, and it is a virtue of logic that it forces us to do so. However, in the case of **HD**, logic can and should do more.[14] The (allegedly) implicit premise (P6) is not simply some "general relevant background information": it bears a specific, *formal* relation to the explicit premise (P2): (P6) can be obtained from (P2) by (SA). Likewise, (P5) has a formal relation to (P4) and to the premises (P1)-(P4) as a whole: $\overset{\leftrightarrow}{\Box}p$ entails (P4) and it is compatible with each of the other premises. As these examples show, formal tools *can* help us clarify at least some of the tacit premises that are at stake.

Such help is indispensable as soon as one considers more complex scenarios, where proper logical calculations will be required to determine which tacit premises are mutually compatible. Note that, as soon as we go to first order predicative languages, there is not even a positive test for joint consistency of the explicit premises with the tacit ones. Moreover, if we want to model the dynamics of reasoning with conditionals, we should be able to accommodate cases where explicit premises are added along the way, as we reason. In such cases, one would have to double-check consistency with previously added tacit premises, change them again, etc. Describing such a procedure in exact terms will result, essentially, in a formalism much like the one we describe below.

---

[13] See e.g. [15] for an introduction to these systems.

[14] The same applies to Goble's work, as he later acknowledged [17, 16]. We believe that a similar argument can be made for the approach of Carmo and Jones, but this is beyond the scope of the current paper.

3.4 Going non-monotonic

In view of the preceding, one should strengthen **HD** by adding certain defeasible rules of inference. There are several ways to do so – see [29] for a reader-friendly introduction to the field. For reasons of space, we will merely give an indication of the type of system we have in mind, leaving its full exploration for future work. In doing so, we borrow terminology from Makinson's [29].[15]

The first weakness of **HD** concerns the inference from $\vec{\Box}\varphi$ to $\overset{\leftrightarrow}{\Box}\varphi$. The obvious solution would be: treat all claims of the type "if φ is a fixed circumstance, then φ is all that is fixed" as default assumptions. So our default assumptions are all members of the following set:

$$\Delta_1 = \{\vec{\Box}\varphi \to \overset{\leftrightarrow}{\Box}\varphi \mid \varphi \in \mathcal{W}\}$$

How can we define a new consequence relation $\vdash$, using **HD** and Δ_1? As a first stab, let $\Gamma \vdash \psi$ iff there are $\tau_1, \ldots, \tau_n \in \Delta_1$ such that (i) $\Gamma \cup \{\tau_1, \ldots, \tau_n\} \vdash_{\mathbf{HD}} \psi$, and (ii) $\Gamma \nvdash_{\mathbf{HD}} \neg \tau_i$ for all $i \in \{1, \ldots, n\}$. So e.g. from $\Gamma_1 = \{\vec{\Box}p\}$ we can infer $\overset{\leftrightarrow}{\Box}p$; however, from $\Gamma_1' = \{\vec{\Box}p, \vec{\Box}q, \neg \mathsf{U}(p \to q)\}$ we can no longer infer $\overset{\leftrightarrow}{\Box}p$ in view of the derived theorem (AO4). This way, we obtain a non-monotonic, but exact and formal criterion for when it is safe to assume $\overset{\leftrightarrow}{\Box}\varphi$ for some φ. One may think of criterion (i) as giving us more inferential power, whereas criterion (ii) makes sure that, whenever the premises require this, inferences are blocked to maintain consistency.

There are two problems with such an approach. The first is well-known from the general study of non-monotonic logic. That is, sometimes criterion (ii) above applies, but there is nevertheless a *disjunction* $\neg \tau_{i_1} \vee \ldots \vee \neg \tau_{i_k}$ that follows from Γ by means of **HD**. In that case, Γ will yield consequences that are jointly incompatible with Γ. Moreover, its consequences will not be closed under classical logic.

Various solutions to this first problem have been developed. One is to quantify over maximal sets of default assumptions compatible with Γ; another is to rephrase (ii) in terms of minimal disjunctions of negations of default assumptions that follow from Γ. The interested reader is kindly referred to [29] where exact definitions of these two solutions and the properties of the resulting logics are discussed.

The second problem is more specific to the current application. Let $\Gamma_2 = \{\vec{\Box}p, \vec{\Box}q\}$. Intuitively, one would expect that only $\overset{\leftrightarrow}{\Box}(p \wedge q)$ is derivable from Γ_2. However, this premise set is also compatible with the formula $\mathsf{U}(p \leftrightarrow q)$. So, as far as Γ_2 is concerned, there is no reason to block the inferences from Γ_2 to $\overset{\leftrightarrow}{\Box}p$ and $\overset{\leftrightarrow}{\Box}q$.

It seems that in order to stay closer to our logical intuitions, a different type of default assumptions should be maximized, *prior to* the assumptions in Δ_1:

---

[15] For readers familiar with *Adaptive Logics* it should be noted that our proposal here can be readily translated into that framework as well. This has the immediate advantage that one obtains a model-theoretic semantics and a dynamic proof theory in the sense of [2] for the resulting logics. See [37] for a study of the relation between Makinson's default assumption consequence relations and adaptive logics.

$$\Delta_2 = \{\neg U(\varphi \leftrightarrow \psi) \mid \varphi, \psi \in \mathcal{W}\}$$

In other words: two propositions are taken to be non-equivalent by default, i.e., unless the premises indicate otherwise.[16] Returning to our example $\Gamma_2 = \{\vec{\Box}p, \vec{\Box}q\}$, we will thus first infer $\neg U(p \leftrightarrow q)$, $\neg U(p \leftrightarrow (p \land q))$ and $\neg U(q \leftrightarrow (p \land q))$. This at once blocks the derivations of $\vec{\Box}p$ and $\vec{\Box}q$ in view of (AO4).

Note that in the previous paragraph, we emphasized that Δ_2 should receive priority over Δ_1. Again, there are well-studied and well-behaved formal accounts of how to impose a priority structure on default assumptions — see e.g. [37] for a framework that accommodates such refinements.

The second weakness of **HD** that we spotted was that it invalidates (SA), making it impossible to apply general conditional oughts to specific circumstances. This suggests that we use a third type of default assumptions:

$$\Delta_3 = \{O(\varphi|\psi) \to O(\varphi|\psi \land \tau) \mid \varphi, \psi, \tau \in \mathcal{W}\}$$

So, for instance, from $\Gamma_3 = \{O(q|p), \overleftrightarrow{\Box}(p \land r)\}$ we first derive $O(q|p \land r)$, after which we apply detachment to derive Oq. This inference is blocked in the case of $\Gamma'_3 = \{O(q|p), O(\neg q|p \land r), \overleftrightarrow{\Box}(p \land r)\}$, since $O(q|p \land r)$ is incompatible with the second premise. From Γ'_3, one can derive $O\neg q$ by our axiom (DET).[17]

At this point some readers may become suspicious about the whole enterprise of holistic detachment. We argued that one should strengthen $\vec{\Box}\varphi$ to $\overleftrightarrow{\Box}\varphi$ whenever this is possible; this inference is necessary in order to obtain the kind of information that is strong enough to license detachment. We also argued that one should have a defeasible form of (SA) in order to allow that general conditionals are applicable in more specific cases. But why then not give up on the requirement of holism, so that (SA) is not required in the first place? Doesn't that make for a much smoother logic?

Two points in defense. First, some defeasible form of (SA) is highly intuitive in itself. Regardless of the specific circumstances we are in, it seems that we can reason about the relation between conditional oughts, even if these are interpreted as defeasible pieces of advice. From "if you are in Sapporo, you should go to a sushi-bar", we are inclined to infer "if you are in Sapporo with friends, then you should go to a sushi-bar". The inference appears to be valid, regardless of where in the world one happens to be.

---

[16] Note that every member of Δ_2 is of the form $\neg U\tau$, and conversely, every formula of the latter form can be equivalently rephrased as a formula in Δ_2 (simply by putting $\varphi = \tau$ and $\psi = \top$). Treating formulas of the form $\neg U\tau$ as default assumptions gives rise to a logic similar to that studied in [3].

[17] Although this paragraph may suggest the opposite, implementing this idea to obtain a formal, well-behaved logic does bring some complications. More specifically, one needs to restrict the logical form of the assumptions from Δ_3 in various ways, in order to overcome so-called *flip-flop problems*. As above, we leave the technical details for a follow-up paper.

Second, in cases where exceptions are explicitly mentioned – such as, "if you are in Sapporo but you are allergic to fish, then you should not go to a sushi-bar" – we do want to be able to draw the correct conclusion regarding our obligations, relative to the circumstances at hand. If I am in Saporro and I happen to be allergic to fish, then I do not want to derive the conclusion that I should go to a sushi-bar. But if one skips the holistic requirement, that conclusion will have exactly the same logical status as the conclusion that I should not go to a sushi-bar: it is a deductive consequence of the premises.

4 Related work

The literature on dyadic deontic logic and detachment is vast. For reasons of space, we focus on work that is directly linked to ours and draw some high-level comparisons. A full study of these relationships is left for future work. We first focus on traditional, possible worlds semantics, after which we consider norm-based accounts (in the sense of [20]) and other more syntactic approaches.

4.1 Possible worlds semantics that validate restricted detachment

As noted in the introduction, the idea of restricted detachment can be found in many accounts of contrary-to-duty paradoxes. Already in his [21], Hansson distinguishes between mere facts and unalterable ones in relation to detachment [21, p. 394]. Greenspan [19] seems to be the first to formalize fixed circumstances by means of a normal modal operator akin to our $\vec{\Box}$. In her account, those circumstances are tied to temporal reasons: e.g. once the time to leave has passed, it is a fixed circumstance that you will not help. However, in [34] it is shown that there are examples of detachment where the circumstances are not fixed due to temporal (or agential) reasons.

In his [13], Feldman argues in favour of restricted detachment and proposes a logic based on this idea. In a back and forth ([10, 14, 9]) Castañeda criticises Feldman's logic, but acknowledges the importance of what he calls "deontic circumstances" for a proper understanding of ought-claims. Such circumstances are contrasted with "deontic foci", i.e. the things that are the subject of obligations and permissions. Our distinction between mere facts and fixed circumstances has some parallels with Castañeda's distinction between deontic circumstances and deontic foci, but there are also some fundamental differences. Most strikingly, Castañeda argues that only *actions* (of a given agent) can be obligatory, not propositions.[18] Circumstances are represented by propositional variables, whereas formulas of the form $O\varphi$ are only well-formed if φ represents an action. Castañeda deems it unacceptable that deontic circumstances are obligatory (as they are in our account of O,

---

[18] Castañeda uses the term "prescriptions" to refer to the former.

cf. our axiom (BO)), especially for cases of "determined, successfully and carefully planned wrongdoing" [10, p. 13]. Treating the defined operator O_a (cf. Section 2.1) as the "proper" operator for obligations is also unacceptable for Castañeda, since this would still imply that these instances of planned wrongdoing are neither wrong nor right [10, p. 13]. Note however that, as we argued in Section 2.1, we make a distinction between the claim that φ is obligatory given the circumstances – which one can express either by O or O_a – and the claim that the circumstances themselves violate a general obligation – e.g. expressed by $O(\neg\varphi|\top) \wedge \vec{\Box}\varphi$.

In a series of articles ([6, 7, 8]) Carmo and Jones develop a theory of "fixedness" or unalterability with regards to conditional obligations. This theory was a direct source of inspiration for our **HD**. There are however a few problems with the specific route taken by Carmo and Jones. First, as shown in [27], their logic (as defined in [8]) validates a specific type of deontic explosion: if φ and ψ are independent, in the sense that neither strictly implies the other, and if $O(\varphi|\top)$, then $O(\psi|\neg\varphi)$. This is i.a. due to the validity of a (restricted) form of strengthening of the antecedent in their logic.[19] Second, as noted in Section 3.3, one needs to add various tacit premises in order to obtain the correct results with Carmo and Jones' system, even in simple cases. As we argued, this drawback can be overcome by extending the logic non-monotonically. The third and more serious drawback is that this logic cannot handle specificity-cases [7, p. 295] (see also [35]).

According to Van Benthem, Liu and Grossi [36], detachment should be modeled by a formula of the type $O(\varphi|\psi) \to [!\psi]O(\varphi|\top)$, which expresses that if φ is obligatory conditional on ψ, then, after the information is received that ψ, φ becomes an unconditional obligation. Note that here, the holism is built in automatically, since the event $!\psi$ is the announcement of ψ and *nothing but* ψ. However, as of yet, the Hansson-style semantics for dyadic deontic logic has not been equipped with an "all and only" modality, in order to model the (defeasible) inferences from a (possibly incomplete) description of the fixed circumstances to the oughts that can be detached in view of all that is fixed.

In the related field of practical, goal-oriented reasoning, Boutilier [5] proposes a way to determine "actual preferences" on the basis of conditional preference statements and a (finite) knowledge base $\mathcal{K}$. When $I(\psi|\varphi)$ represents that, conditional on φ, ψ is true in all preferred states, the agent has an actual preference for ψ iff $I(\psi| \bigwedge \mathcal{K})$ holds. Note however that this type of inference is not modeled at the object level of the logic: there is no operator for "all the agent knows", or for "what is preferred given all the agent's knowledge". Boutilier's article is mainly focussed on the logic of the conditional preference operator.

Van der Torre [38] suggested that Boutilier's proposal could be turned into a

---

[19]In fact, one can prove an even stronger fact: in the logic of Carmo and Jones, $O(\varphi|\top), \neg U \neg(\neg\varphi \wedge \psi) \vdash O(\psi|\top)$. So whenever ψ is compatible with the violation of a general obligation that φ, then ψ is obligatory conditional on $\neg\varphi$.

principle of deontic reasoning, using Levesque's "all and only" modality. Van der Torre calls the resulting rule "exact factual detachment" (EFD); it is formally identical to our holistic detachment. He only discusses this rule, but develops no semantics or axiomatization for the resulting logic. He then goes on to argue that EFD leads to the validity of the truth schema $O\varphi \to \varphi$ and concludes that "if EFD is accepted, then the relation between facts and absolute obligations is identical to the relation between antecedent and consequent of the conditional obligations" [38, p. 88]. This is however incorrect, at least as long as one can distinguish between mere facts and knowledge (or in our interpretation: what is fixed). Indeed, as we argued in Section 2.2, $O\varphi \land \neg\varphi$ is perfectly satisfiable, even in a logic that validates holistic detachment.

4.2 Norm-based, syntactic accounts

There is also a wide range of accounts that do not attempt to reduce the truth or applicability of conditional norms to some external reality such as a preference relation or deontic function defined over possible worlds. Instead, these accounts take a set N of conditional norms as primitive, and use that N to construct an operational, rule-based semantics for deontic logic. Technically, N is just a set of pairs (ψ, φ) in a formal language.

One account of this type is developed in Horty's [24]. Here, norms are represented as default rules, and facts are conceived as the triggers of unconditional oughts. Conditional oughts $O(\varphi|\psi)$ are interpreted in terms of what follows from a given default theory when adding ψ to the factual information. In Horty's framework, detachment is treated as a defeasible principle: one applies it as long as the result remains consistent with the given facts; moreover, no distinction is made between mere facts and what is considered fixed.

Input/output logic [30, 31] is another well-known class of systems in which conditionals (as syntactic entities) are primitive. Here again, one can define various input/output relations that, when given a set of input F and a set of conditionals N, fix an output $out(F, N)$, corresponding to our situational obligations. The input roughly plays the role of our fixed circumstances, with the difference that the input is not necessarily included in the output (contrary to our axiom (BO)). In contrast to Horty's account, input/output logic (at least in its original form [30, 31]) cannot handle specificity-cases.

In [33], Parent and van der Torre discuss exact factual detachment (cf. supra) as a property of I/O-logics. The idea here is that, if $(\varphi, \psi) \in N$, then ψ is in the output of N under the input φ. In other words, if one has *exactly* the input φ, and if there is a conditional norm to the effect that ψ is obligatory if φ is the case, then ψ is indeed obligatory. In the original I/O-logics, and in the specific system $\mathcal{O}_3$ proposed by Parent and van der Torre in [33], exact factual detachment is a derived property; also there, specificity-cases cannot be handled in the way **HD** does.

Straßer [35] develops a non-monotonic logic which features specific expressions of the type "the obligation to do φ, conditional on ψ, is overridden", denoted by $\bullet\mathsf{O}(\varphi|\psi)$.[20] The latter phrase receives a purely syntactic definition, in terms of the existence of other obligations and circumstances. In these systems, detachment is represented by a rule of the following type:

$$\psi, \mathsf{O}(\varphi|\psi), \neg\bullet\mathsf{O}(\varphi|\psi) \vdash \mathsf{O}\varphi$$

By adding a suitable set of axioms that govern the behavior of $\bullet\mathsf{O}(\varphi|\psi)$, and by adding a defeasible mechanism that validates the inference from $\mathsf{O}(\varphi|\psi)$ to $\neg\bullet\mathsf{O}(\varphi|\psi)$, one can then ensure that specificity-cases and contrary-to-duties are adequately handled. In this framework, no operators for "fixed circumstances" are used; also, no semantics for the $\bullet$-operator is provided.

In more recent work, Beirlaen and Straßer [4] have used structured argumentation frameworks to model deontic reasoning on the basis of conditional obligations and "fixed facts", making use of a normal modal operator to model the latter. Here, one looks at all possible arguments that can be built for a given claim (e.g. $\mathsf{O}p$) and defines various attack relations between such arguments. These attack relations in turn allow one to determine a "grounded extension" of the premises, which can be seen as a set of "safe arguments" that in turn deliver the situational obligations. As Beirlaen and Straßer show, specificity and other parameters can be readily built into this framework by adopting an appropriate definition of the attack relation.

The mentioned accounts are arguably richer and more fine-grained than our **HD**. Still, one argument in favour of more traditional approaches (such as our own), and contra norm-based or other purely syntactic approaches such as the ones mentioned above, is that we get one unified theory in which we can not only reason about what should hold given some F and N. That is, using principles such as our (ATT), we can also derive general rules $\mathsf{O}(\varphi|\psi)$ from premises that merely state what is fixed, and what we consider the "correct" normative judgement in that situation. It should however be admitted that from this viewpoint, **HD** is just a preliminary "toy logic". A more elaborate, insightful semantics and formal language should be developed in order to flesh out and solidify this argument.

5 Concluding remarks

What we hope to have shown is that one can model holistic detachment in deontic reasoning in a monotonic, specificity-sensitive way, making use of an "all and only"-operator. We do not reject non-monotonicity altogether, but move it more upwards in the inference chain that leads to situational obligations. In a slogan: not the rule of detachment, but its premises are defeasible.

---

[20]In fact, Straßer distinguishes two ways in which an obligation can be overridden, resulting in two operators $\bullet_p$ and $\bullet_i$, and two different rules of detachment.

A lot of work remains to be done in order to develop **HD** and its nonmonotonic extensions into a full-fledged, formal theory. Let us just mention three general lines of research. First, more refined accounts of holistic detachment should be studied. One may for instance require that all and only those fixed circumstances that are *relevant* should be taken into account. To explicate such a principle formally, it may be useful to apply techniques from the study of relevant or hyperintensional logics. Second, we hope to be able to work out richer, more insightful semantics for our deontic conditionals, following the traditional accounts cited in Section 4. Third, as noted in the Introduction, we believe there are distinct types of conditional norms, ranging from mere (defeasible) advice to strict legal stipulations. Logically speaking, such conditionals display very different behavior; a general theory of dyadic deontic logic should take this into account.

References

[1] Lennart Åqvist. *Deontic Logic*, volume 8 of *Handbook of Philosophical Logic*, chapter 4, pages 147–264. Kluwer Academic Publishers, 2nd edition, 2002.

[2] Diderik Batens. A universal logic approach to adaptive logics. *Logica Universalis*, 1:221–242, 2007.

[3] Diderik Batens and Joke Meheus. The adaptive logic of compatibility. *Studia Logica*, 66(3):327–348, 2000.

[4] Mathieu Beirlaen and Christian Straßer. A structured argumentation framework for detaching obligations. In Olivier Roy, Allard Tamminga, and Malte Willer, editors, *Deontic Logic and Normative Systems*, pages 32–48. College Publications, 2016.

[5] Craig Boutilier. Toward a logic for qualitative decision theory. In *Proceedings of the Fourth International Conference on Principles of Knowledge Representation and Reasoning*, KR'94, pages 75–86. Morgan Kaufmann Publishers Inc., 1994.

[6] José M. C. L. M. Carmo and Andrew J. I. Jones. A new approach to contrary-to-duty obligations. In D. Nute, editor, *Defeasible Deontic Logic*, pages 317–344. Synthese Library, 1997.

[7] José M. C. L. M. Carmo and Andrew J. I. Jones. Deontic logic and contrary-to-duties. In Dov M. Gabay and Franz Guenther, editors, *Handbook of Philosophical Logic*, volume 8, pages 147–264. Kluwer Academic Publishers, 2nd edition, 2002.

[8] José M. C. L. M. Carmo and Andrew J. I. Jones. Completeness and decidability results for a logic of contrary-to-duty conditionals. *Journal of Logic and Computation*, 23(3):585, 2013.

[9] Hector-Neri Castañeda. Moral obligations, circumstances, and deontic foci (a rejoinder to Fred Feldman). *Philosophical Studies*, 57(2):157–174, 1989.

[10] Hector-Neri Castañeda. Paradoxes of moral reparation: Deontic foci vs. circumstances. *Philoshophical Studies*, 57(1):1–21, 1989.

[11] Brian F. Chellas. *Modal Logic: An Introduction*. Cambridge University Press, 1980.

[12] Roberto Ciuni. Conditional doxastic logic with oughts and concurrent upgrades. In Alexandru Baltag, Jeremy Seligman, and Tomoyuki Yamada, editors, *Logic, Rational-*

ity, and Interaction, pages 299–313. Springer, 2017.

[13] Fred Feldman. *Doing the Best We Can: An Essay in Informal Deontic Logic*. Philosophical Studies Series in Philosophy 35. Springer Netherlands, 1986.

[14] Fred Feldman. Concerning the paradox of moral reparation and other matters. *Philosophical Studies*, 57(1):23–39, 1989.

[15] Lou Goble. A logic for deontic dilemmas. *Journal of Applied Logic*, 3:461–483, 2005.

[16] Lou Goble. Prima facie norms, normative conflicts, and dilemmas. In Dov Gabbay, John Horty, Xavier Parent, Ron van der Meyden, and Leendert van der Torre, editors, *Handbook of Deontic Logic and Normative Systems*, volume 1, pages 241–351. College Publications, 2013.

[17] Lou Goble. Deontic logic (adapted) for normative conflicts. *Logic Journal of the IGPL*, 22(2):206–235, 2014.

[18] Valentin Goranko and Solomon Passy. Using the universal modality: Gains and questions. *Journal of Logic and Computation*, 2(1):5–30, 1992.

[19] P. S. Greenspan. Conditional oughts and hypothetical imperatives. *The Journal of Philosophy*, 72(10):259–276, 1975.

[20] Jörg Hansen. Reasoning about permission and obligation. In Sven Ove Hansson, editor, *David Makinson on Classical Methods for Non-Classical Problems*, pages 287–333. Springer Netherlands, 2014.

[21] Bengt Hansson. An analysis of some deontic logics. *Noûs*, 3(4):373–398, 1969.

[22] Sven Ove Hansson. Situationist deontic logic. *Journal of Philosophical Logic*, 26(4):423–448, 1997.

[23] Risto Hilpinen and Paul McNamara. Deontic logic: A historical survey and introduction. In Dov Gabbay, John Horty, Xavier Parent, Ron van der Meyden, and Leendert van der Torre, editors, *Handbook of Deontic Logic and Normative Systems*, volume 1, pages 3–136. College Publications, 2013.

[24] John F. Horty. *Reasons as Defaults*. Oxford University Press, 2012.

[25] I. L. Humberstone. The modal logic of "all and only". *Notre Dame J. Formal Logic*, 28(2):177–188, 1987.

[26] Albert Jonsen. Casuistry: An alternative or complement to principles. *Kennedy Institute of Ethics Journal*, 5(3):245, 1995.

[27] Bjørn Kjos-Hanssen. A conflict between some semantic conditions of Carmo and Jones for contrary-to-duty obligations. *Studia Logica*, (1):173–178, 2017.

[28] Hector J. Levesque. All I know: A study in autoepistemic logic. *Artificial Intelligence*, 42(2-3):263–309, 1990.

[29] David Makinson. *Bridges from Classical to Nonmonotonic Logic*, volume 5 of *Texts in Computing*. King's College Publications, London, 2005.

[30] David Makinson and Leon van der Torre. Input/output logics. *Journal of Philosophical Logic*, 29:383–408, 2000.

[31] David Makinson and Leon van der Torre. Constraints for input/output logics. *Journal of Philosophical Logic*, 30:155–185, 2001.

[32] Eric Pacuit, Rohit Parikh, and Eva Cogan. The logic of knowledge based obligation. *Synthese*, 149(2):311–341, 2006.

[33] Xavier Parent and Leendert van der Torre. "Sing and dance!". In Fabrizio Cariani, Davide Grossi, Joke Meheus, and Xavier Parent, editors, *Deontic Logic and Normative Systems*, pages 149–165. Springer, 2014.

[34] Henry Prakken and Marek Sergot. Contrary-to-duty obligations. *Studia Logica*, 57(1):91–115, 1996.

[35] Christian Straßer. A deontic logic framework allowing for factual detachment. *Journal of Applied Logic*, 9:61–80, 2011.

[36] Johan van Benthem, Davide Grossi, and Fenrong Liu. Priority structures in deontic logic. *Theoria*, 80(2):116–152, 2014.

[37] Frederik Van De Putte. Default assumptions and selection functions: A generic framework for non-monotonic logics. In Felix Castro, Alexander Gelbukh, and Miguel Gonzalez, editors, *MICAI*, volume 8265 of *Lecture Notes in Computer Science*, pages 54–67. Springer, 2013.

[38] Leendert van der Torre. *Reasoning About Obligations*. PhD thesis, Erasmus Universiteit Rotterdam, 1997.

www.ingramcontent.com/pod-product-compliance
Lightning Source LLC
Chambersburg PA
CBHW050124170426
43197CB00011B/1704